JOURNEY

AN ILLUSTRATED HISTORY OF THE WORLD'S GREATEST TRAVELS

JOURNEY

AN ILLUSTRATED HISTORY OF THE WORLD'S GREATEST TRAVELS

Foreword Simon Reeve
Editorial consultant Father Michael Collins
Contributors Simon Adams, R.G. Grant, Andrew Humphreys

DK

DK LONDON
Senior Editor Angela Wilkes
Senior Art Editor Gadi Farfour
Editors Andy Szudek, Kate Taylor, Anna Fischel
Designers Vanessa Hamilton, Renata Latipova, Nicola Rodway
Picture Researcher Sarah Smithies
Managing Editor Gareth Jones
Senior Managing Art Editor Lee Griffiths
Producer, Pre-Production Gillian Reid
Senior Producer Mandy Inness
Jacket Designer Surabhi Wadhwa
Design Development Manager Sophia M.T.T.
Jacket Editor Claire Gell
Associate Publishing Director Liz Wheeler
Art Director Karen Self
Publishing Director Jonathan Metcalf

DK DELHI
Senior Art Editor Ira Sharma
Project Art Editor Vikas Sachdeva
Art Editor Roshni Kapur
Assistant Art Editors Rohit Bhardwaj, Devika Khosla
Project Editor Antara Moitra
Editor Madhurika Bhardwaj
Assistant Editor Sugandha Agarwal
Managing Editor Soma B. Chowdhury
Managing Art Editor Arunesh Talapatra
Cartography Coordinator Rajesh Kumar Mishra
Cartography Manager Suresh Kumar
Picture Researcher Aditya Katyal
Picture Research Manager Taiyaba Khatoon
Senior DTP Designer Harish Aggarwal
DTP Designers Rajesh Singh Adhikari, Syed Md Farhan, Pawan Kumar
Production Manager Pankaj Sharma
Pre-production Manager Balwant Singh
Jacket Designers Suhita Dharamjit, Tanya Mehrotra
Managing Jackets Editor Saloni Singh

This edition published in 2022
First published in Great Britain in 2017 by
Dorling Kindersley Limited
DK, One Embassy Gardens, 8 Viaduct Gardens,
London, SW11 7BW

The authorised representative in the EEA is
Dorling Kindersley Verlag GmbH. Arnulfstr. 124,
80636 Munich, Germany

10 9 8 7 6 5 4 3 2 1
001–312782–Mar/2022

A CIP catalogue record for this book is available from the British Library.
ISBN: 978-0-2413-6363-8

Printed in the UAE

For the curious
www.dk.com

CONTENTS

1 THE ANCIENT WORLD

EDITORIAL CONSULTANT
Father Michael Collins is a native of Dublin, Ireland. He has written several books with DK and publishes articles about his travels which have a focus on archeology, ancient cultures, and civilizations. *Journey* is based on Michael's original idea about the history of humanity's travels.

LEAD CONTRIBUTOR
Andrew Humphreys is a journalist, author, and travel writer who has written or co-written more than 35 books for DK, Lonely Planet, *National Geographic*, and *Time Out*. His journalism often involves travel with a historical slant and has appeared in *The Financial Times*, *The Telegraph*, and *Condé Nast Traveller*. He is also the author of two books on the golden age of travel in Egypt. He wrote Chapters 4, 5, 6, and 7 of *Journey*.

CONTRIBUTORS

Simon Adams worked as an editor of children's reference books before becoming a full-time writer 25 years ago. Since then, he has written and contributed to more than 60 books, specializing in history, travel, and exploration. A keen jazz fan, he now lives and works in Brighton. He wrote Chapter 1 of *Journey*.

R.G. Grant has written extensively on history, military history, current affairs, and biography. His publications include *Flight: 100 Years of Aviation*, *A Visual History of Britain*, and *World War I: The Definitive Visual Guide*. He was consultant editor on DK's *The History Book* published in 2016. He wrote Chapter 2 and part of Chapter 3 of *Journey*.

FOREWORD

Simon Reeve is an adventurer and *New York Times* bestselling author who has travelled to more than 120 countries making award-winning documentaries for the BBC. They include *Caribbean, Sacred Rivers, Indian Ocean, Tropic of Cancer, Equator, Tropic of Capricorn, Pilgrimage, Greece, Turkey,* and *Australia*. Simon has received a One World Broadcasting Trust Award for "an outstanding contribution to greater world understanding" and the prestigious Ness Award from the Royal Geographical Society. His unique brand of documentaries, combining travel and adventure with global environmental, wildlife, and conservation issues, has taken him across jungles, deserts, mountains, and oceans, through some of the most remote and beautiful regions of our planet.

4

THE AGE OF EMPIRES

5

THE AGE OF STEAM

6

THE GOLDEN AGE OF TRAVEL

7

THE AGE OF FLIGHT

Foreword

In a museum in the Italian town of Bolzano lies the leathery body of a time-traveller. Otzi the iceman, as he is known, was found frozen on a ridge in the Alps in 1991 by German ramblers. Such was the workmanship on his clothes and equipment that the authorities initially thought he was a lost hiker from the early 1900s. Only after carbon-dating did scientists realise that Otzi was truly ancient, and had been travelling through the Alps some 5,300 years ago, before Stonehenge was built. His discovery was a revelation.

Otzi's sophisticated travel equipment was made of at least 15 types of wood, each used for a specific purpose. His clothing included leggings, insulated deer hide shoes, and a bearskin hat. He carried a backpack, embers for lighting fires, a medicine kit, drills, scrapers, and weapons. His axe was cast and sharpened in a way that would be difficult even today. Copper for the blade, which was fixed to the yew haft in a manner designed to achieve the perfect leverage, came from central Italy. Different craftsmen had made his arrows, and his flint tools had been quarried near Verona. Otzi was better equipped than many modern hikers.

Otzi is evidence of long hikes and ancient trade at a time when Europe was sparsely populated with stone-age tribes, the Egyptian pyramids were just being built, and the Mesopotamians were inventing writing. And as this marvellous book makes clear, humans have been on the move ever since. The consequences have been profound. Innovation, brutal conquest, trade, settlement, and love flow from our innate urge to explore. Our journeys today can be a quest to

distant lands or local ramblings around unknown hills that are close to home. Near or far, humans have always wanted to travel and discover.

Within these pages are the voyages of the early Polynesians and ancient Greeks, and the dangers and delights of medieval pilgrimage. There is the search for Prester John, the horrific journeys of slaves, the travels of Wilfred Thesiger – the first to thoroughly explore the desert of the fabled Arabian Empty Quarter – and the extraordinary tale of Annie Londonderry, the first woman to cycle around the world in the 1890s.

Few now need to spend months trekking to a destination because most journeys today are safe and swift, eased by cheap and efficient airlines. With a single budget flight, for example, I was able to visit the magnificent palace at Knossos, in Crete, the birthplace of European civilization, from where ancient Minoans traded around the eastern Mediterranean, a story told in Chapter 1. Yet there is still a thrill to modern travel, and, with just a few moments of planning, we can create personal journeys and memories that will truly linger.

Whether seeking inspiration for a personal odyssey or immersing yourself in the past, leaf through the pages of this book and wonder at the epic travels of our ancestors, and of pioneers who discovered and mapped our planet. Travellers helped to create our civilization and forge our modern world. Their endless adventures are evidence that going on a journey is fundamental to our species. Journeys, clearly, are the story of us.

SIMON REEVE

Introduction

Humans have always been wanderers. Our distant ancestors were nomads for thousands of years before they founded any settlements. Beyond the many practical motives for travel – the pursuit of trade, warfare, pilgrimage, the search for new lands to settle or conquer – there has always been a more primitive human urge, the impulse to find out what lies over the hill, trace a river to its source, sail an unfamiliar coastline, or explore ever further into the unknown.

Looking back from the present hi-tech age, the scale of the journeys once made by people who travelled only on foot, on horseback, or in small boats propelled by sails or oars is simply astonishing. The soldiers of Alexander the Great marched all the way from Greece to northern India, while the Mediterranean sailors of the ancient world ventured as far as the southern tip of Africa. Without so much as a compass to guide them, the Vikings sailed from Scandinavia to the shores of North America, and Polynesian sailors set off across vast expanses of the Pacific Ocean in search of new islands to colonize.

Long journeys to distant places were not the preserve of a bold minority of adventurers. In medieval times, thousands of pilgrims from Christian Europe undertook arduous journeys by land and sea to visit the holy places of Jerusalem, as did Muslims to Mecca, and merchants transported goods in caravans across the arid wastes of the Sahara Desert or along the mountainous Silk Road from China to Europe.

About 500 years ago, the oceanic voyages of Christopher Columbus and other European sailors made it possible for cartographers to draw up broadly accurate maps of the world for the first time. But for centuries after these voyages of discovery, many areas of the world remained a mystery. Even in the 20th century, people were still foraying across uncharted deserts or jungles, or to places where no human had set foot before, such as the Arctic and Antarctica.

As steamships, railways and, later, aircraft made long-distance journeys more common, the romance of travel was sustained by luxury and novelty, from the Orient Express to flying boats. More recently, the oceans and outer space have become targets for exploration, and the lust for adventure has been sated by treks and journeys that challenge human endurance. More recently still, the COVID-19 pandemic has reminded us that travel is a privilege, not a right, while the climate crisis is presenting new challenges and opportunities for sustainable eco-travel and conservation-centric experiences.

▷ **Desceliers's world map**
This map was created in around the 1530s by French chartmaker Pierre Desceliers. Made in the style of a sea chart, it has compass roses and navigation lines, but was clearly a work of art rather than for use at sea.

“ I travel not to go anywhere, but **to go**. I travel for **travel's sake**. The **great affair is to move**. ”

ROBERT LOUIS STEVENSON, *TRAVELS WITH A DONKEY IN THE CEVENNES*

LA LIGNE:
LA ZONA TORRIDA
AMERIQUE
TROPIQUE
DE:
MER AUSTRALLE:
Riuiere de plate
Destroict de magellan

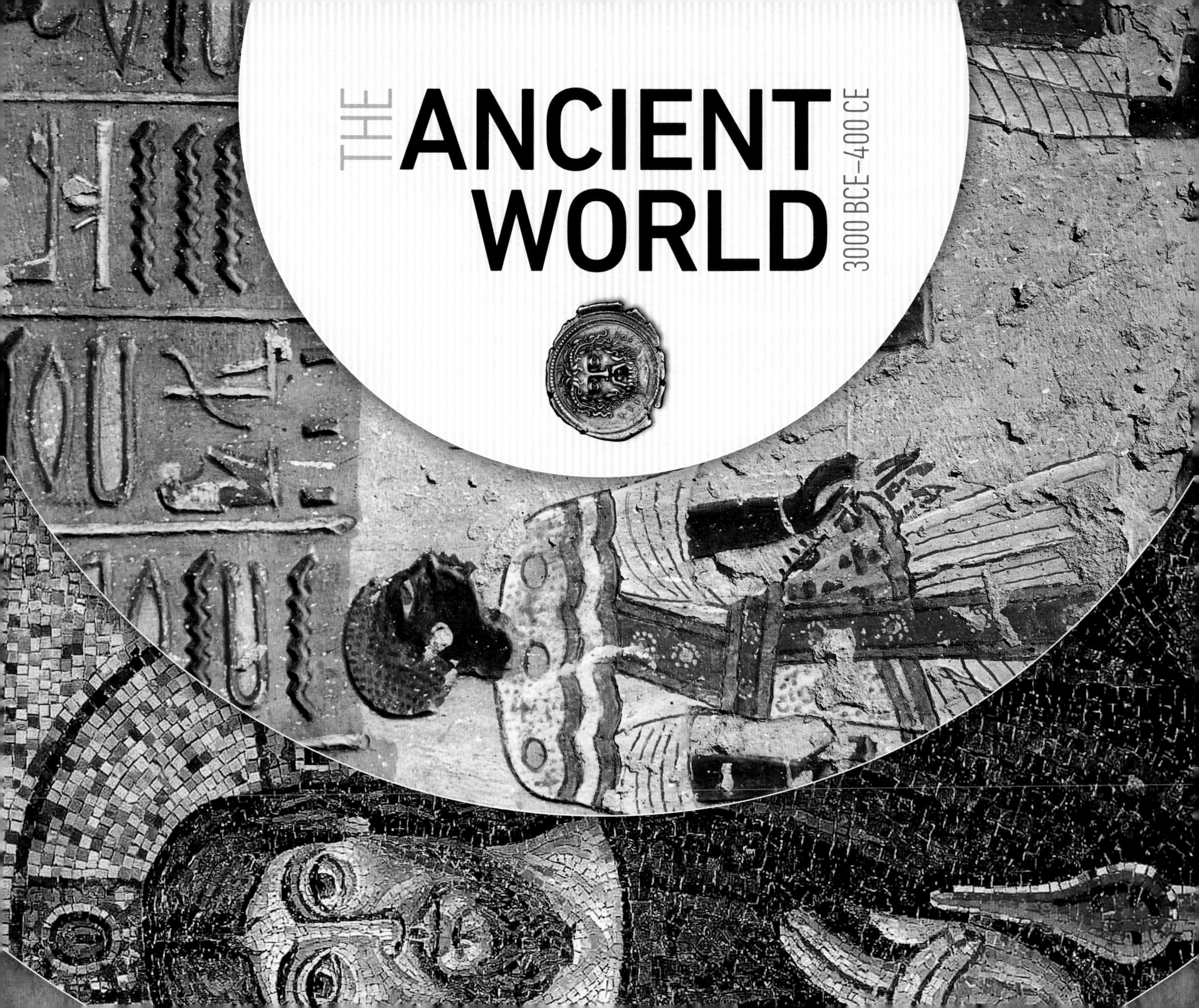
THE
ANCIENT
WORLD
3000 BCE–400 CE

THE ANCIENT WORLD, 3000 BCE–400 CE

Introduction

The ability to travel swiftly over long distances is common to many species, but bipedalism, coupled with a slightness of build, hunter-gatherer instincts, and an apparently insatiable curiosity, are all hallmarks of *Homo sapiens* – a creature for whom no stone, it seems, can be left unturned. Exploration – that restless need to shine light on the unknown – is the lifeblood of the species. Thanks to our ability to conceptualize, we can imagine better worlds and search for them – sometimes cooperating, sometimes warring along the way.

The first travellers

We do not know the name of the first traveller, or what journey they made, but no doubt they were a hunter-gatherer, searching for shelter, food, or water. Foraging and returning to the fold may have been a daily, weekly, or monthly routine. With the advent of farming and the domestication of animals, shepherds travelled with their flocks from winter to summer pastures and back again. Nomads moved in search of water and grazing lands, while armed warriors moved across the lands in search of booty and conquest.

International trade developed when the first cities were founded in Mesopotamia in the third millennium BCE. These cities traded goods and services with their neighbours, and their merchants travelled great distances by land and sea in search of markets. Echoes of long-distance trade around the Indian Ocean can be heard in the *Epic of Gilgamesh*, dating to around 2100 BCE. Likewise, in the Mediterranean, the Minoans of the second millennium BCE traded goods across the inland sea. Meanwhile, the

AN ASSYRIAN RELIEF c.800 BCE DEPICTS CEDARWOOD BEING SHIPPED FROM LEBANON

A ROMAN MOSAIC PORTRAYS THE GREEKS' ARCHETYPAL TRAVELLER, ODYSSEUS

A 7TH-CENTURY FRESCO SHOWS CHINESE DIPLOMAT ZHANG QIAN TRAVELLING IN CENTRAL ASIA

> “ A man who has been through **bitter experiences** and travelled far **enjoys even his sufferings after a time**. ”
>
> HOMER, THE *ODYSSEY*

Egyptians bartered with the fabulous land of Punt, and the Phoenicians, who sold dye, reconnoitred the entire coast of Africa. By the first millennium BCE, the Polynesians were already voyaging across the vast expanse of the Pacific Ocean – not for trade, but to find new islands.

The classical world

The Greeks of the 5th century BCE produced the first map of the known world. Herodotus, the “father of history”, recorded voyages to distant lands in the 4th century, while Pytheas travelled north from his beloved Mediterranean to the unknown and icy Thule (possibly Iceland) in around 325 BCE. Couriers ran an efficient long-distance postal service throughout the Persian Empire. Alexander the Great conquered vast lands, spreading Greek culture, but also absorbing much from the lands which he and his armies invaded. In China, diplomacy sent Zhang Qian out into the world in the 2nd century BCE, preparing the ground for that international artery of trade, the Silk Road, which ran between China and Europe.

In their organized and efficient way, the Romans revolutionized travel. Their road system made movement around their vast empire relatively straightforward, and travellers were aided by road maps and roadside inns. Likewise, the seafarers of the Indian Ocean made use of detailed guides to the ports and coastlands along their routes, while Strabo and Ptolemy established geography as an academic subject. Slowly, the world was becoming more familiar to its inhabitants, and the possibilities of travel, not just for trade but also for adventure and pure exploration, were multiplying accordingly.

THE ROADS BUILT BY THE ROMANS INCREASED TRAVEL THROUGHOUT THEIR EMPIRE

PTOLEMY’S WORLD MAP OF c.150 CE REMAINED AUTHORITATIVE UNTIL THE RENAISSANCE

THE PURPLE DYE SOLD BY THE PHOENICIANS WAS LOVED BY EMPERORS SUCH AS JUSTINIAN I

Journeys from the cradle

In around 2340 BCE, the world's first empire arose in Sumer in Mesopotamia. Ruled by King Sargon of Agade, it traded food, ceramics, metals, and precious stones with its many neighbours in the region.

The empire created by Sargon of Agade emerged in the fertile lands between the Tigris and Euphrates rivers, in what is now southern Iraq. The rivers provided water to drink and irrigation for crops, as well as a means of travel. Apart from water, however, the raw materials required for the military campaigns and conquests of the empire were in short supply. All that its sandy soil could produce was little more than grain and other basic crops, as well as the necessary mud and straw for building houses.

All the necessities of an empire, therefore, had to be imported. Wood to construct ships and wagons came from the cedar trees of Lebanon to the west, while tin and copper to make tools, weapons, and utensils came from Anatolia to the north, Sinai to the south, or Iran to the east. Stone for building also came from Anatolia, and ceramics from India. Gold, silver, and precious gemstones came from across the region. Pearls were sourced from Bahrain in the Persian Gulf, turquoise and ivory from India, and deep blue lapis lazuli, used for jewellery and incorporated into the panels of the Standard of Ur, came all the way from distant Afghanistan.

▷ **Standard of Ur**
The Standard of Ur, a wooden box of unknown purpose, dates to around 2600 BCE. Its sides are covered with intricate, inlaid mosaics depicting scenes of war and peace.

The first travellers

Such goods were traded by merchants along recognized trade routes, the products passing from merchant to merchant as they were carried by pack animals such as donkeys, or hauled in wagons towards Sumer. Bulk goods were shipped by boat up and down the Euphrates, the Tigris being less amenable to travel. The merchants would reach a market town where they would then sell on their goods to the next merchant before returning home. It is these traders who were the journeymen of Sumer, the travellers of the empire. Most would have been foreigners bringing their goods and products for sale in Sumer. They spoke the many different languages of the region, and

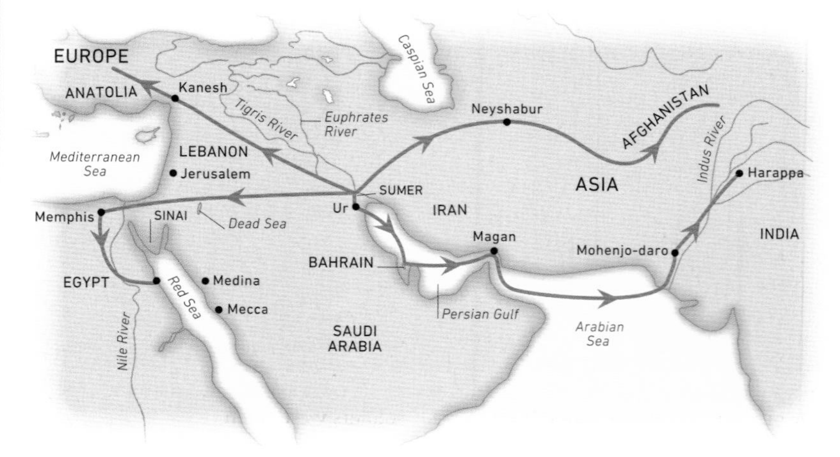

▽ **The Sumerian Empire**
Sargon of Agade conquered the city-states of Mesopotamia to create a single unified empire. Trade routes spread out to India in the east, and to Europe and Africa in the west.

IN CONTEXT
Sargon of Agade

The exact location of Agade (or Akkad) is still unknown, but archaeologists presume it was in southern Mesopotamia, on the banks of one of the branches of the Euphrates River. The ancient *Sumerian King List*, a stone tablet written in Sumerian that records all the kings of Sumer, names Sargon as the son of a gardener and then the cup-bearer to King Ur-Zababa of Kish, who reigned for around six years before Sargon deposed him. Sargon then ruled his new kingdom from around 2340 to 2284 BCE, although some sources suggest his reign only lasted 37 years. During this time, he conquered the whole of Mesopotamia and extended his power to the Mediterranean coast and into modern-day Turkey and Iran.

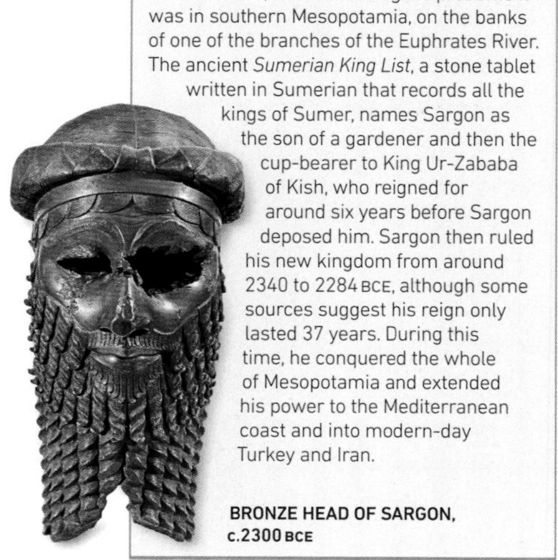

BRONZE HEAD OF SARGON, c.2300 BCE

◁ **Bill of sale**
Clay tablets inscribed in a Sumerian cuneiform script have been found in the city of Shuruppak. Dating to around 2500 BCE, they record the goods bought and sold in the city, including shipments of silver for the city's governor.

although few would have been literate, their knowledge of the known world was immense.

Around the known world

The scale and scope of the journeys these merchant-travellers undertook was extensive. The resin found in the royal tomb of Queen Puabi, who ruled Ur in around 2600 BCE, came from Mozambique on the southeast coast of Africa, and clay seals recording commercial transactions have been found bearing the Indus Valley script of the distant Indus Empire in India. A detailed record of this long-distance trade is found in *The Epic of Gilgamesh*, a lengthy epic poem written about Gilgamesh, King of Uruk, who ruled around 2100 BCE.

The other travellers of the empire were the soldiers of Sumer, who sallied forth around the region for conquest and plunder. Wagons similar to those they would have used feature on the Standard of Ur, which depicts, among many things, the Sumerian army on the move. These soldiers were often abroad for many months, travelling across new lands, before returning home when their victories were won. The empire that Sargon and his soldiers created lasted for around 200 years until, in 2154 BCE, it collapsed altogether after the invading Gutians, a nomadic people from the Zagros Mountains in western Iran, overran the empire.

▽ **Sargon on campaign**
This 20th-century illustration shows Sargon leading his army in Syria. The Mediterranean gave him easy access to Europe and North Africa.

△ **Ship procession**
In this fresco from Akrotiri – a Bronze Age settlement on the island of Thera – a Minoan ship is rowed past a coastal town. The fresco has kept its strong colours, despite its great age.

Minoan seafarers

Europe's first civilization took root on Crete around 3000 BCE. It was established by the Minoans, whose voyages around the eastern Mediterranean coast made them among the first travellers in Europe.

Until the late 19th century, historians and archaeologists were unsure about the ancient history of Europe. Existing historical records dated the first European Iron Age civilization to the 8th or 7th century BCE, but the Greek myths of Homer suggest that an earlier, bronze-using civilization must have flourished somewhere. In 1870, the German archaeologist Heinrich Schliemann discovered the Bronze Age site of Troy in modern-day Turkey, and then went on to find gold at Mycenae on the Greek mainland. But the most important find took place in 1900, when British archaeologist Arthur Evans discovered the magnificent palace at Knossos, on Crete. He dated the site to around 2000 BCE, and in doing so established the island as the birthplace of European civilization. As Knossos was the legendary birthplace of King Minos of Crete, this newly discovered civilization became known as the Minoan.

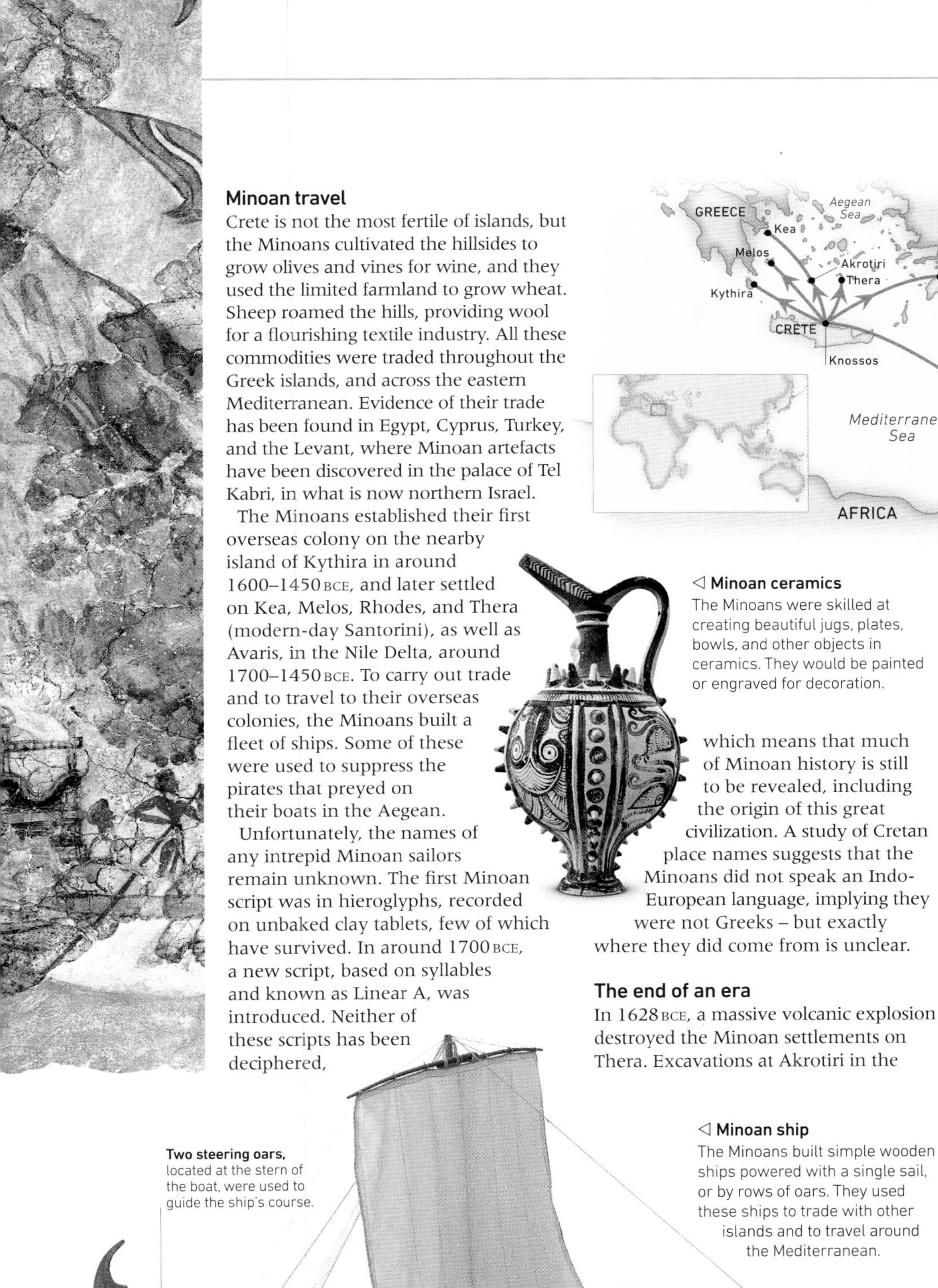

◁ **The Minoan civilization**
Minoan kingdoms were established on Crete in around 2000 BCE, and Knossos became the dominant power in around 1700 BCE. The Minoans then extended their power north into the islands of the Aegean.

Minoan travel

Crete is not the most fertile of islands, but the Minoans cultivated the hillsides to grow olives and vines for wine, and they used the limited farmland to grow wheat. Sheep roamed the hills, providing wool for a flourishing textile industry. All these commodities were traded throughout the Greek islands, and across the eastern Mediterranean. Evidence of their trade has been found in Egypt, Cyprus, Turkey, and the Levant, where Minoan artefacts have been discovered in the palace of Tel Kabri, in what is now northern Israel.

The Minoans established their first overseas colony on the nearby island of Kythira in around 1600–1450 BCE, and later settled on Kea, Melos, Rhodes, and Thera (modern-day Santorini), as well as Avaris, in the Nile Delta, around 1700–1450 BCE. To carry out trade and to travel to their overseas colonies, the Minoans built a fleet of ships. Some of these were used to suppress the pirates that preyed on their boats in the Aegean.

Unfortunately, the names of any intrepid Minoan sailors remain unknown. The first Minoan script was in hieroglyphs, recorded on unbaked clay tablets, few of which have survived. In around 1700 BCE, a new script, based on syllables and known as Linear A, was introduced. Neither of these scripts has been deciphered, which means that much of Minoan history is still to be revealed, including the origin of this great civilization. A study of Cretan place names suggests that the Minoans did not speak an Indo-European language, implying they were not Greeks – but exactly where they did come from is unclear.

◁ **Minoan ceramics**
The Minoans were skilled at creating beautiful jugs, plates, bowls, and other objects in ceramics. They would be painted or engraved for decoration.

The end of an era

In 1628 BCE, a massive volcanic explosion destroyed the Minoan settlements on Thera. Excavations at Akrotiri in the 1960s and '70s, however, revealed a large palace with many fine frescos that survived. Volcanic fallout damaged the palaces on Crete, but they were restored, only to be wrecked again in around 1450 BCE. No-one knows the nature of that second catastrophe – it may have been an earthquake or the arrival of the Mycenaeans – but the palace of Knossos alone remained standing.

◁ **Minoan ship**
The Minoans built simple wooden ships powered with a single sail, or by rows of oars. They used these ships to trade with other islands and to travel around the Mediterranean.

Two steering oars, located at the stern of the boat, were used to guide the ship's course.

A single mast supports a broad sail woven from linen or Egyptian papyrus.

IN CONTEXT
Bronze Age grandeur

The palace of Knossos dates from around 1900 BCE. It consists of a series of multi-storey buildings containing state rooms, shrines, archives, workshops, and numerous grain and oil stores, all arranged around a central courtyard. The palace had a sophisticated drainage and sewage system. Discovered by Arthur Evans in 1900 CE, the palace has been partially restored, revealing the splendour of its decoration and richness of its contents. The palace was taken over and occupied by the Mycenaeans in around 1450 BCE until it was destroyed by fire and finally abandoned by around 1200 BCE.

KNOSSOS PALACE, RESTORED TO SHOW SOME OF ITS ORIGINAL COLOUR

Travel in ancient Egypt

The ancient Egyptians were great travellers who sailed up the Nile and across the seas in sturdy wooden boats. They also used two-wheeled war chariots, both as weapons and for transport.

△ **Relief of Harkhuf**
Harkhuf was born on Elephantine, an island on the Nile, and became governor of Upper Egypt. Most of what we know about him comes from the inscriptions on his tomb at Aswan.

The River Nile was the lifeblood of ancient Egypt, providing water for drinking and cooking, and irrigation for the crops that grew along its banks. The annual flood fertilized the fields. Above all, the Nile was the highway of Egypt, the main artery of communication that held the country together. People travelled up and down the Nile in wooden boats driven by large sails, while cargo ships carried grain, livestock, and even huge blocks of granite to build new temples and palaces. The Egyptians named their ships, just as we do today; one commander started off aboard a ship called *Northern* and was later promoted to captain *Rising in Memphis*.

At first, the Egyptians restricted their voyages to the Nile, but by 2700 BCE, they were exploring both the Red Sea and the Mediterranean Sea. During the reign of Sneferu (2613–2589 BCE), a huge fleet of 40 sea-going ships sailed north across the Mediterranean to Byblos, in what is now Lebanon, to collect cedar wood. During the reign of Queen Hatshepsut (1478–1458 BCE), another large trading fleet sailed down the Red Sea and into the Indian Ocean to the land of Punt (see pp.22–23).

The travels of Harkhuf

A lot is known about Egyptian travellers due to the *Autobiography of Harkhuf*, a series of inscriptions on a tomb at Qubbet el-Hawa, one of a number of rock-cut tombs on the western bank of the Nile, opposite modern-day Aswan. The inscriptions record that Harkhuf, who was born nearby, was appointed governor of Upper Egypt by King Merenre I (r. 2287–2278 BCE) and oversaw the trading caravans that headed south to Nubia.

Harkhuf was tasked with increasing trade with Nubia and building alliances with local Nubian leaders, to prepare the ground for Egyptian expansion into their country. He led at least four expeditions into Nubia, and brought back on one trip someone he described to the young pharaoh Pepi II as a "dwarf" – most probably a pigmy.

On one of these expeditions, Harkhuf travelled to the land of Iyam. It is not certain where Iyam was, but it was probably on the fertile plain that opens out south of modern-day Khartoum, where the Blue Nile meets the White Nile, although some historians believe it lay in the Libyan Desert to the west of Egypt.

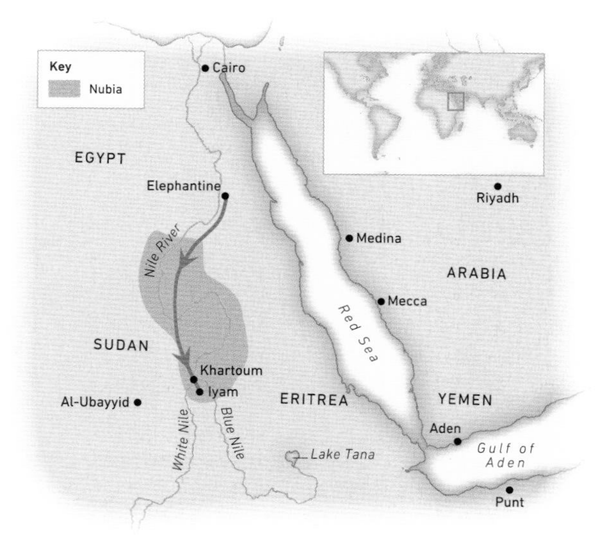

◁ **Harkhuf's travels**
Harkhuf led at least four major expeditions south into Nubia. On one of these journeys, he visited Iyam, a place many now believe is located south of modern Khartoum in Sudan.

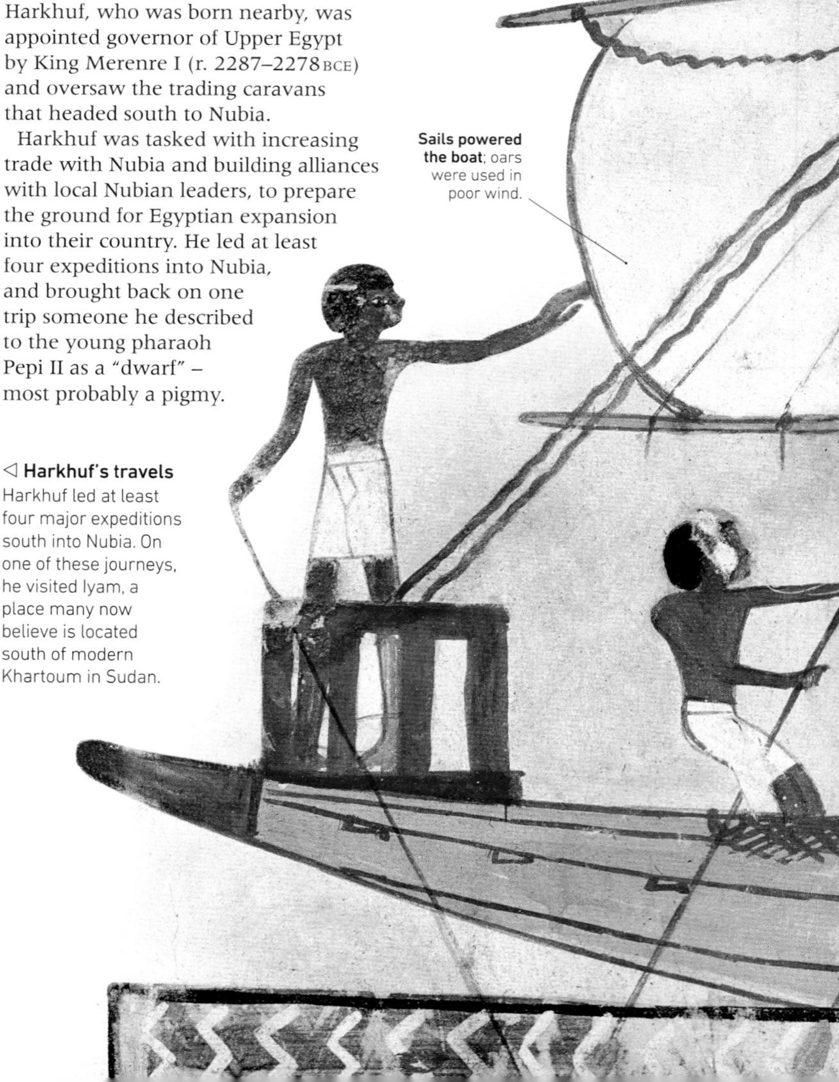

Sails powered the boat; oars were used in poor wind.

Travelling by chariot

The second mode of transport used by the ancient Egyptians was the chariot. Both the chariot and the horse were introduced into Egypt by the Hyksos invaders of the 16th century BCE. The chariot itself was a lightly built, wooden affair with two spoked wheels, manned by two soldiers and pulled by two horses.

These chariots were incredibly effective weapons of war. The chariot served as a mobile platform from which soldier-archers could fire at the enemy. During the Battle of Kadesh in 1274 BCE, near modern-day Homs in Syria, Pharaoh Ramesses II (r. 1279–1213 BCE) led about 2,000 war chariots into battle against the Hittites, who opposed him with at least 3,700 war chariots of their own. Evidence of these chariots can be seen in the tomb paintings of the pharaoh Tutankhamun (r. 1332–1323 BCE) and in the many paintings of Ramesses II at war. Away from the battlefield, chariots were used in the civilian world to transport people and goods quickly and efficiently from place to place.

◁ **Nubian traders**
The Nubians lived along the upper reaches of the Nile to the south of Egypt. They traded gold, ebony, ivory, copper, incense, and wild animals with the Egyptians. During the Middle Kingdom (2040–1640 BCE), the Egyptians expanded into Nubia and gradually took over the country.

◁ **On wheels**
The war chariot was used as a means of transport. Here, a scribe is driven by a charioteer who controls two horses using long reins.

A rear steering oar kept the boat on course.

◁ **Watercraft**
The Egyptians built small riverboats and larger seagoing boats out of cedar planks nailed together. The gaps between the planks were sealed with pitch.

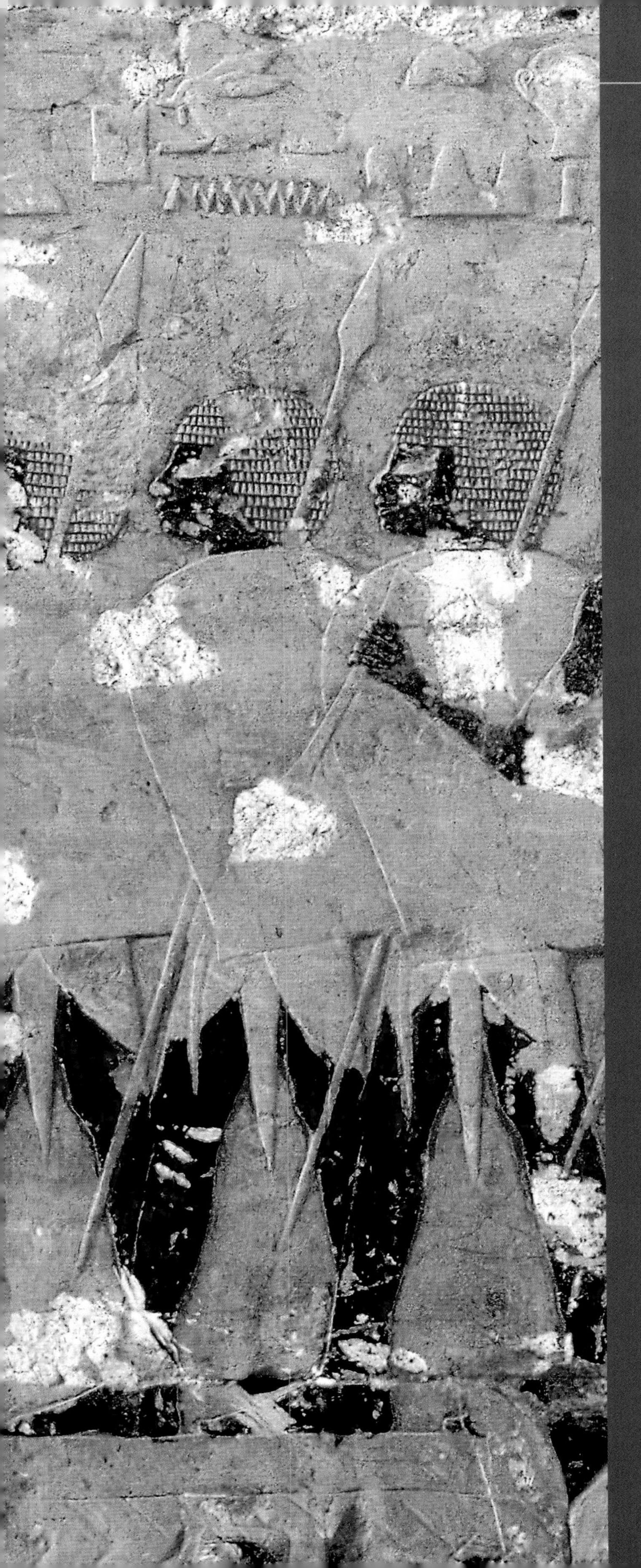

The expedition to Punt

About 3,500 years ago, Queen Hatshepsut of Egypt sent an expedition to the mysterious land of Punt. A record of the voyage survives, carved in stone.

The location of the country that the ancient Egyptians called Punt is an unsolved puzzle. The best guess is that it lay on the east coast of Africa, perhaps in modern-day Somalia or Eritrea. Punt had long been the source of highly valued incense used in Egyptian religious ceremonies, but it had remained an unknown, semi-mythical country. The story of the expedition is told in a bas-relief at the temple that was erected at Hatshepsut's burial site at Thebes (now Luxor), after her death in 1458 BCE. An oracle of the god Amun apparently instructed Hatshepsut to send the mission to Punt. Five boats crammed with Egyptian officials, soldiers, servants, and trade goods sailed south down the Red Sea, under the command of Hatshepsut's chancellor, Nehasi. Each boat had a square-rigged sail and 30 oarsmen. They kept close to the coast, as they were fearful of the dangers of the open sea. Arriving at Punt, they marched inland across a series of hills. The temple bas-relief depicts Punt as a place with lush vegetation, cattle grazing, and domed houses on stilts, accessed by ladders.

The unexpected visitors received a warm and respectful welcome from Punt's rulers, King Parahu and Queen Ati, who expressed amazement that they had found "this country unknown to men". They exchanged jewellery and weapons for a host of exotic products: elephant tusks, leopard skins, rare woods, and, above all, the trees from which frankincense and myrrh were extracted. These were carried back to Egypt and planted to produce their own incense. The bas-relief shows the Egyptians also returning with live animals, including pet baboons on leads.

The expedition sailed back northwards along the Red Sea coast, then journeyed overland by donkey caravan to the Nile Delta, before setting sail again and travelling up the Nile to Thebes. There the incense trees were planted in a garden dedicated to the god Amun. The existence of the bas-relief in Hatshepsut's temple testifies to the importance the Egyptians attached to the success of this bold mission beyond the boundaries of their known world.

◁ **Egyptian soldiers on the expedition**
These Egyptian soldiers taking part in the Punt expedition are depicted on a wall of Hatshepsut's mortuary temple. The ruined building is at Deir el-Bahri on the west bank of the Nile, opposite Luxor.

Polynesian navigators

Some of the most amazing journeys in the history of navigation were made by the people who left Asia thousands of years ago to sail across the Pacific and settle on the islands of Polynesia.

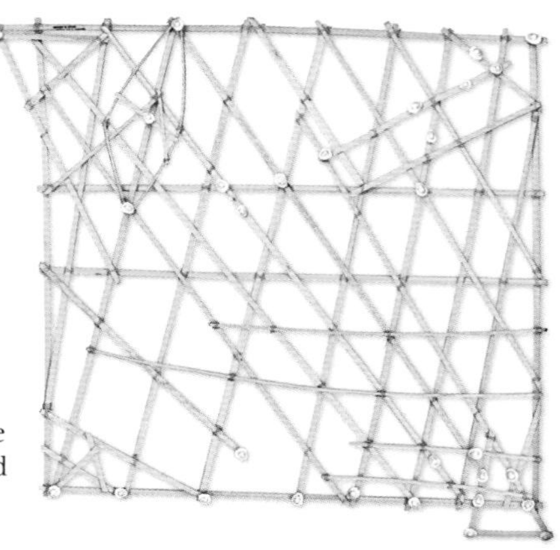

△ **Stick map**
Polynesian navigators made maps out of sticks and lengths of thread, which represented the swells and currents around islands. Shells were used to depict islands, although the precise symbols varied from one mapmaker to another.

The story of Polynesian navigation begins more than 2,000 years ago, when people, probably originally from islands in Southeast Asia, sailed across the northern Pacific and began to settle in the western islands of Micronesia. From there, they spread further, making sea journeys to settle much further south, on islands such as Samoa and Tonga. The dates of these migrations are uncertain, because the people left no written records, but it is likely that the settlement of the southern islands took place between around 1300 and 900 BCE. By the 13th century CE, all of Polynesia, bounded by Hawaii in the north, Easter Island in the east, and New Zealand in the south, had probably been settled.

Wind, sea, and sky

The people who made these perilous journeys travelled in outrigger canoes (canoes with stabilizing floats) made of wood. They had no navigational instruments, and early historians often explained their journeys by suggesting that they were blown off-course by storms. However, modern Polynesian sailors can navigate in the traditional way, without instruments, and their ancestors no doubt did the same. They worked out which way they were going using a combination of techniques: observing the stars, watching the ocean, and following the migration routes of birds. Polynesian sailors use similar methods today, but benefit from a knowledge of the positions of islands that their ancestors could not have had.

Like all early navigators, they took clues from the sky. The position of the rising and setting sun gave the compass directions, and stars provided other key points to refer to. Cloud formations also provided useful information. For example, certain types of cloud often formed above islands, and clouds moving in a V-formation sometimes

▷ **Bora Bora**
This view of the island of Bora Bora, one of the Society Islands, shows one of its peaks, which is part of an extinct volcano. The island was probably first settled by Polynesians around the 4th century CE.

indicated land – a phenomenon caused by clouds reacting to heat rising from an island's surface. It was even possible to work out something about the type of land by the colour of the cloud, since clouds tend to appear darker above forests and lighter above sand.

Migration routes and swells

Experience at sea also gave sailors an understanding of swell patterns. Swells are waves that are generated by distant winds and weather systems, and they can form regular patterns, which help sailors to navigate. The Polynesians would keep their canoes at the same angle to the swells they observed. Navigators would then pay close attention to any sudden changes in the motion of the canoe to indicate that they were veering off-course.

The movements of birds could also give navigators information about the position of land. Some Pacific birds migrate over long distances, and so could have helped sailors navigate to distant lands. For example, the long-tailed cuckoo might have helped people sail from the Cook Islands to New Zealand – and the Pacific golden plover follows a route between Tahiti and Hawaii.

▷ **Long-tailed cuckoo**
The Pacific long-tailed cuckoo is shown here with its young, and a grey warbler. It is one of the birds whose migrations follow a route between the Cook Islands and New Zealand.

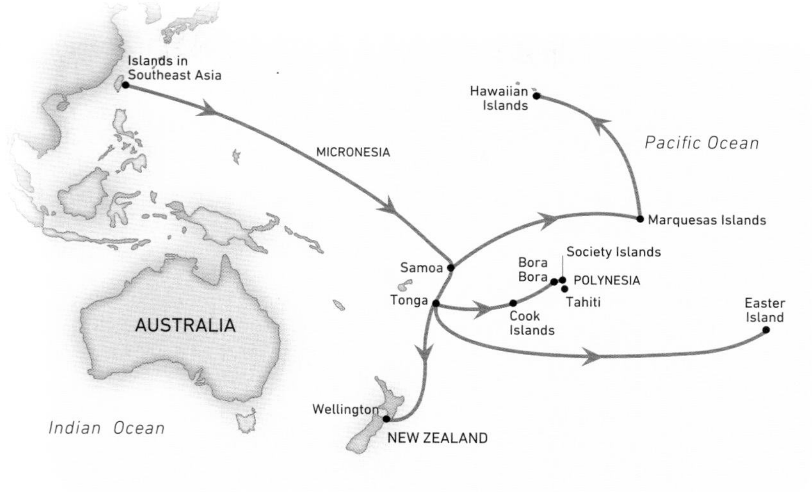

◁ **Polynesian expansion**
Polynesian navigators conquered a vast area of the Pacific known today as the Polynesian Triangle. Hawaii, Easter Island, and New Zealand form the three points of the triangle.

Inherited information

As navigators encountered islands that were new to them, they were added to their mental atlas, and their position memorized in relation to the stars. They then passed on this knowledge to their children by word of mouth and drawn charts, including diagrams made of sticks and shells. By the time of the first western contact, local sailors' knowledge was vast. British explorer Captain James Cook (see pp. 172–75) reported that the navigator Tupaia, who was from Raiatea in the Society Islands, and who helped Cook on his first expedition in the area, knew some 130 islands within a 3,200-km (1,988-mile) radius of his home.

IN CONTEXT
Voyaging canoes

The Polynesians were skilled boat builders. They used small canoes with dug-out hulls and outriggers for short journeys, but for long voyages a much larger, double-hulled vessel evolved. This typically had two plank-built hulls, connected by beams lashed together with plant fibres, and sometimes fitted with a deck between the hulls. The timbers were thoroughly caulked, probably with tree sap. Although fitted with large sails made of a fibre such as *hala*, the craft could also be paddled. Voyaging canoes could be up to 30 m (98 ft) long, and had enough space to carry several families, plus their belongings and supplies.

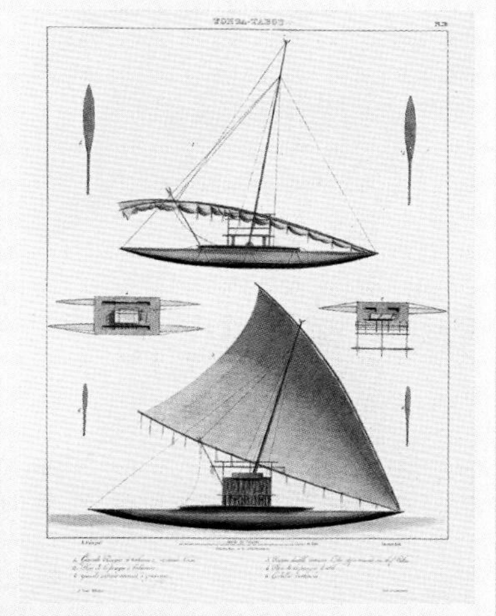

AN 1820s DRAWING OF TWIN-HULLED AND OUTRIGGER SAILING CANOES FROM TONGA

Around Africa

The Phoenicians traded across a vast maritime empire based around the Mediterranean Sea between 1500 and 300 BCE. They even sailed around Africa – the first people to accomplish this remarkable feat.

△ **Imperial purple**
The purple dye discovered by the Phoenicians was highly prized as its colour did not fade. It was later restricted to being used only by the Byzantine imperial court, hence its name.

The Phoenicians lived along the eastern coastline of the Mediterranean Sea, in the areas that are now Lebanon, Syria, Israel, Palestine, and Turkey. Their first capital was at Byblos, Lebanon, but was moved further south, to Tyre, in 1200 BCE. In 814 BCE, however, they transferred their capital and centre of operations across the Mediterranean to Carthage, in what is now Tunisia. The Phoenicians did not create a single, unified empire with vast tracts of land. Instead, they established their expanding number of ports, which were scattered around the southern Mediterranean coastline, as independent city-states. The name of this vast empire, Phoenicia, was coined by the ancient Greeks, to refer to the "land of purple," as the Phoenicians were renowned for their trade in the purple dye derived from the Murex snail. They also developed a royal blue dye from a related snail.

Trade and culture

The Phoenicians lived by trade. They sold wine to the Egyptians in terracotta jars, and bought Nubian gold in return. They also traded with the Somalis of eastern Africa. With the Greeks, they traded wood and glass, as well as slaves and the fabled purple cloth. They acquired silver from Sardinia and Spain, which they traded with King Solomon of Israel. Tin was obtained from Galicia on the Atlantic coast of Spain, and even from distant Britain. They mixed this metal with copper from Cyprus to create the harder bronze alloy.

The Phoenicians sailed the Mediterranean in tub-shaped merchant ships. As they traded on the seas, they spread the use of their alphabet throughout the region. The Greeks adopted this alphabet and then passed it on to the Romans, creating the Roman alphabet that is still in use around the world today.

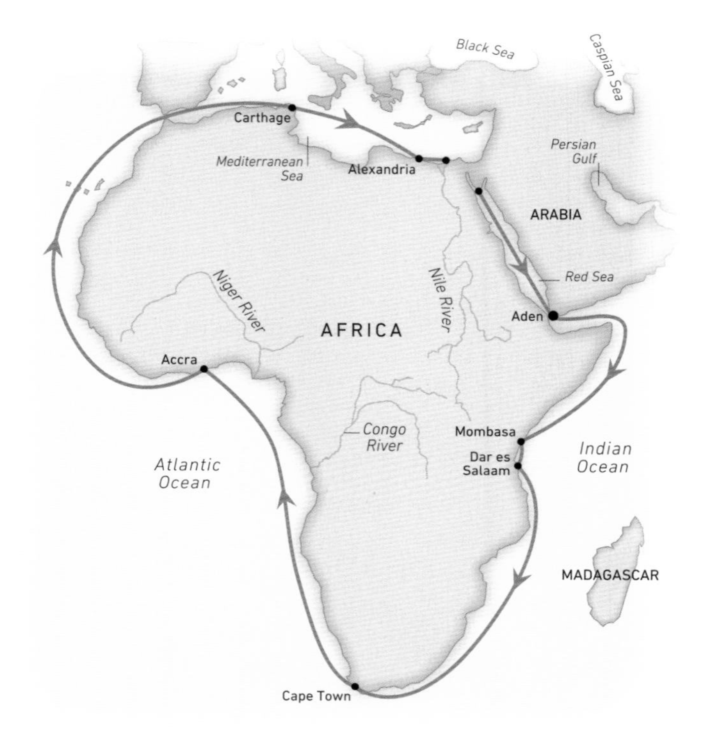

△ **Route around Africa**
The exact route taken by the Phoenicians is unknown, but it is thought they sailed south into the Indian Ocean before heading west around Africa. They then went north up the Atlantic Ocean, and east across the Mediterranean Sea.

▷ **Silver coin**
This silver coin depicts the merchant ships that brought such great wealth to the Phoenicians. It also shows a mythical sea horse.

Phoenician exploration

Although there is a lot of archaeological evidence of the Phoenician's travels, there is little in the way of written evidence. Ancient Gaelic myths recount a Phoenician and Scythian expedition to Ireland led by the king of Scythia, Fénius Farsa. A Greek *periplus* manuscript listing ports and coastal landmarks also records that in the 6th or 5th century BCE, Hanno the Navigator set sail from Carthage with 60 ships to explore northwest Africa. He set up seven colonies in what is now Morocco, and possibly reached as far south as Gabon on the Equator.

The most fascinating account of Phoenician travel occurs in Herodotus' *The Histories*, in which he records that Pharaoh Necho II of Egypt, (r. 610–595 BCE), sent a Phoenician expedition down the Red Sea around 600 BCE. For three years, the fleet sailed down the Indian Ocean and

The Greek world

Over the course of many centuries, the ancient Greeks developed one of the most advanced civilizations in the world. It was a centre of philosophy, learning, ingenious architecture, and maritime exploration.

The Greeks began to build their remarkable civilization in the 8th century BCE. The old Mycenaean script had been forgotten, so the Greeks took the Phoenician alphabet and adapted it to create a new Greek alphabet. The first written records appear around this time. By 680 BCE, they had introduced a system of coinage, indicating the growth of a merchant class. The country was divided into small, self-governing communities, a development imposed by geography – mountain ranges and the sea cut off access to neighbouring lands. At first, these communities were governed by aristocratic families, many of whom were later overthrown by populist tyrants. One such tyrant, Hippias, was deposed in Athens in 510 BCE, after which the Athenians set up the world's first democracy to govern themselves.

City-states

By the 6th century BCE, four city-states had emerged as the dominant forces in ancient Greece. They were namely Athens, Sparta, Corinth, and Thebes, each of which fought with the others for dominance. Athens and Corinth became major maritime and trading powers, while Sparta became a militarized state in which every male was a soldier, and slaves (known as helots) worked the land for the state.

However, when danger faced the country, rivalries were set aside. In 492, 490, and 480 BCE, the Persians invaded, forcing the Greeks to unite to fight the enemy. National institutions were also maintained, the most famous being the Olympic Games, established in 776 BCE. A rapidly increasing population and a shortage of arable land in the 8th and 7th centuries BCE forced many Greeks to set up colonies overseas, particularly on the coasts of the Mediterranean and the Black Sea. Although independent, these colonies retained commercial and religious ties to their founding cities in Greece. They generated enormous amounts of wealth from commerce »

△ **Navel of the world**
Delphi, in central Greece, was the seat of an oracle that the Greeks consulted when they had to make important decisions. They considered it the omphalos, or navel, of the world.

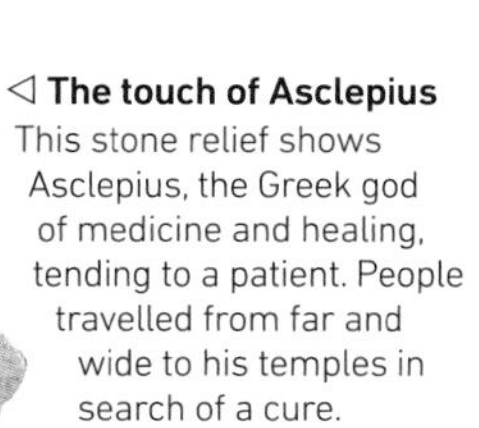

◁ **The touch of Asclepius**
This stone relief shows Asclepius, the Greek god of medicine and healing, tending to a patient. People travelled from far and wide to his temples in search of a cure.

▷ **Ancient Greece**
The first Greeks lived on the mainland and islands of what is now modern-day Greece. From 750 BCE, they established colonies in Italy, France, Spain, Turkey, Cyprus, North Africa, and the south coast of the Black Sea.

» and manufacturing, which only strengthened the Greeks' hold on the Mediterranean and the lands around it.

Sailing the world

As a maritime people, the Greeks were great travellers. They mapped the world they knew and explored as far as they could. In the 6th century BCE, Anaximander, a philosopher and geographer who lived in Miletus (in present-day Turkey) produced the first known map of the world. The map was circular, with the Aegean Sea at its centre and the three known continents of Europe, Asia, and Libya (Africa) arranged around it. Surrounding all three was a large ocean.

The main boat of the Greeks was the three-tiered trireme, which was used for both transportation and war. In this mighty vessel, they sailed the Mediterranean and beyond. The historian Herodotus (484–c.425 BCE) records that, during the 6th century BCE, a ship from the island of Samos was blown off course through the Straits of Gibraltar and arrived at the city of Tartessos on the Atlantic coast of Spain. Friendly relations were soon established with Arganthonios, the king of Spain, who encouraged the Greeks to trade with his country. On doing so, the Greeks then learned of trade routes to Britain, the River Elbe in Germany, and as far away as the Shetland Isles north of Scotland.

△ **Olympic sports**
This Greek vase shows an athlete throwing a discus. Discus throwing was one of the five sports in the Pentathlon competed for in the original Olympic Games, held in Olympia every four years from 776 BCE.

> " The barbarians pointed out... **the place where the sun lies down**... the **night** is **very short**... "

PYTHEAS, AS RECORDED BY GEMINUS OF RHODES, 1ST CENTURY BCE

Pytheas explores

One of the most remarkable travellers of the ancient world was Pytheas, a Greek from the colony of Massalia (modern-day Marseilles) in southern France. Pytheas was that rare traveller who journeys not for trade but for scientific interest – although he was certainly interested in the possibilities of trade. In around 325 BCE, he set sail for northern Europe, avoiding the Straits of Gibraltar, which the Carthaginians had closed to all but their own shipping. His route possibly took him up the River Aude and then down to the mouth of either the Loire or Garonne into the Bay of Biscay. From there, he explored Brittany and sailed across the English Channel to Britain, where he visited the tin mines of Cornwall and commented on the amber trade with Scandinavia. He then sailed north between Ireland and Britain, describing the latter as roughly triangular in shape and surrounded by many islands. In particular, he identified the isles of Wight, Anglesey, and Man, and the Hebrides, the Orkneys, and the Shetlands.

Pytheas was fascinated by all he saw. He noted that the tidal ranges of the north were much higher than those of the Mediterranean, and he was the

◁ **The Lighthouse of Alexandria**
The Pharos, or Lighthouse of Alexandria, was one of the Seven Wonders of the World. Built between 280 and 247 BCE, it stood more than 120 m (394 ft) tall, and was for many centuries the tallest man-made structure in the world.

first to suggest that the tides were caused by the Moon. He was also the first Greek to describe the north's long periods of winter darkness – reports of which had reached the Mediterranean centuries before, but had never been confirmed. To complement this, he provided accurate notes on the shortening of the summer nights before the 24-hour days, or "midnight sun", around the summer solstice.

Pytheas also reported a country of perpetual ice, and the existence of icebergs and other Arctic phenomena. He described the relationship between the Pole Star and the Guard Stars, and used a simple gnomon, or sundial, to calculate his latitude. It is unclear how far north he actually travelled, since he called his northernmost stopping point Thule, which could possibly be Iceland, or one of the smaller northern islands. Unfortunately, his own account of his voyage has not survived, although excerpts have been quoted or paraphrased by later authors, most notably Strabo in his *Geographica* (see pp.42–43).

Seven Wonders

The Greeks had many reasons for travelling. Some did so to visit shrines and temples, particularly the temples of Asclepius, the god of medicine, where healing might be found. Others travelled for work, be they lawyers, scribes, actors, or craftsmen. A few travelled for adventure or just for pleasure – the Seven Wonders of the World being the highest on a tourist's list. These were the places that both Herodotus and the scholar Callimachus of Cyrene decided, while working at the Museum of Alexandria, were the greatest man-made sites in the world.

The number seven was important because it represented plenty and perfection, and was also the number of known planets (then five), plus the Sun and the Moon. The original Seven Wonders were the Great Pyramid of Giza, the Hanging Gardens of Babylon, the Statue of Zeus at Olympia, the Temple of Artemis at Ephesus, the Mausoleum of Halicarnassus, the Colossus of Rhodes, and the Lighthouse of Alexandria. Of these, only the Great Pyramid at Giza still exists today.

▽ The face of Zeus
This ancient coin bears the face of Zeus, the Greek god of the sky and thunder who ruled as king of the gods on Mount Olympus. He was married to Hera, but had many romantic encounters, fathering, among others, Athena, Artemis, Helen of Troy, and the Muses.

IN CONTEXT
Greek passports

It is not known when the first passport was issued, but an early reference to the practice can be found in the Book of Nehemiah in the Hebrew Bible, dating to around 450 BCE. It states that Nehemiah, an official working for King Artaxerxes I of Persia, asked for permission to travel to Judea. In response, the king granted him leave and gave him a letter addressed "to the governors beyond the river", requesting safe passage for Nehemiah as he passed through their lands. The ancient Greeks had their own form of passport. This was a clay tablet stamped with the owner's name, and was used by travellers or messengers between military headquarters. The one shown here belonged to Xenokles, a border commander in the 4th century BCE.

CLAY PASSPORT BELONGING TO XENOKLES, c.350 BCE

◁ The trireme
The most important Greek ship was the trireme – a long, thin galley powered by three banks of oars. Fast and agile, it was the main warship of Greece, and played a crucial role in the wars against the Persians.

The travels of Odysseus

For ten years, Odysseus, the legendary king of Ithaca in Greece, travelled the length and breadth of the Mediterranean Sea, desperate to return home. His mythical adventures are related in Homer's epic poem the *Odyssey*.

△ **Monstrous Scylla**
This fragment of a Greek terracotta vase from c.300 BCE shows Scylla, a sea monster who lived on the rocks opposite another monster, Charybdis, in the *Odyssey*.

Since the earliest times, people have regaled each other with tales of extraordinary journeys, in which heroes have battled with monsters and demons in order to reach their destination. Homer's *Iliad* and *Odyssey* tell such tales, based on a masterful blend of fact and fiction.

The myths of ancient Greece relate that in about 1260 BCE, Zeus, king of the gods, held a banquet on Mount Olympus. Eris, the goddess of strife, was not invited, and in revenge threw into the banquet the golden Apple of Discord, inscribed with the word *kallistei*, meaning "for the fairest".

The three most beautiful goddesses, Hera, Athena, and Aphrodite, each claimed the apple, and so Zeus, who refused to choose between them, asked Paris, son of King Priam of Troy, to judge who was the fairest. As bribes, Hera offered him land and Athena tried to tempt him with skills. Aphrodite, however, offered him Helen, the most beautiful woman on Earth. Paris chose Aphrodite, who duly made Helen fall in love with him.

However, Helen was already married to King Menelaus of Sparta, so Paris had to raid Menelaus's house and steal her away to Troy. The Greeks were furious and sent an army to besiege Troy and recover Helen. The war lasted for 10 years, and only ended when the Greeks broke into Troy by hiding inside a giant wooden horse. Many heroes fought and lost their lives in the war, but Odysseus, king of Ithaca, survived. The Greek poet Homer records the last few weeks of the war in the *Iliad*, a tale about warrior values, then, in the *Odyssey*, he tells of the long, exciting voyage of Odysseus, and the personal quests and ordeals he had to endure in order to return home safely.

The *Odyssey*

Homer's epic begins after the Trojan War has ended. Odysseus and his men sail from Troy in 12 ships but are driven off-course by storms. They visit the Lotus-Eaters, a race who only eat the lotus plant, and are then captured by the Cyclops Polyphemus, from whom

▽ **Off-course in the Mediterranean**
A 16th-century map charts Odysseus's convoluted journey around the islands of the Mediterranean after the Trojan War.

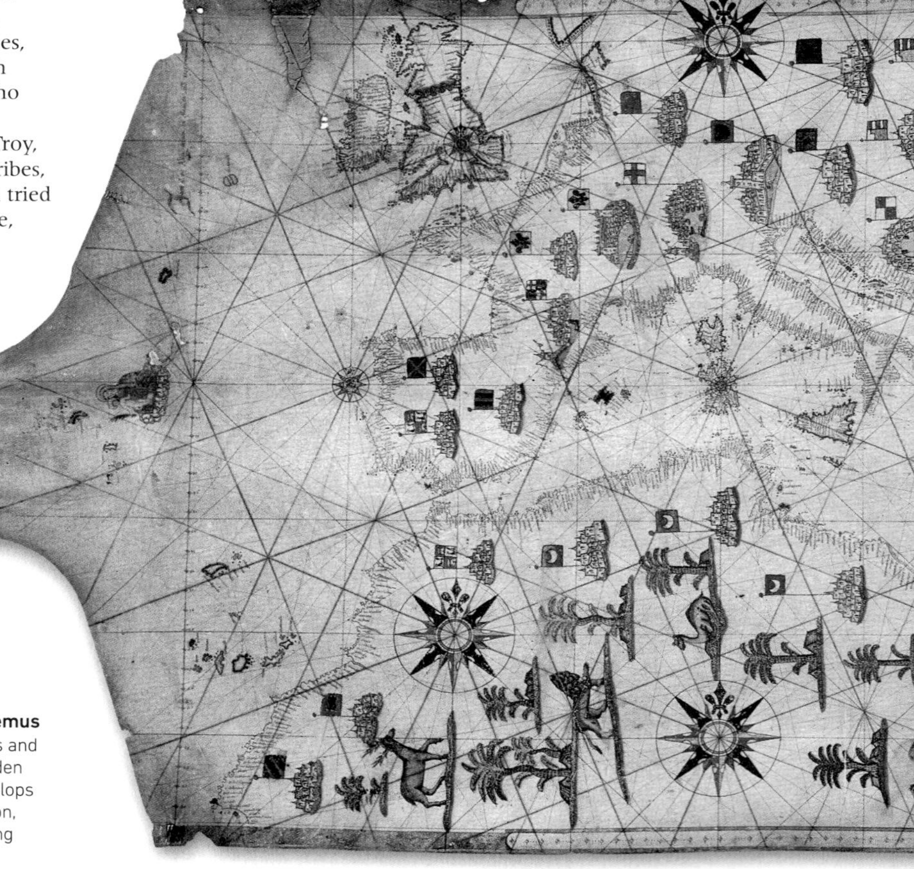

◁ **The blinding of Polyphemus**
This krater shows Odysseus and his men driving a long wooden stake into the eye of the Cyclops Polyphemus, son of Poseidon, god of the sea, who is holding them captive.

◁ **The Sirens' song**
This Roman mosaic shows Odysseus strapped to the mast of his ship to prevent him following the deathly bidding of the Sirens' song.

they escape after blinding him with a wooden stake. Foolishly, Odysseus tells the Cyclops who he is, and the Cyclops tells his father, Poseidon, god of the sea. Poseidon then condemns Odysseus to spend 10 years wandering the oceans.

Further misfortunes soon befall Odysseus. He meets the Laestrygonian cannibals and visits the witch-goddess Circe, who turns his men into swine. On her instructions, Odysseus crosses the ocean and reaches the western edge of the world, where he encounters the spirits of his dead family and friends. Odysseus and his men then avoid the bewitching call of the Sirens – beautiful but dangerous creatures who try to lure them to shipwreck on the rocks – and steer their tortuous way between the six-headed monster Scylla and the whirlpool Charybdis. Finally, the sailors are shipwrecked again and everyone apart from Odysseus is killed.

Washed ashore on the island of Ogygia, Odysseus is held captive by the nymph Calypso for seven years. Calypso falls in love with him, but as he is already married to Penelope, he spurns her. Eventually, Calypso is ordered by Zeus to release him. Odysseus is returned to Ithaca by the Phaeacians, where he is reunited with his son, Telemachus, and rids the islands of the tyrannical suitors who have been competing for Penelope's hand.

The legacy of the *Odyssey*

Although a fictional account based on myths passed down orally from generation to generation, the *Odyssey* may be partly based on events that actually took place. It also does more than recount a journey – it gives journeying itself a meaning. To go on an "odyssey" is to go on a quest – to travel in order to find oneself and come back strengthened. More than this, it means to become a hero – one who prevails against the slings and arrows of fate. Importantly, though, the *Odyssey* moves beyond the warrior-song of the *Iliad*, and focuses instead on the final goals of peace, family, and home.

IN PROFILE
Homer

The ancient Greeks believed that Homer was the author of the *Iliad* and the *Odyssey*, and that he was a blind poet born in Ionia in Anatolia (in modern-day Turkey), between 1102 BCE and 850 BCE. However, it is unclear if Homer actually existed. While one group of modern scholars does believe that a single man wrote most of the *Iliad* and possibly the *Odyssey*, another claims that both poems are the work of many contributors and that "Homer" is a label for their efforts. What is generally accepted is that the poems were originally composed and handed down orally before they were written down during the 8th century BCE.

REPLICA BUST OF HOMER MADE IN THE ANCIENT ROMAN STYLE

" **Tell me**, Muse, of the **man of many devices**, who **wandered far and wide** after he had sacked Troy's **sacred city** and saw the towns of **many men** and **knew their mind**. "

HOMER, WRITING ABOUT ODYSSEUS IN THE *ODYSSEY*

MACEDONIAN KING, 356–323 BCE

Alexander the Great

One of the greatest empires the world has ever seen was created by a young Macedonian warrior who travelled the known world in search of fame and empire and left a legacy that still survives today.

In spring 333 BCE, Alexander of Macedon stood in the city of Gordium in central Turkey. Here, the Gordian knot bound the yoke of a ceremonial wagon to a tall pole, its ends tied so that it was impossible to untie. It was believed that whoever untied the knot would become the master of Asia. Alexander took out his sword and slashed the knot with a single blow. "What difference does it make how I loose it?" he said. He was master of Asia now.

Everything about Alexander is extraordinary. Born into the royal family of Macedon in northern Greece, he became king in 336 BCE at the age of 20. His life's mission was to carry out his father's plan to attack the mighty Persian Empire in retaliation for its invasion of Greece in 480–479 BCE.

Conqueror of the world

Over the course of the next five years, he did just that. In 334 BCE, he crossed the Hellespont that separated Greece from Persian-controlled Asia. With his highly trained, professional army, he won three decisive victories over the Persians, conquering them by 331 BCE. Then the Persian emperor Darius III was killed by his cousin Bessus, who declared himself Darius's successor. Alexander's pursuit of Bessus took him across the Hindu Kush and into central Asia, where he finally killed his foe near what is now Samarkand, in Uzbekistan, in 329 BCE.

However, Alexander was keen to journey on. Like many Greeks of the time, he believed that the world was surrounded by a great ocean, so once he had reached its shores he would have conquered the known world. To find it, he crossed the Hindu Kush again in 327 BCE and invaded India. Here, he won a great victory against a vast Indian army, but he lost the backing of his troops. He tried to persuade them that the great ocean was close, but they were not convinced.

Reluctantly, Alexander led his men through the Gedrosian Desert of southern Persia to Babylon, where, ill with fever, he died aged just 32. He had conquered a vast empire and created one that was even bigger. He established cities bearing his name that still exist today, and spread Greek culture throughout the known world. Not for nothing is he known as Alexander the Great.

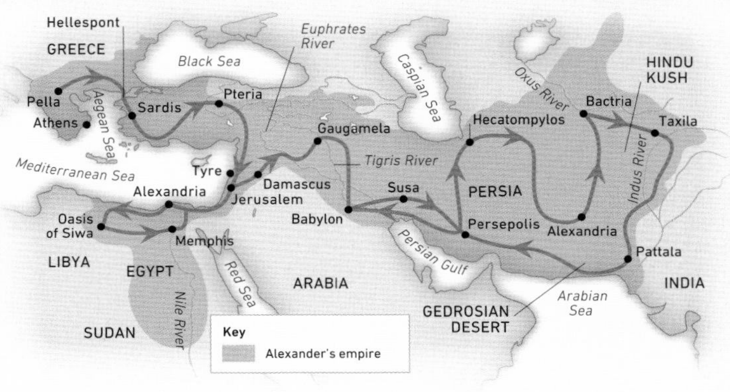

△ **Alexander's journeys**
From 334 BCE until his death in 323 BCE, Alexander travelled the breadth of the great Persian Empire, crossing into central Asia, India, and Egypt.

◁ **Alexander the Great portrait**
This detail from the *Alexander Mosaic*, c.100 BCE, shows Alexander at the Battle of Issus, where he fought his greatest foe, Darius III of Persia, and won a great victory despite having a smaller army.

◁ **Alexander and the oracle**
This bas-relief depicts Alexander dressed as a pharaoh. In 331 BCE, Alexander consulted the oracle of Amun at the Siwa Oasis. He may have believed that he was the son of Amun.

KEY DATES

- **356 BCE** Born in Pella, Macedonia.
- **343** Tutored by Aristotle.
- **336** Becomes king of Macedon after the assassination of his father, Philip.
- **335** Defeats the Thebans and takes control of Greece.
- **334** Crosses into Asia to attack the Persian Empire and wins a great battle at the River Granicus.
- **333** Defeats the Persians again at Issus and captures Palestine.

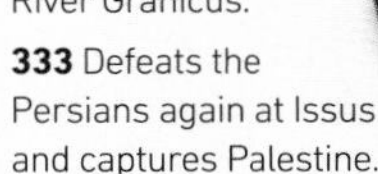

GREEK COIN OF ALEXANDER AT WAR IN INDIA

- **332** Invades Egypt and becomes Pharaoh.
- **332** Founds Alexandria and consults the oracle of Amun, claiming to be a god.

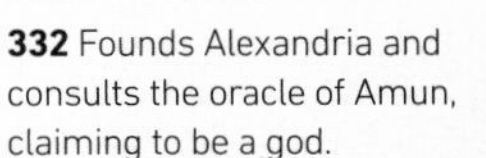

- **331** Conquers the Persian Empire.
- **329** Kills Bessus, who slayed Darius III.
- **327** Invades India.
- **325** Leaves India and marches back through the deserts of southern Persia.
- **323** Arrives in Babylon, where he dies.

ALEXANDER IN CHINA, A PLACE HE NEVER VISITED, 4TH CENTURY BCE

The travels of Zhang Qian

Isolated from the rest of the world, China first opened its eyes to the West when the diplomat Zhang Qian led an expedition in 138 BCE that led to the establishment of the Silk Road.

At the time of Emperor Wu, China knew almost nothing about the countries beyond its borders. Wu wanted to find out about them and to establish commercial ties, but he was prevented from reaching them by the hostile nomadic Xiongu tribes. They encircled the Han dynasty to the west and controlled what is now Mongolia and western China. Wu therefore needed to form an alliance with the friendly Yuezhi people, who lived beyond the Xiongu. He chose Zhang Qian, a military officer and a member of his court, to lead a diplomatic mission to negotiate with the Yuezhi.

◁ **Persian box**
This silver box from Persia was found in the tomb of Emperor Zhao Mo, who ruled southern China and northern Vietnam from 137 BCE until his death 15 years later. It is thought to be the earliest product imported into China.

Into the unknown

In 138 BCE, Zhang Qian left the capital Chang'an with about 100 men, including Ganfu, a Xiongu guide who had been captured in war, and headed west. Unfortunately, he and his party were soon captured by the Xiongu and held captive for 10 years. During this period of imprisonment, Zhang Qian was given a wife by the Xiongu, who bore him a son. He also managed to gain the trust of the Xiongu leader. In 128 BCE, he, his wife, son, and Ganfu managed to escape and made their way along the north side of the inhospitable Tarim Basin. From there, they went to Dayuan in the Ferghana Valley (now Uzbekistan), where Zhang Qian first saw the powerful Ferghana horses, and then south into Yuezhi territory. The peaceful Yuezhi people had no wish to fight the Xiongu, so instead, Zhang Qian spent his time studying their culture and economy before moving west into Daxia, the Greco-Bactrian kingdom that had declared independence from the Seleucid Empire in 250 BCE. Here, he learned about Shendu (India) to the south and Anxi or Parthia (Persia) and Mesopotamia to the west, as well as the nomadic kingdoms on the steppes to the north.

△ **Zhang Qian**
This 7th-century fresco in the Mogao caves on the Silk Road in central China depicts Zhang Qian and his party taking leave of Emperor Wu as they set out on their first expedition in 138 BCE.

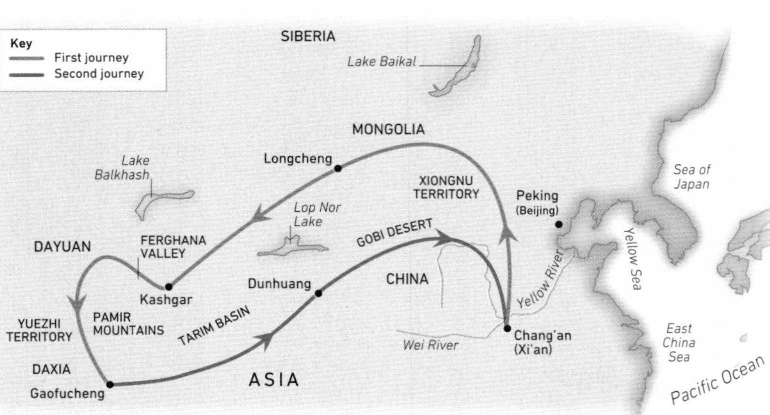

▷ **The Silk Road**
Zhang Qian's epic journey to Daxia was the first step in the establishment of the Silk Road – a network of routes that linked China to Central Asia, and eventually to the ports of the Mediterranean Sea.

Recaptured by the Xiongu

In 127 BCE, Zhang Qian and his party decided to return to China, this time travelling along the southern edge of the Tarim Basin. Once again, they were captured by the Xiongu, but their lives were again spared, as the Xiongu people appreciated Zhang Qian's sense of duty and his fearlessness in the face of death.

Two years later, the Xiongu leader died and, in the midst of the chaos and subsequent infighting for the succession, Zhang Qian, his wife, his son, and Ganfu managed to escape and return to China. Emperor Wu welcomed Zhang Qian's report that sophisticated civilizations lay to the west that China could trade with. They valued Han merchandise and wealth, and could also supply goods, such as Ferghana horses, that China wanted. Zhang Qian's report appears to have produced immediate results, as in 114 BCE, the trade routes from China into central Asia were organized into the Silk Road.

IN CONTEXT

The Ferghana horse

The Chinese called the Ferghana Valley horses "divine", for they were the finest mounts they had ever seen. After Zhang Qian reported on these horses on his return from his epic journey in 125 BCE, the Han began to import them in such large numbers that Ferghana ended the trade. In response, Emperor Hu sent an army in 113 BCE to capture some horses, but it was defeated. In 103 BCE, he sent another army of 60,000 men, who managed to acquire 10 horses, as well as an undertaking that Ferghana would supply two heavenly horses each year. Chinese military supremacy was assured.

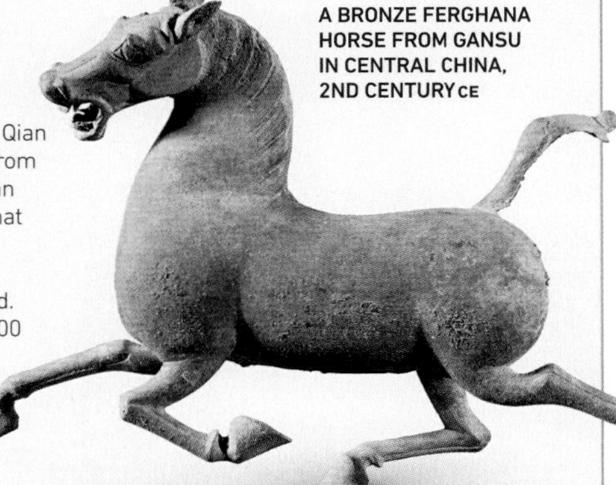

A BRONZE FERGHANA HORSE FROM GANSU IN CENTRAL CHINA, 2ND CENTURY CE

The Silk Road

The road from China to the west was now open for business, and China was no longer isolated by hostile tribes. Although Zhang Qian had been unable to develop commercial ties with the countries he had visited himself, he did try to develop trading links with India and, in 119–115 BCE, he succeeded in establishing a trading relationship with the Wusun to the far northwest of China, and thus with distant Persia. The trading future of China was now secure.

Across the Alps

In 218 BCE, Carthaginian general Hannibal Barca marched an army, including war elephants, from Spain to Italy, making an epic crossing of the Alps.

The North African city of Carthage was a bitter rival of the Roman Republic. Inheriting command of the Carthaginian army in Spain from his father, Hamilcar, Hannibal planned a surprise overland attack on Rome. Mostly recruited from the tribes of Spain, his forces also included Libyan infantry, Numidian horsemen, and 37 elephants – of a small African breed often used in battle in the ancient world. This army, more than 100,000 strong, set off from New Carthage (Cartagena) in southeast Spain in late spring. By the time it reached Roman Gaul, it was late summer and the force had dwindled to 60,000, partly because Hannibal wanted only the best troops with him for a rapid advance with lightweight baggage.

The major natural obstacle in Gaul was the Rhône. Hannibal's army, including the panicking elephants, crossed the broad river on rafts. There was some fighting with local tribes and with a Roman army sent to intercept the Carthaginians, but by November, Hannibal had reached the foothills of the Alps. His exact route over the mountains remains a matter of dispute. Mountain tribesmen harassed the Carthaginians on their ascent over the rough terrain, causing the loss of men and animals, and it took nine days to reach the snowbound crest of whichever pass was chosen.

The demoralized soldiers only agreed to start the daunting downward leg of the crossing after much exhortation from Hannibal. Horses, mules, and elephants struggled to find a footing on icy tracks bordered by steep precipices. The soldiers from warmer climes suffered in the unaccustomed freezing cold. At one point, a landslide carried away the path and a new one had to be laboriously constructed amid the ice and snow. The exhausted force finally reached flat ground after a crossing probably lasting three to four weeks.

When he arrived on the north Italian plain, Hannibal began preparations for his attempt to conquer Rome. He would campaign in Italy for 15 years and inflict terrible defeats on Roman armies, but he never achieved the total victory he sought. Eventually, it was Rome that destroyed Carthage. Hannibal committed suicide in 183/182 BCE, a fugitive from Roman power.

◁ **Hannibal and his army**

This 16th-century painting by Jacopo Ripanda depicts Hannibal and his army entering battle in suitably exotic outfits. Hannibal might have mounted an elephant's back to survey the battlefield, but the animals' main use was as a shock force in the vanguard of an attack.

GEOGRAPHER, 64/63 BCE–24 CE

Strabo

The author of one of the first books of geography ever written, Strabo provided the world with an invaluable source of information about the people and places of the Roman Empire and its neighbours.

Like many prominent people from the ancient world, Strabo is a figure about whom little is certain. What is known is that he was born into a wealthy Greek family in Amaseia in Pontus (Turkey), in either 64 or 63 BCE. Pontus had only recently been incorporated into the Roman Empire, following the death of its military leader, Mithridates VI, and Strabo's family had held prominent positions in the former government. It is also known that in 44 BCE Strabo moved to Rome, where he lived for at least 13 years and studied philosophy and grammar under the Greek poet Xenarchus and Tyrannion of Amisus, a grammarian and noted geographer.

Exploring the world

Around this time, Strabo also began to travel. In 29 BCE, he visited Corinth and the small Greek island of Gyaros. He sailed up the Nile to the temples of Philae in the Nubian kingdom of Kush in around 25 BCE, and at some point travelled even further south to Ethiopia. He also visited Tuscany, in Italy, and explored his homeland of Asia Minor. At the time, the Roman Empire under Emperor Augustus (27 BCE–14 CE) was peaceful, which enabled Strabo to travel widely.

Some time around 20 BCE, Strabo wrote his first book, *Historical Sketches*, which covered the history of the known world from the conquest of Greece by the Romans in the second century BCE. Unfortunately, only a fragment of this work survives, despite its popularity at the time (it is mentioned by several classical authors). Nevertheless, the book was only a precursor to his main work, *Geographica* – a geographical history of the world that is still in print today. It is unclear when he started work on the book, but some historians date it to around 7 BCE – others place it as late as 17 or 18 CE. The latest passage in the text that can be accurately dated refers to the death of Juba II of Mauretania, who died in 23 CE. What is clear is that Strabo worked on the book for many years and revised it as he went along.

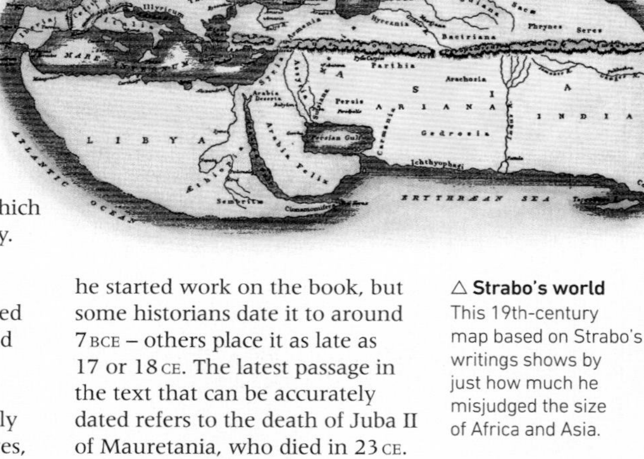

△ **Strabo's world**
This 19th-century map based on Strabo's writings shows by just how much he misjudged the size of Africa and Asia.

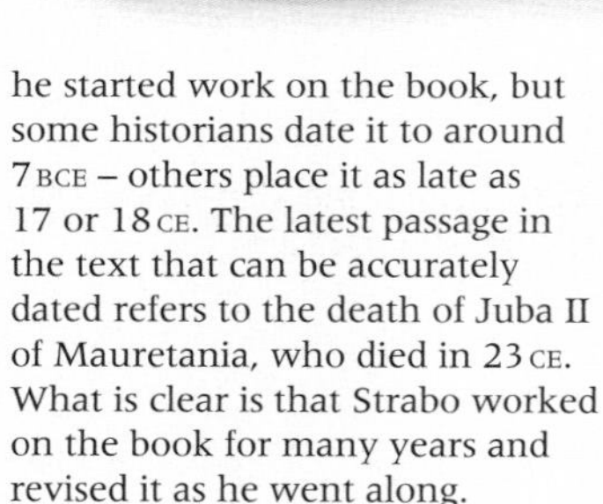

The shifting Earth

In *Geographica*, Strabo cites Eratosthenes, the Greek "Father of Geography", who worked out the circumference of the Earth, and Hipparchus, who discovered the precession of Earth's equinoxes. However, their work was purely scientific – Strabo's aim was to take such ideas and incorporate them into a practical guide for travellers, particularly politicians and statesmen. As such, the book is full of insights, many of which are original. For example, Strabo was the first person to describe the formation of fossils and the effects of a volcanic eruption on a landscape. Most importantly, he also discussed the question of why marine shells are buried in the earth so far above and away from the sea. He concluded that it was not the change in sea levels that made these shells move, but shifts in the land itself, which had slowly risen up above the sea. For this thought, which anticipates plate tectonics, Strabo therefore deserves to be known not only as a great traveller, but also as one of the founders of modern geology.

◁ **The Great Pyramids**
In around 25 BCE, Strabo sailed up the Nile, passing the Great Pyramids on his way to the temples of Philae, which now lie beneath Lake Aswan in southern Egypt.

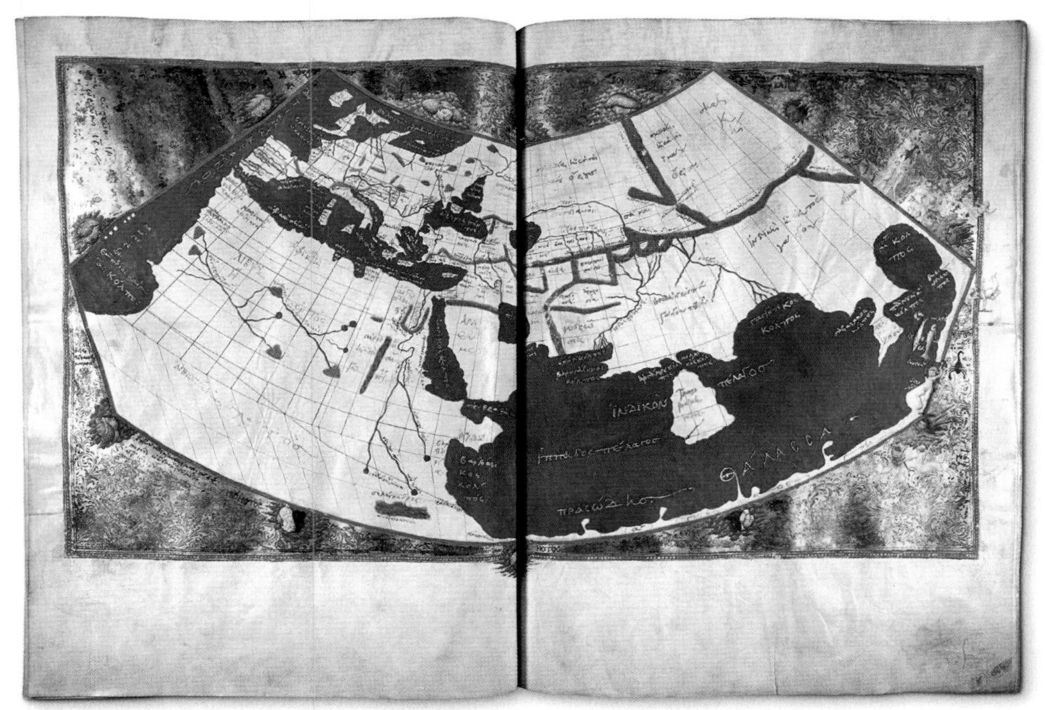

◁ **The known world**
Engraved in 1482 by Johannes Schnitzer, this map of the world shows Ptolemy's understanding of the *ecumene* – the known or inhabited world. Europeans were as yet unaware of the existence of the Americas and Australasia, the Pacific Ocean was still unknown to them, and European sailors had yet to round the southern tip of Africa.

ON TECHNOLOGY
The quadrant

This early 16th-century engraving shows Ptolemy, a quadrant in his hand, taking measurements with the assistance of Urania, the Greek muse of astronomy. The quadrant is a simple instrument that can measure angles up to 90°, making it possible to measure the altitude of a heavenly body. Ptolemy would not, however, have had a quadrant like this, which was not invented until the early 1700s. He did, however, build an early, primitive type of a quadrant with which he could take daily measurements of the height of the Sun at noon.

PTOLEMY GUIDED IN HIS WORK BY URANIA, THE GREEK MUSE OF ASTRONOMY

" I know that I am **mortal**... but when I trace... the **windings to and fro**... of the **heavenly bodies**, I no longer touch **earth** with my **feet**. "

PTOLEMY, IN THE *ALMAGEST*, 150 CE

drew up a catalogue of 48 major constellations. He also recorded all the data that was necessary to work out the positions of the Sun, the Moon, and the planets, the rising and setting of the stars, and the eclipses of the Sun and the Moon.

Ptolemy's attention to detail can be seen throughout his major work, the *Geographia*. He based it on an earlier treatise by the Greek geographer Marinus of Tyre, a work that has since been lost. Ptolemy's own book consists of three sections divided among eight books. Book I concerns cartography, while Book II to the start of Book VII is a gazetteer providing the longitude and latitude of every place in what was then the known world. Ptolemy measured latitude from the equator in Africa, but stated it in terms of hours: the equator was set with 12 hours of midsummer daylight, the Arctic with 24. The prime meridian, from which longitude is measured, runs through the Fortunate Islands, in the Atlantic Ocean, which are thought to be the Canary Islands. The rest of Book VII provides details on the three different projections or perspectives needed to construct a map of the world. Finally, Book VIII contains a series of regional maps.

◁ **The Earth-centred sphere**
In this 15th-century painting by Justus van Gent and Pedro Berruguete, Ptolemy is shown holding an armillary sphere, a model of the objects in the sky with the Earth at its centre, as he believed it to be.

The legacy of Ptolemy

Ptolemy's maps were more accurate than any others in existence at the time, although his calculations of longitude stretched the world considerably from east to west. Calculating longitude, however, remained problematic to work out until Galileo tackled the problem in the 17th century. The translation of *Geographia* into Arabic in the 9th century, and then into Latin in 1406, ensured that Ptolemy's influence lasted for a thousand years. His ideas were not superseded until the 16th century.

LA DIN

TRADE AND
CONQUEST
400–1400

TRADE AND CONQUEST, 400–1400

Introduction

In medieval times, all long-distance travel was an adventure, undertaken at risk of life and fortune. Contemporary maps were full of blank spaces – unknown territories that were marked as sites of mythical wonders. To a Christian, it was credible that if they left Europe, they might stumble upon the location of the Earthly Paradise or the kingdom of the legendary Prester John. Travellers' tales, among the most avidly read literature of the time, inextricably mixed fantastic invention with eyewitness reports.

Yet, despite the limits of formal geographical knowledge, the world was knitted together with trade routes that stretched across continents. Emperor Charlemagne in western Europe exchanged gifts with the caliph of Baghdad; the Roman popes sent ambassadors to the Mongol rulers of Central Asia. A remarkable number of people took to the primitive roads and the perilous seas in search of profit or adventure, conquest or salvation.

Faith and trade

Throughout the age, thousands of people travelled for their religion. In the 7th century, the Chinese Buddhist Xuanzang embarked on the long journey to India in search of the roots of his faith. The creed of Islam, established in the same century, obliged its believers to make the hajj, or pilgrimage to Mecca. As Islam spread, annual caravans of hajji travelled towards the Red Sea on routes across Asia and northern Africa. Christians expanded their own tradition of pilgrimage to sacred sites, drawing the faithful to a wide range of holy places near and far. The greatest wish for many Christians was to make the long journey to Jerusalem.

A 13TH-CENTURY ARAB MANUSCRIPT SHOWS A TYPICAL MERCHANT'S DHOW OF THE TIME

AS TODAY, MUSLIMS WERE OBLIGED TO MAKE THE HAJJ AT LEAST ONCE IN THEIR LIVES

THE LATEST NAVIGATIONAL AIDS ARE TESTED IN *THE TRAVELS OF SIR JOHN MANDEVILLE* (1357)

> " It **behoves a man** who **wants to see wonders** sometimes to **go out of his way**. "
>
> *THE TRAVELS OF SIR JOHN MANDEVILLE*, c.1350

Other voyagers were not so peaceful. None travelled more widely than the Vikings, whose thirst for adventure matched their lust for land and plunder. Viking warbands eventually sailed beyond the limits of the known world to Iceland, Greenland, and the edge of the Americas. The European Crusader knights also voyaged with sword in hand, wresting Palestine from the Muslims in 1099. The establishment of Crusader states in the Eastern Mediterranean became another stimulus to travel and trade, making the fortunes of the Italian maritime cities of Venice and Genoa.

A golden age

In the 13th century, the conquests of the Mongols under Genghis Khan and his successors imposed a measure of political unity on an area stretching from the Middle East to China, utilizing the ancient Silk Road trade routes. In the same period, camel caravans conducted a busy trade across the Sahara Desert to the source of gold in the mysterious kingdom of Mali. Seaborne commerce, benefiting from the introduction of the astrolabe and the compass as aids to navigation, linked the ports of the Red Sea to East Africa, India, Indonesia, and, ultimately, China. The Venetian Marco Polo was far from unique in making the journey from Europe to Mongol-ruled East Asia. The Muslim traveller Ibn Battuta, driven by sheer curiosity, was able to ramble the world from North Africa to Beijing. However, by the end of the 14th century, this golden age of Eurasian travel was fading as war and insecurity imposed new barriers on travel. The next great journeys would be made across the oceans.

KING (AND SOON SAINT) LOUIS IX OF FRANCE, WHO DIED ON THE EIGHTH CRUSADE IN 1270

GENGHIS KHAN, WHOSE CONQUESTS UNIFIED AN AREA STRETCHING FROM THE MIDDLE EAST TO CHINA

VENETIAN TRAVELLER MARCO POLO USED THE SILK ROAD TO REACH MONGOL-RULED EAST ASIA

△ **The Mahabodhi Temple, Bodh Gaya**
This 19th-century watercolour shows the Mahabodhi Temple in India. Here, while meditating under a peepul (fig) tree, the Buddha found the answers to his questions about suffering and achieved the state of enlightenment.

Xuanzang's journey to India

Chinese scholar Xuanzang's expedition to India was one of the greatest overland journeys of its time. He discovered much about a little-known country and took back to China hundreds of texts that helped to spread Buddhism in his homeland.

The Buddhist scholar Xuanzang was born in Goushi (in what is now Henan province), China, c.602. From an early age, he read widely, especially religious texts, and by the age of 20, he had become a Buddhist monk. His reading convinced him that the Buddhist scriptures that were available in China were either incomplete, or poor translations of the Indian originals. So, in 629, he decided to travel across Asia to India to find authentic copies of the Buddhist scriptures that he could take back to China and translate. He did this in spite

▷ **Xuanzang**
The scholar and traveller is depicted on the wall of the Dacien Temple, in China's Shanxi province, which was built to house scriptures he brought back from India.

> " I would **rather die** trying to take the **last step westward than** try to make it **back to** the **East alive**. "

VOW MADE BY XUANZANG

of the fact that China's borders were officially closed, meaning that he risked being a fugitive in his country when he returned.

Xuanzang's journey turned out to be an epic 17-year expedition. First, he had to travel across northern China. Taking the Silk Road across the Gobi Desert, he followed the Tian Shan mountains into dangerous territory where he narrowly escaped being robbed. He then travelled along the Bedel Pass into Kyrgyzstan, where he met the great Khan of the Göktürks. The Göktürks were a league of nomadic Turkic peoples who had been at war with Tang China, but now had peaceful relations with the Chinese. Xuanzang then continued southwest through Uzbekistan to its capital, Tashkent, pushing on westwards to Samarkand, where there were many abandoned Buddhist temples. Travelling further south, he came to Termez, where there was a monastery that was home to more than 1,000 Buddhist monks.

Reaching India

In Kunduz, Xuanzang was advised to make a diversion westwards to Afghanistan, where he encountered another large Buddhist community. Here he began to collect Buddhist texts and found many Buddhist relics. One local monk, Prajñakara, helped him study some of these early scriptures and went with him to Bamyan, where he found a vast number of monasteries as well as the famous large Buddha statues carved into a rock face.

Travelling on to Gandhara (Kandahar), Xuanzang took part in a religious debate and met Hindus and Jains for the first time. Here, near the border of modern Pakistan, Xuanzang felt himself to be in the Indian world, and close to his goal at last. Leaving Gandhara, he undertook a long journey around India, meeting local rulers, staying in monasteries, talking to Buddhist monks, and collecting Buddhist texts.

Among the highlights of Xuanzang's period in India were his visits to major north Indian sites associated with the Buddha himself. This included a journey to what is now Bangladesh, where he found some 20 monasteries devoted to both the Hinayana and Mahayana traditions of Buddhism. He also stayed at the monastery at Nalanda, eastern India, a renowned centre for the study of Mahayana Buddhist scriptures. At Nalanda, Xuanzang studied various subjects, especially Sanskrit, and found an inspirational master in the monastery's abbot, the renowned philosopher Silabhadra.

Xuanzang returned to China in 645. He had travelled some 16,000 km (9,942 miles) by horse and camel, collected 657 Buddhist texts, been triumphant in debates with Buddhist monks and Hindu teachers, and even, on one occasion, converted a group of criminals to Buddhism. He spent the rest of his life translating the texts he had collected into Chinese, making it possible for Buddhism to spread widely across the country. Emperor Taizong, impressed by his great achievements, not only pardoned him for making his journey but also appointed him a court adviser.

◁ **Buddhas of Bamyan, Afghanistan**
Carved into a cliff face in Bamyan, central Afghanistan, in the 6th century, the largest of these statues was 53 m (174 ft) tall. It was destroyed by the Taliban in 2001, but a reconstruction project is underway.

IN CONTEXT
Buddhist scriptures

The principal Buddhist scriptures originated in India and consisted either of the Buddha's words, or those of his close disciples. However, many of the texts available in China were incomplete or poor translations of the Sanskrit originals. Xuanzang felt compelled to travel to India, which was the source of scriptures such as the important group called *Prajñāpāramitā* (*Perfection of Wisdom*), and to translate them anew. When he returned to China, Emperor Taizong helped him set up a translation centre in the capital Chang'an, where Xuanzang, together with a team of scholars and students, set about translating the material he had brought back from his journey.

CHINESE TRANSLATION OF *PERFECTION OF WISDOM*, BY CALLIGRAPHER ZHAO MENGFU

Journey to the West

Journey to the West is a novel published in China in the 16th century, often attributed to Wu Cheng'en. The story is based very loosely on Xuanzang's great journey across Asia.

Although based on the Buddhist scholar Xuanzang's journey to India, *Journey to the West* is not a realistic account but incorporates many mythological and religious elements into the tale. The Buddha himself is at the heart of the narrative. Concerned by the sinful behaviour of the people of China, he singles out Xuanzang as the person best suited to go to India to bring back the Buddhist scriptures needed to instruct the people how to live moral lives.

In the story, Xuanzang is accompanied on his journey by four disciples. These are mythical figures who are persuaded or inspired by the Chinese Buddhist deity Guanyin to travel with Xuanzang and protect him, enduring the dangers of the journey as a way of atoning for their sins. They take animal form but display human frailties. The most prominent of the four disciples is the violent and cunning Sun Wukong, or Monkey, a shape-shifting immortal who led a rebellion in heaven. The other three are: Zhu Bajie, a rapacious pig; Sha Wujing, a quiet and reliable river ogre; and Yulong, the Dragon King of the Western Sea, who takes the form of the white horse that carries Xuanzang on his travels.

The main section of the narrative describes the trip across Asia and consists of a series of adventures and mishaps in which the five travellers encounter monsters, evil magicians, and challenges such as flaming mountains. They emerge from these hazards largely unscathed and eventually reach India. There, they have more adventures before finally reaching their goal, a site called Vulture Peak, where they receive the scriptures from the Buddha himself and make their way back to China.

Containing both prose and poetry, *Journey to the West* combines comedy, adventure, and religious content in a unique way. This mixture of materials has made it one of the best-loved classics of Chinese literature. It has also helped to keep alive the memory of the great Xuanzang (see pp.58–59) and his real-life journey.

▷ **Pilgrimage to India**
This mural depicts Xuanzang being carried by the Dragon King, Yulong, who has taken the form of a white horse. Behind them are the Monkey King, Sun Wukong, and Zhu Bajie, a half-human, half-pig monster.

△ ***Navicella***
This 1628 oil painting is based on a mosaic by Giotto di Bondone, which is located on the façade of St Peter's Basilica. It depicts Jesus walking on water through a storm on Lake Galilee to join his disciples on a boat.

Kingdom of Heaven

Travel played a key role in the Gospels, from Mary and Joseph's trip to Bethlehem for a census, to the visit of the Magi from the East. Journeys were also crucial to the early spread of Christianity.

The Gospel accounts depict Jesus as a wandering Jewish preacher who travelled throughout Palestine, largely on foot, with a band of disciples. He was crucified by the Roman authorities in around 33 CE, but his teaching was continued by his followers who travelled enormous distances around the world by sea and by land.

The spread of early Christianity was facilitated by the Roman road system. As Rome expanded and conquered new territories, well-patrolled roads enabled people to travel longer distances than ever before. Commerce followed, leading to prosperity and economic stability. Emperor Augustus (r. 27 BCE–14 CE) boasted that the world had never known such peace until his reign.

△ **Constantine I**
Roman Emperor from 306–337 CE, Constantine reversed centuries of imperial persecution, and decreed tolerance for Christians, allowing them to move freely throughout the Roman Empire.

Christianity initially put down roots in towns and cities, rather than in rural areas. However, practising Christianity was forbidden by Emperor Nero (r. 54–68 CE) and until the reign of the first Christian emperor, Constantine I, nearly 300 years later, Christians were persecuted throughout the Roman Empire. Furthermore, like Judaism, Christianity was a religion of sacred scriptures, and many ancient Greek and Roman authors disputed elements of the developing creed. Yet despite external threat and internal conflict and divisions, Christianity continued to spread rapidly, and in 301 CE, Armenia became the first nation to officially embrace the religion.

The spread of Christianity

Pilgrimage soon became a distinctive part of Christianity. Constantine built several churches over the tombs of saints in Rome, which became a focus for religious travellers. His mother, Helena, made a pilgrimage to Palestine in around 325 CE. Another notable pilgrim was a woman called Egeria who, in a letter known as the *Itinerarium Egeriae* (*Travels of Egeria*) recounted her visits to many sites in the Holy Land.

As a result of Roman traders travelling in the 1st century, Christianity spread rapidly through the British Isles from the 2nd to the 5th centuries. Monasteries were established by monks who were noted for their learning and writing.

The Western Roman Empire collapsed in the late 5th century, partly because of numerous barbarian invasions, but a Roman emperor continued to rule the East from Constantinople. This meant that roads were no longer maintained as they had been and aqueducts fell into disrepair. Many routes became dangerous and the ports and seas were under threat of piracy.

IN PROFILE

Saint Paul

Paul of Tarsus (c.5–67 CE) was one of the earliest Christian missionaries to travel long distances on Roman roads, preaching about Jesus. Greek, Latin, and Semitic languages were the dominant tongues in the area of the Mediterranean basin. As a citizen of the Roman province of Cilicia, St Paul was fluent in Hebrew and spoke the Greek *koine* dialect made popular by the armies of Alexander the Great. Translators were also available in all the ports and could be easily hired in large towns.

St Paul recorded his journeys across Asia Minor and Europe in the *Acts of the Apostles*. The cause of his death is uncertain, but it is thought that he was executed in Rome at Emperor Nero's command.

MOSAIC DETAIL SHOWING ST PAUL, ST SOPHIA'S CATHEDRAL, KIEV

◁ **Rylands Library Papyrus**
In order to spread Christianity, the gospels, the letters of St Paul, and other Christian writings were copied by hand and then disseminated. The Rylands Library Papyrus is a fragment of a copy of St John's Gospel from around the 2nd century CE.

Christianity in Europe

During the 6th and 7th centuries, monks from Britain travelled throughout Europe, founding monasteries and centres of learning. They also re-established literacy, which had lapsed in some places. This self-imposed exile from home was called "white martyrdom".

By the 8th century, Christianity had taken hold in Europe. Many countries were ruled by Christian monarchs, and Greek missionaries, such as Methodius and Cyril, had created the Cyrillic alphabet to translate the Bible. This same alphabet still forms the basis of modern-day Russian.

▽ **Hagia Eirene**
From the mid-4th century, Christians began to build churches as public places of worship. Like Hagia Eirene in Istanbul, they were rectangular in shape with a rounded end, similar to the Roman basilicas used as law courts.

" Remember **Christ's disciples**. They **rowed** their **heavy ships** to shore, then abandoned everything to **follow Christ**. "

ÆLFRIC OF EYNSHAM, ENGLISH ABBOT

The spread of Islam

In 632, a new religion emerged in eastern Asia. Within a few years, the armies of Islam had established an empire that stretched from the borders of India in the east to the Atlantic Ocean in the west.

Muhammad, the founder of Islam, was born in the western Arabian city of Mecca in around 570. He was orphaned at an early age and raised by his uncle. He used to hide away in a mountain cave for nights on end to pray, and in around 610 he began to receive visits from the angel Gabriel, who revealed to him the first verses of what became the Qur'an, Islam's holy book. As the revelations continued, Muhammad began to preach their message, stating that he was a prophet and messenger from God. After being warned of a plot to assassinate him, Muhammad and his followers fled Mecca for Medina in 622. This *Hijra*, or flight, marks the start of the Islamic calendar.

In Medina, Muhammad united all the tribes under his leadership, and in 630 he returned to Mecca with an army of 10,000 Muslim converts. By the time of his death in 632, most of the Arabian peninsula had converted to Islam.

Across the world

Muhammad's successor, Abu Bakr (r. 632–34), completed the religious and political union of the various Arab tribes, but under the next two caliphs – Umar and Uthman – a rapid expansion began that changed the face of the region. Muslim armies took Damascus in 635, and Jerusalem in 638, and defeated the Persian Sasanian Empire in 642. To the west, Muslim armies conquered Egypt in 640, and Tunisia by 680, reaching the Atlantic coast of Morocco by 683. In 711, Muslim armies attacked and conquered Spain. By now, the Islamic empire stretched from the Atlantic coast to the borders of India, creating the largest empire the world had even seen.

The limits of its expansion became clear when, between 670 and 677, and again between 716 and 717, Muslim armies failed to capture Constantinople, the capital of the Byzantine Empire. Expansion north over the Pyrenees into France was halted by the Franks at Poitiers in 732, although success over the Chinese at the battle of the River Talas in 751 opened up Central Asia to Muslim control.

△ **The Qur'an**
Muslims believe that the Qur'an was revealed by God to Muhammad through the angel Gabriel. The revelations were collected into one book by Muhammad's successor, Abu Bakr, in 634. The Qur'an consists of 114 *surah* (chapters) divided into numerous *ayah* (verses).

▽ **The Islamic world**
Soon after its birth in 632, Islam spread throughout Arabia, reaching Palestine in 634, and Persia in 643. Westwards, it swept across Egypt and North Africa, crossing over into Spain and southern France in 711.

Key
Territories under Islam
Atlantic Ocean
EUROPE
FRANCE
THE PYRENEES
Mediterranean Sea
Caspian Sea
Aral Sea
CENTRAL ASIA
CHINA
Black Sea
Talas River
SPAIN
TUNISIA
Córdoba
Tangier
Fez
MAGHREB
MOROCCO
LIBYA
EGYPT
Cairo
Jerusalem
Damascus
PERSIA
Indus River
Persian Gulf
INDIA
Medina
Mecca
SAUDI ARABIA
Red Sea
Arabian Sea
AFRICA

◁ ***Mihrab***
A semi-circular niche in the wall of a mosque, a *mihrab* shows the direction of the holy Kaaba, or "House of Allah," in Mecca. This highly decorated *mihrab* is inside the late 15th-century Bara Gumbad Mosque in New Delhi, India.

Early inventors

The early astrolabes combined a planisphere (a star chart that used a rotating, adjustable disc to work out which stars were visible that day) and a dioptra, or sighting tool, to measure angles. By such primitive means, the angle of the Sun, Moon, and stars, and their position in the heavens relative to the ground, day or night, could be worked out. Armed with this knowledge, travellers could then plot their latitude to find out where they were on Earth.

It is not completely clear who invented the first mechanical astrolabe. Credit is usually given to Hypatia of Alexandria, due to a letter written to her by one of her students, Synesius of Cyrene. However, this seems unlikely, given the fact that Hypatia's father, Theon of Alexandria, had already written a treatise – sadly now lost – on the working of the astrolabe. It is also claimed that Ptolemy had used an astrolabe in his calculations for *Tetrabiblos*, his book on astrology, some two centuries earlier. Apollonius of Perga, a Greek astronomer who lived in what is now southern Turkey, is thought to have invented an astrolabe around 350 years before Ptolemy.

All the individuals attributed to the invention of the astrolabe were Greeks living in the Roman Empire, and the Greeks continued to use and develop astrolabes during the succeeding Byzantine Empire. John Philoponus, a Byzantine philosopher from Alexandria, wrote the earliest extant treatise in Greek on the instrument in around 530. By the mid-600s, Severus Sebokht, a scholar and bishop who lived in Mesopotamia, wrote a major treatise on the astrolabe, and described it as being made of brass, indicating that by then it was an instrument of some complexity.

△ **Celestial globe**
Made by Ibrahim ibn Said al-Sahli in Valencia, Spain, in around 1085, this globe of the heavens is believed to be the oldest such object in the world.

Muslim advances

In the 640s, Muslim armies from Arabia occupied most of the Byzantine Empire, and knowledge of the astrolabe passed to the Arabs. They further developed it, adding new dials and discs to make it more complex and accurate. In its most advanced form, the astrolabe consisted of a *mater*, or base plate, marked with a degree scale around its outer rim. On top of this, two rotating plates and a ring were aligned to work out the angles of the known stars, and thus determine latitude. If the latitude was already known, the astrolabe could be used in reverse to calculate the time.

While the Arabs used the astrolabe as a navigational tool, they also discovered a more specific purpose for the instrument. When praying, Muslims have to face the holy city of Mecca, the birthplace of Muhammad. By using the astrolabe, Muslims could calculate *qibla* – the direction of Mecca. Used both as a secular and a religious instrument, the astrolabe soon became indispensable for travellers and worshippers alike.

> " It uses as its **servants geometry** and **arithmetic**, which it would not be improper to call **a fixed... truth**. "
>
> SYNESIUS OF CYRENE (c.373–c.414 CE), WRITING TO HIS FRIEND PAEONIUS

▽ **The Ptolemaic universe**
This illustration from 1708 shows Ptolemy's geocentric view of the universe, in which Earth is orbited by the Sun, Moon, and all the planets.

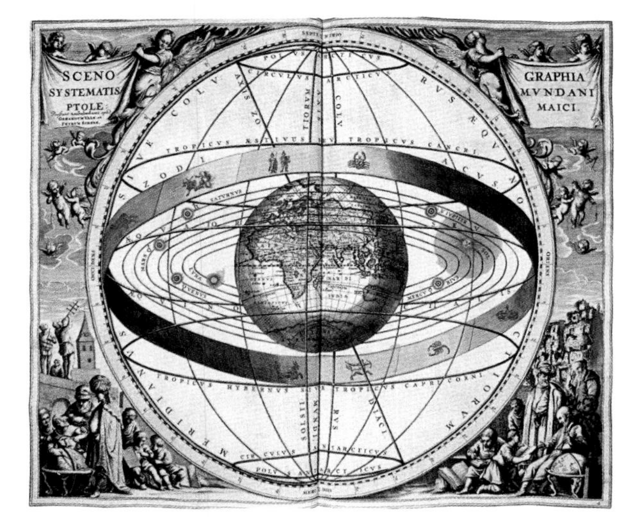

Voyages of the Vikings

The most wide-ranging voyagers of medieval Europe, Viking bands travelled as far as Iceland, Constantinople, and North America, driven by a hunger for land, plunder, trade, and a thirst for adventure.

▽ **Viking shields**
Viking warriors carried round shields made of hand-painted wooden planks riveted together. They fought with a shield in one hand and an axe, sword, or spear in the other.

The Vikings hailed from Denmark, Sweden, and Norway. In the 8th century, these were turbulent pagan areas of Europe that stood in contrast to relatively orderly Christian states such as Anglo-Saxon England or Charlemagne's empire in France and Germany. Scandinavian society was dominated by warbands – groups of warriors following a leader distinguished by his strength, courage, and success in battle. These warbands fought one another both at home and overseas. Shipbuilders in Scandinavia had developed the longship – a fast, shallow-draught war vessel capable of sailing on both seas and rivers. In 793, a Viking warband crossed the North Sea and carried out a ferocious raid on the monastery at Lindisfarne, an island off the coast of northeastern England. Two years later, Vikings sailed around the north of Scotland, raiding monasteries at Iona in the Hebrides, and others off the coast of Ireland. They also occupied the Shetland and Orkney Islands, which then became permanent bases for further raids that were carried out on the British Isles.

▽ **Viking travels and trade routes**
Vikings ventured across the North Sea to the British Isles and down Europe's Atlantic coast; south along the rivers of Russia and Ukraine to the Black Sea; and across the Atlantic to Iceland, Greenland, and America.

From around 830, the scope of Viking activity expanded dramatically. Their raiding parties, of increasing size, struck around the coast of northwest Europe and advanced deep inland along major rivers such as the Rhine, Seine, and Loire. A permanent Viking settlement on the island of Noirmoutier, off the mouth of the Loire, provided the base for a famous expedition launched in 859 by the warband leaders Björn Ironside and Hastein. Sailing their longships southwards along the Atlantic coast of the Iberian Peninsula, the Vikings entered the Mediterranean through the Straits of Gibraltar. There they made a prolonged stay, plundering both Muslim and Christian states in Spain, North Africa, the south coast of France, and Italy, before returning in triumph to the Loire weighed down with booty, in 862.

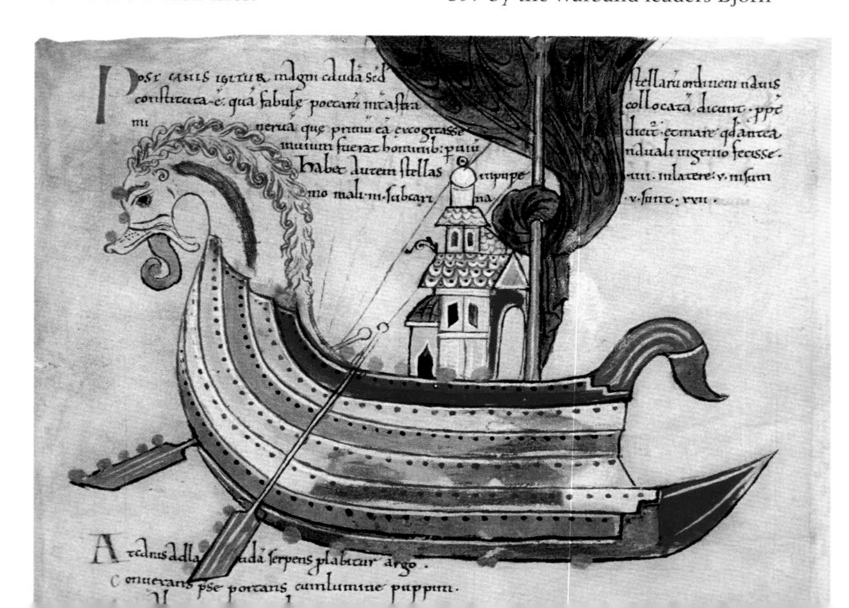

▷ **Ocean voyages**
This 11th-century codex illustration shows a Viking longship. Due to the shape of the prow, these ships were called "dragonships" by the Franks.

Expanding horizons

By that time, the Vikings were raising their ambitions from hit-and-run raids to conquest and settlement. In 865, a substantial force invaded eastern England and embarked on a sustained campaign of conquest. After 14 years of warfare, Alfred the Great, the Anglo-Saxon King of Wessex, stemmed the Viking advance, but Danish settlers became a permanent presence in north and eastern England. Vikings also settled in Ireland and Scotland, and as far north as the Faroe Islands. Vikings, led by their ruler, Rollo, settled in northern France. By the early 10th century, longships crossing the North Sea were no doubt

◁ **The Vikings attack Paris**
This 19th-century coloured lithograph depicts the besieged Frankish city of Paris, France. In 845, and again in 885–886, the Vikings sailed inland up the Seine. However, the Vikings never captured the city.

outnumbered by knarrs, the broader, deeper-hulled ships used to transport livestock, stores, and trade goods.

The Vikings who voyaged to the British Isles and western Europe chiefly came from Denmark and Norway. In eastern Europe, it was primarily the Swedes who forged new trade routes and embarked on raids. Crossing the Baltic in the course of the 9th century, they established control of an area stretching from Lake Ladoga, south to Kiev, where they became known as "Rus" or "Varangians". The Rus Vikings travelled along the great rivers – the Vistula, the Dnieper, the Dniester, and the Volga – down to the Black Sea and the Caspian. This brought them into contact with two key civilizations substantially wealthier and more advanced than those of Western Europe – the Christian Byzantine Empire and the Muslim Abbasid Caliphate. Viking traders penetrated as far as the Abbasid capital Baghdad in modern-day Iraq. Thousands of Arabic silver coins, found in buried Viking treasure hoards in Sweden, attest to the importance of the trade established with the Muslim world. »

" **Never before**... was it thought that such an **inroad from the sea** could be made. "

LETTER FROM ALCUIN TO KING ETHELRED, 793

ON TECHNOLOGY
Viking longship

The famous Viking longships, used for raiding and long-distance voyages, had a sleek hull that was clinker-built – that is, constructed of overlapping planks. Lightweight and with a shallow draught, the longships travelled fast and could easily be sailed up rivers or landed on beaches. They had a square sail, but were rowed by their crew on still days. The longships employed on coastal raids were exceptionally long and thin, those intended for ocean voyages were somewhat broader and shorter. Most of the examples of longships that survive today were unearthed from ship burials, the style of interment adopted for high-status Viking warriors.

9TH-CENTURY GOKSTAD BURIAL SHIP IN THE VIKING SHIP MUSEUM, OSLO, NORWAY

△ **Viking coin**
Around the 10th century, the Vikings adopted silver coinage in imitation of the more civilized countries that they raided. The images on these coins provide evidence of the design of Viking ships.

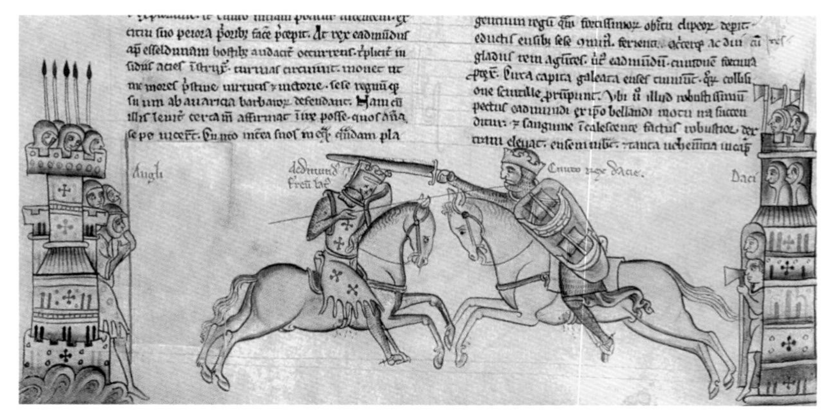

▷ **King Cnut the Great**
A medieval illustration shows Cnut, king of Denmark, defeating Anglo-Saxon king Edmund Ironside in battle. Cnut became ruler of all England in 1016, and later added Norway to his North Sea Empire.

» Viking ambitions were rarely limited to trade alone – plunder and conquest were always a temptation. In the summer of 860, some 200 Rus Viking longships commanded by the warriors Askold and Dir sailed down the Dnieper into the Black Sea, bent upon seizing the Byzantine capital, Constantinople. They were frustrated by the city's formidable fortifications, but Viking armies returned to attack Constantinople repeatedly over the following century. A more cooperative relationship with the Byzantine Empire developed from 988, when the ruler of Kievan Rus, Vladimir I, converted to Christianity. As a gesture of peace, Vladimir sent a group of Rus warriors to enter the service of the emperor in Constantinople. These warriors founded the Varangian Guard, which became an elite mercenary element of the Byzantine army – Viking ferocity placed at the service of the Empire.

The Vikings' most extraordinary voyages of exploration were those made across the North Atlantic. The Vikings first landed in Iceland when a sea captain heading for the Faroes was blown off-course and returned to tell of a "land of snow". This inspired Danish warrior Garðar Svavarsson to make an exploratory voyage that circumnavigated Iceland in around 860. The prospect of empty lands in the north attracted Norwegian warband leaders in search of remote locations where they could live untrammelled by authority. The first attempt at a settlement was made by Flóki Vilgerðarson, but he returned declaring the country uninhabitable. The first durable Viking Icelandic settlement was established by Ingólfr Arnarson, in around 874.

Exploring North America

By the second half of the 10th century, land in Iceland was becoming scarce. In 983, Erik the Red, a notably unruly immigrant to Iceland from Norway, became probably the first European to see the ice cliffs of Greenland. He returned from an exploratory voyage to organize a fleet of colonists, leading 25 knarr transport ships with settlers and supplies back to Greenland. In or around 1000, the final step in the Vikings' extraordinary adventure was taken. Travelling west from Iceland, a certain Bjarni Herjólfsson got lost in fog and missed Greenland, ending up sailing along an unknown coast. Inspired by Bjarni's report of a forested

▽ **Lindholm Høje**
The Viking burial site at Lindholm Høje, in Denmark, has graves marked with stones laid out in the shape of a boat. These "stone ships" reflect the importance of sea travel in the Viking world.

land, Leif Erikson, first son of Erik the Red, sailed with 35 men in search of the new country. He sailed to Newfoundland, the Gulf of St Lawrence, and beyond to the south, although exactly where remains disputed. He built a camp at a place he called Vinland because his men found grapes growing there. It is speculated that this may have been somewhere between present-day Boston and New York. Subsequent voyages attempted to establish a permanent Viking settlement in North America, but the effort failed, probably because of vigorous resistance by indigenous peoples. Yet Leif Erikson's men had almost certainly been the first Europeans to land in North America.

By that time, the golden age of Viking voyages was coming to an end. In the Scandinavian homelands, kingdoms evolved that established their rule over the warbands, putting an end to their raids. By the early 11th century, a North Sea Empire had taken shape, ruled by King Cnut of Denmark, who was also ruler of England and Norway. Outside Scandinavia, Viking warriors and settlers became integrated into their host societies. The Slav population of eastern Europe absorbed the Rus Vikings. In northern France, the Vikings adopted French language and customs, intermarried with local people, and became the Normans. With the collapse of the North Sea Empire after Cnut's death in 1035, the Viking era effectively came to an end.

◁ **Icelandic sagas**
Viking voyages were recorded in sagas written by Icelanders in the 13th and 14th centuries. They were based on family histories passed down by word of mouth.

◁ **Gotland stone**
The lower half of this carved stone from the Swedish island Gotland shows a Viking longship. The upper half depicts a scene from Norse mythology – Viking warriors who have died in combat entering Valhalla.

IN CONTEXT
L' Anse aux Meadows

The remains of the Viking settlement at L' Anse aux Meadows in Newfoundland, Canada, were identified in 1960. They provided the first material evidence to support the story of Leif Erikson's landing in America as related in the Icelandic sagas. Sited in a sheltered cove, the settlement consisted of eight buildings made of turf over a wooden frame. These would probably have accommodated some 90 people, as well as housing a forge and facilities for ship repair. Radiocarbon dating has confirmed that the settlement was established around 1000, a date that fits with the evidence from the sagas. The settlement seems to have been abandoned within 20 years of its foundation.

AN AERIAL VIEW OF A RECONSTRUCTED LONGHOUSE AT L' ANSE AUX MEADOWS

" Leif **set sail** as soon as he was ready... and **lighted upon lands** of which before he had **no expectation**. "

THE SAGA OF ERIK THE RED, c.1265

△ **Jerusalem siege**
A 13th-century image of the Crusaders' siege of Jerusalem in 1099 shows Christian soldiers defending the Church of the Holy Sepulchre, the traditional site of Christ's crucifixion.

The Crusades

Between the 11th and 13th centuries, tens of thousands of Christian soldiers from Europe made the 5,000-km (3,100-mile) journey to the eastern Mediterranean to fight for their faith against the warriors of Islam.

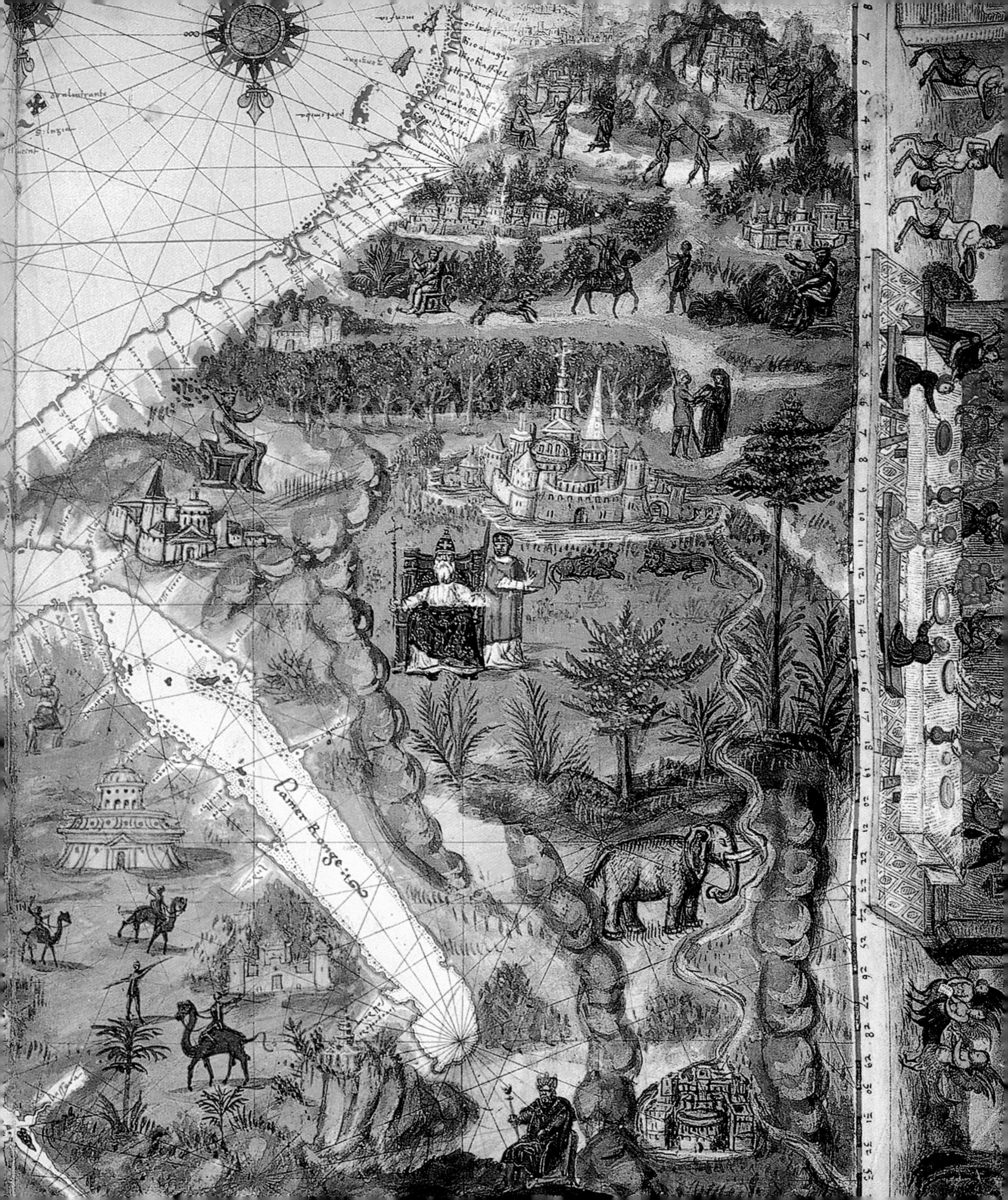
Lamer Rouge

Medieval pilgrimages

Pilgrimage was the most common motive for long-distance travel in medieval times. Each year, thousands of believers left the safety of their homes to embark upon perilous journeys to distant sacred places.

△ ***Reception of Pilgrims and Distribution of Alms***
Domenico di Bartolo's fresco, painted in 1442, is in a hospital in Siena that was dedicated to caring for pilgrims on their way to Rome. In this painting, pilgrims are being offered food and medical care.

Although all religions have had some tradition of travel to holy sites, medieval Christians were especially drawn to pilgrimage. The religious importance of visiting Palestine, revered as the setting for the life of Jesus, was established as early as the 4th century CE, when Helena, mother of the Roman Emperor Constantine, made a well-publicized trip to Jerusalem, searching for relics of Christ's crucifixion, and had the Church of the Holy Sepulchre built on the alleged site of Jesus's tomb. Christian pilgrimage remained a relatively small-scale affair until the 10th century, after which it rapidly expanded and diversified.

Jerusalem was never surpassed as a goal for the Christian pilgrim, but other pilgrimage sites proliferated

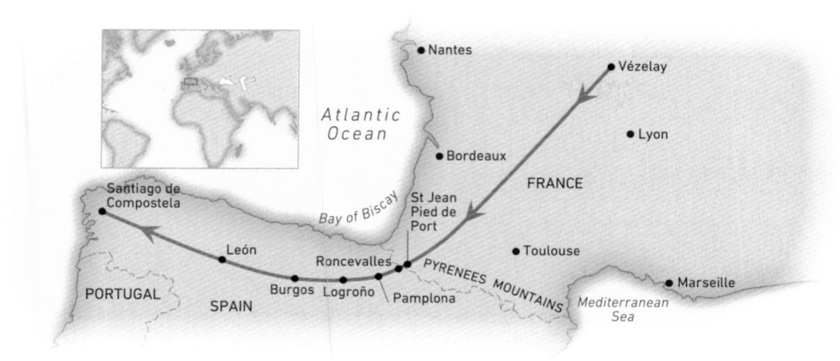

△ **The way to Santiago**
Most of the pilgrims who travelled to Santiago de Compostela followed established routes across France and northern Spain, from assembly points at Vézelay, Paris, and Le Puy.

through the Middle Ages. Rome claimed the tombs of the apostles Peter and Paul, but many lesser shrines housing a sacred relic or the corpse of a saint also built up large followings. Pilgrims flocked to the Shrine of the Three Kings in Cologne Cathedral, Germany, which was said to contain the bones of the Magi from the biblical nativity story, or to the city of Santiago de Compostela in northern Spain, the alleged burial place of St James the Apostle. In England, Canterbury became the principal pilgrimage site after the murder of the saint and martyr Archbishop Thomas Becket in the cathedral in 1170.

Pilgrims came from all levels of society and their motives were many and various. The journey might be a form of penance undertaken in expiation of a sin or crime, or even a legal punishment imposed on a wrongdoer. The pilgrim might be seeking a miraculous cure for an illness or fulfilling a vow to a saint who had been asked to intercede at some moment of peril. Some had deep religious motivation, while others, more like modern tourists, primarily wanted to see the world. For most people, the journey to Jerusalem – long, expensive, arduous, and dangerous – could only be a once-in-a-lifetime experience, but a pious Christian might make pilgrimages to more local sites several times a year.

Catering to pilgrims

Pilgrims are usually represented with a coarse garment, staff, purse, and scallop shell, but in reality, most wore their ordinary clothes and carried a spare pair of shoes. An easy prey for robbers and wild animals, they travelled in groups for safety and mutual support. Pilgrims bound for Santiago de Compostela, for example, traditionally gathered at the Benedictine abbey of Vézelay in central France. From there, the route across the Pyrenees and through northern Spain was dotted with pilgrim hostels established by pious local landowners or by the Benedictine monastic order, which was highly active in fostering pilgrimage. These hostels might provide a range of services, from the repair of shoes broken on the stony roads to hospital beds and barbers. Other pilgrims travelled to northern Spain by sea, hiring ships for the return journey from England or Germany, rather in the manner of today's charter flights.

△ **Badge of a pilgrim**
This pilgrim's cloak is bedecked with scallop shells, the emblem of St James, whose shrine drew pilgrims to Santiago de Compostela.

Once the pilgrims reached their final destination, they would usually take part in a carefully stage-managed ritual designed to satisfy both their religious urge and their desire for mystery and wonder. Approaching the popular English shrine of the Virgin Mary at Walsingham in Norfolk, for example, pilgrims were instructed to take off their shoes and walk the last "Holy Mile" barefoot, singing religious songs. Reaching the shrine itself, they »

IN CONTEXT
The Canterbury Tales

Written by the English courtier Geoffrey Chaucer in the late 14th century, *The Canterbury Tales* is a collection of verse stories purportedly narrated by a group of pilgrims bound for the shrine of Thomas Becket in Canterbury Cathedral. The pilgrims range from a chivalrous knight and his son to a bawdy miller, a friar, and a raunchy widow.

As well as offering a panorama of English medieval society, Chaucer's work shows the variety of motives and attitudes found among pilgrims, some of whom are far from godly folk. The impression created is of travellers who have embarked on a jaunt, rather than an act of penance.

A WOODCUT OF THE KNIGHT FROM THE 1490 EDITION OF *THE CANTERBURY TALES*

◁ **Shrine of the Magi**
Cologne Cathedral in Germany was built to house a sarcophagus that is said to contain the remains of the biblical Magi. It still attracts hosts of pilgrims today.

> “ The pilgrim should carry with him **two sacks** – one right **full of patience**, the other containing **200 Venetian ducats**. ”
>
> PIETRO CASOLA, *PILGRIMAGE TO JERUSALEM IN THE YEAR 1494*

▷ **Stained glass**
This window in Canterbury Cathedral shows pilgrims at the shrine of Archbishop Thomas Becket, murdered there in 1170. His death was followed by miraculous cures that guaranteed the site's popularity.

were led through a chapel where they would kiss the finger bone of St Peter, past a well, and into another dimly lit chapel containing the site's most sacred relic, drops of the Virgin Mary's breast milk.

Although pilgrims benefitted from the charity of the Church and pious laymen, mass pilgrimage inevitably became a source of profit for many enterprising individuals who provided services to the religious throng. Pilgrims bought travel guides that offered advice on the journey and detailed information on the sites. Among the most famous of these, manuscripts of *Mirabilia Urbis Romae* (*The Marvels of Rome*), a guide to the monuments of Rome, remained in circulation from the 12th to the 16th century. All pilgrims wanted to return with souvenirs of their journey. These were readily provided by hawkers at pilgrimage sites, and ranged from badges to display on their clothing to replica relics with alleged healing powers. The cathedrals or abbeys controlling the shrines and the towns in which they were sited prospered on the necessary spending and voluntary offerings of the pilgrims.

A dangerous road

The pilgrimage to Jerusalem was the most difficult to undertake, both because of the distance from Christian Europe and the presence of Muslims, either in lands around the Holy City or actually in possession of it. The overland journey to the Levant through the Balkans and Anatolia was always fraught with dangers. Increasing Muslim hostility eventually made the land route effectively impossible. Sea passages became an essential part of journeys to Palestine, and from the 13th century, the maritime city of Venice developed a virtual monopoly of the pilgrim trade. However, crossing the Mediterranean on board a Venetian galley brought its own hardships and dangers. The ships were overcrowded, insanitary, infested with rats, and poorly provisioned. They were also liable to interception by Muslim corsairs. Once in the Holy Land, visitors were keen to

△ **Rome travel guide**
This woodcut appeared as an illustration in *Mirabilia Urbis Romae*, a medieval guide to the wonders of Rome that was bought by many generations of pilgrims visiting the papal city.

▷ **The *mahmal* leaves Cairo**
This 19th-century lithograph depicts the *mahmal* – a ceremonial palanquin in which the Sultan of Cairo was borne to Mecca on the annual hajj. It was first employed by the Mamluk Sultan Baybars in the 13th century.

make a thorough tour of the key sacred sites, including the Holy Sepulchre, the Mount of Olives, Bethlehem, the Pond of Bathsheba, and Mount Zion. When the Crusaders were in control of the area, pilgrims could count on the protection of the Knights Templar and the ministrations of the Knights Hospitaller. But in periods of Muslim domination, they could expect much harassment and considerable expenditure on bribes and fees.

The Muslim pilgrimage

Muslims had their own tradition of pilgrimage, although it differed from that of the Christians. The hajj to Mecca was an obligation for the pious Muslim, but pilgrimages to lesser sacred sites played a relatively small part in Muslim religious practice. The hajj was highly organized, with caravans of thousands of pilgrims departing annually from gathering points at Cairo, Damascus, and Basra. Water, food, and lodging were provided along the way. The perils and hardships associated with Christian pilgrimage were certainly not unknown to Muslims, however. Caravans were preyed upon by Bedouin tribes in the desert and taxes were levied on pilgrims by the rulers of the Mecca region.

The Muslim pilgrimage has not only survived, but has expanded until the present day. The Christian tradition of pilgrimage, by contrast, went into severe decline from the 16th century. The rise of the Muslim Ottoman Empire made the Holy Land almost inaccessible to European Christians. The veneration of saints was rejected by the Protestant Church as a superstition and even the Catholic hierarchy eventually became suspicious of popular faith in the miraculous power of relics and shrines.

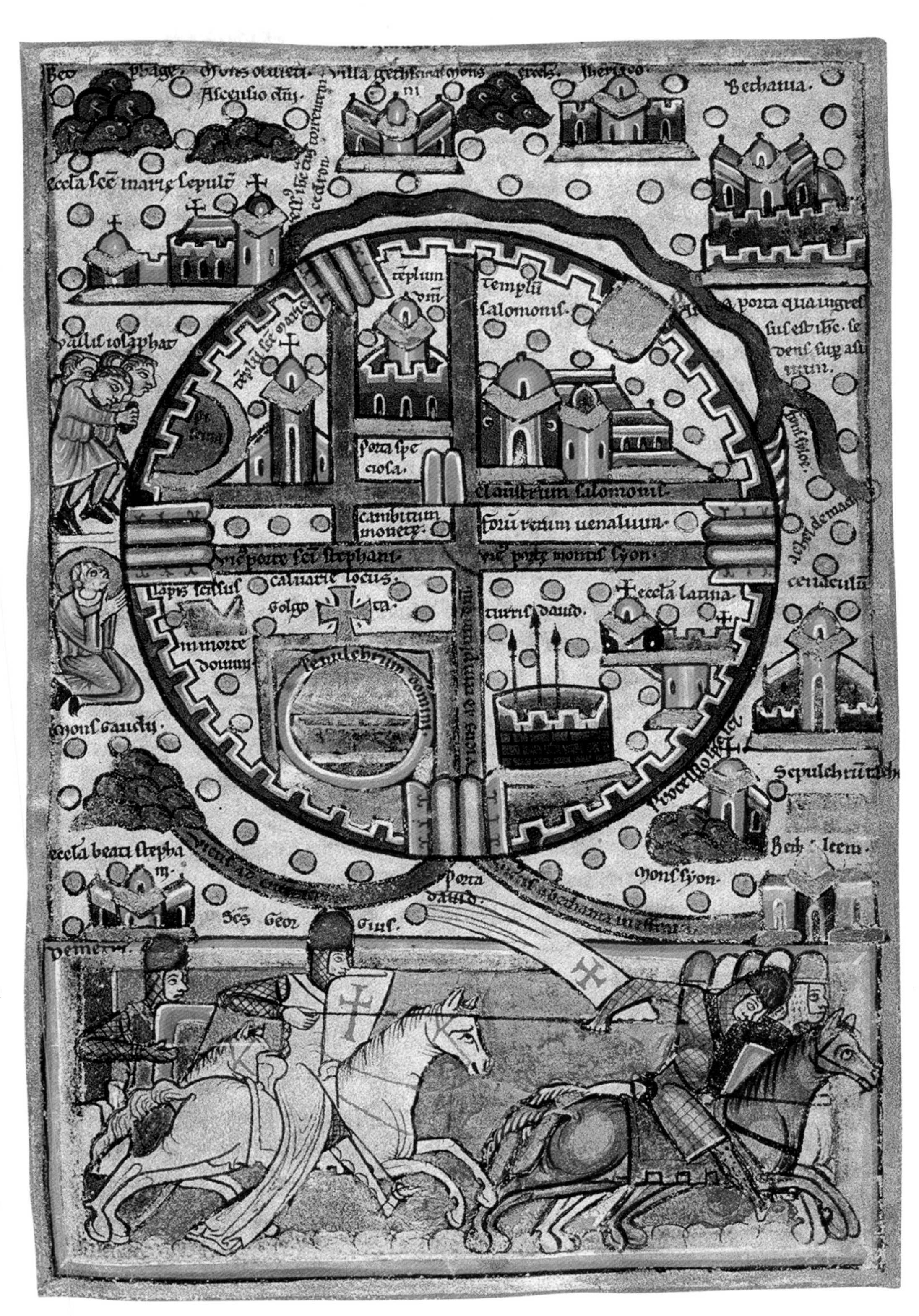

▷ **Medieval plan of Jerusalem**
A stylized plan of the Holy City, dating from around 1200, is accompanied by a scene of battle between Crusader knights and Muslims. The Crusades were partly launched in order to guarantee Christian pilgrims access to Jerusalem.

Medieval travel accounts

Medieval Europeans revelled in picturesque accounts of foreign travels. Ignorant of the wider world, they could not distinguish fact from fantasy, credulously consuming tales in which imaginary monsters appeared in otherwise accurate portraits of exotic lands.

The most influential of medieval travel writers was a man who faked his identity and who may never have travelled at all. Appearing in the mid-14th century, *The Travels of Sir John Mandeville* was one of the most widely read books in Europe. The author describes himself as a knight from St. Albans, England, but researchers have found no trace of such a person. Authorship of the book has been speculatively ascribed to, among others, a physician from Liège and a monk from Ypres, but no-one really knows Mandeville's identity. Some believe the author had at least travelled to the Holy Land, for he describes Jerusalem in extensive detail, but no-one thinks his supposed eyewitness accounts of India, Ethiopia, and China are authentic. For example, Mandeville describes Ethiopians as having only one leg, which they use to shelter themselves from the sun. He also claims to have passed through lands where people have the heads of dogs and dwarves thrive on the smell of apples. Yet Mandeville's text includes many not wholly inaccurate second-hand accounts of the customs of distant countries. Christopher Columbus read it, and discovered the idea that a man who travels in a straight line will end up back where he started – because the world is a sphere.

◁ **Irish boatmen**
An illustration from Gerald of Wales's *Topographia Hibernica* shows two Irishmen paddling a coracle. Gerald describes the Irish as "suffering their hair and beards to grow enormously in an uncouth manner".

" I, **John Mandeville**, **Knight**... have been **long time** over the sea, and have seen **many diverse lands**, and many... **kingdoms** and **isles**... "

THE TRAVELS OF SIR JOHN MANDEVILLE, c.1350

Approaching authenticity

Popular travel accounts were not necessarily of journeys to far distant places. The widely circulated *Topographia Hibernica*, written by Gerald of Wales in the 12th century, was a description of Ireland. Gerald found the Irish "a rude people... living like beasts", their hair too long and their clothes too rough. Although much of his account was first hand, it still contained imaginary marvels, such as an encounter between a priest and an articulate werewolf.

As well as the famous Marco Polo (see pp.88–89), a number of Europeans wrote more or less accurate accounts of journeys to the East. The earliest was by Franciscan friar Giovanni da Pian del Carpine, who travelled as a papal ambassador to the court of the Mongol Khan at Karakorum in 1245–47. He was followed by another friar, William

◁ **Medieval astronomers**
This tinted version of a picture attributed to John Mandeville shows astronomers on the summit of Mount Athos, Greece, using various astronomical instruments.

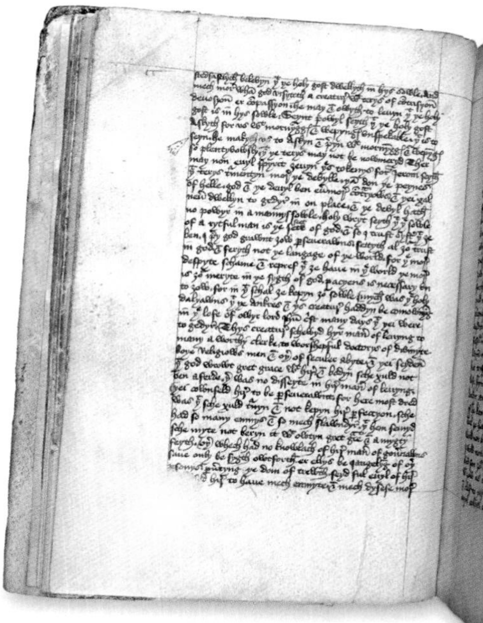

△ **Autobiographical manuscript**
The manuscript of the autobiography of 15th-century Englishwoman Margery Kempe, which relates her travels to holy sites. It was only rediscovered in 1934 after centuries of oblivion. It now resides in the British Library.

◁ **Odoric of Pordenone**
A medieval illustration of the narrative of Odoric's travels in Asia shows the Franciscan friar about to embark on his mission. He was beatified as a saint in 1755, four centuries after his death.

of Rubruck, who travelled to Mongolia and back in 1253–55. William wrote the most level-headed medieval travel account of Asia – but, being frankly sceptical about the existence of monsters and human freaks, it was not widely read.

Yet another Franciscan, Odoric of Pordenone, embarked on a journey through Asia in 1318, returning to Italy 12 years later, having visited Persia, India, Indonesia, and China. The account of his travels contains details of unquestionable veracity – such as the Chinese use of cormorants for fishing – and was highly influential, much of it plagiarized by Mandeville. For information about India, Europeans could turn to the *Mirabilia* of Bishop Jordanus Catalani, who lived on the subcontinent some time between 1321 and 1330. Jordanus is reliable on Indian customs, but resorts to marvels when writing of other parts of Asia and Africa.

An interesting contrast to these often dubious travel accounts is provided by the autobiography of Margery Kempe. Written in the 1430s, this vividly relates the fears and ecstasies of a woman from King's Lynn in Norfolk, England, during her often stressful journeys to holy sites in Palestine and Europe. Kempe's work, little known in its day, conveys the experience of medieval travel far more authentically than the more famous books of marvels.

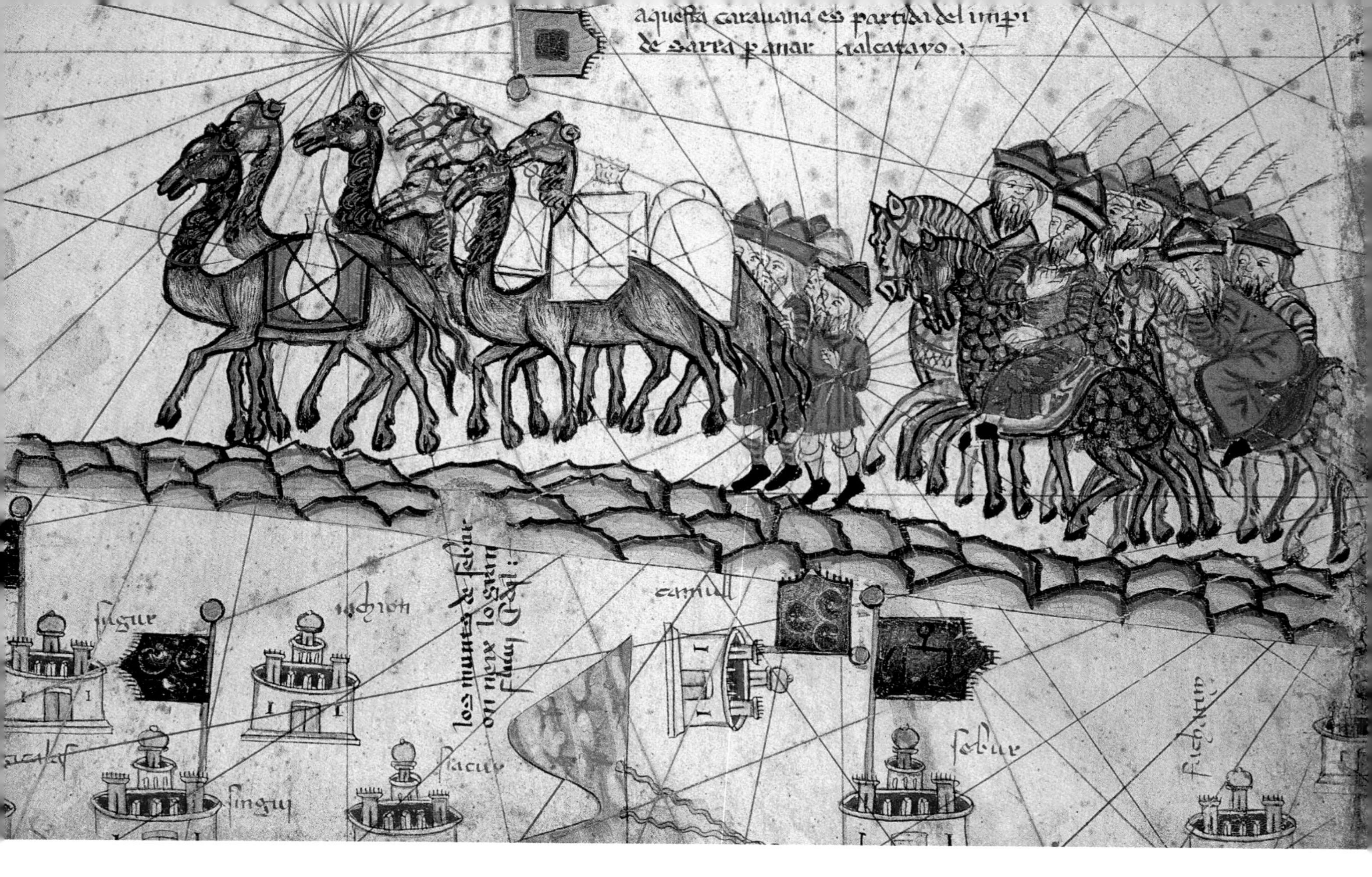

△ **Silk Road caravan**
An image from the 14th-century *Catalan Atlas* shows Marco Polo on the Silk Road crossing the rugged Pamir Mountain range. Merchants ride on horseback and their baggage is carried by camels.

The Silk Road

In the 13th century, the Mongol khans created a vast land empire that stretched across the breadth of Asia. Under their rule, travel flourished along the Silk Road linking China to the Muslim Middle East and Europe.

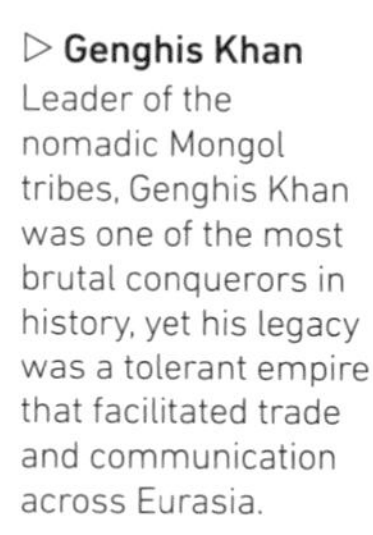

▷ **Genghis Khan**
Leader of the nomadic Mongol tribes, Genghis Khan was one of the most brutal conquerors in history, yet his legacy was a tolerant empire that facilitated trade and communication across Eurasia.

By the time the Mongol Empire was founded, the Silk Road already had a history stretching back over a thousand years – the wealthy elite of ancient Rome had worn silk imported across Asia from Han-dynasty China. But for centuries, the trade route had only operated partially and intermittently, disrupted by wars, banditry, and the predations of greedy local rulers. When Genghis Khan united the Mongol tribes of Central Asia in 1206 and launched them on far-flung campaigns of conquest, the initial effect on trade was disastrous. Famous cities along the Silk Road, from Baghdad and Balkh to Samarkand and Bukhara, were left in ruins after Mongol attacks. However, by the second half of the 13th century, in control of an empire stretching from Persia to China, the Mongol rulers had learned to appreciate the value of well-maintained communication routes and the wealth that trade could bring. The peace that they imposed and the religious tolerance that they practised led to a steep rise in the number of foreign merchants on the Silk Road.

For merchants from Christian Europe, the main starting points for the journey east were Constantinople and the Black Sea coast, while Muslim traders set off from Cairo or Damascus. The travellers

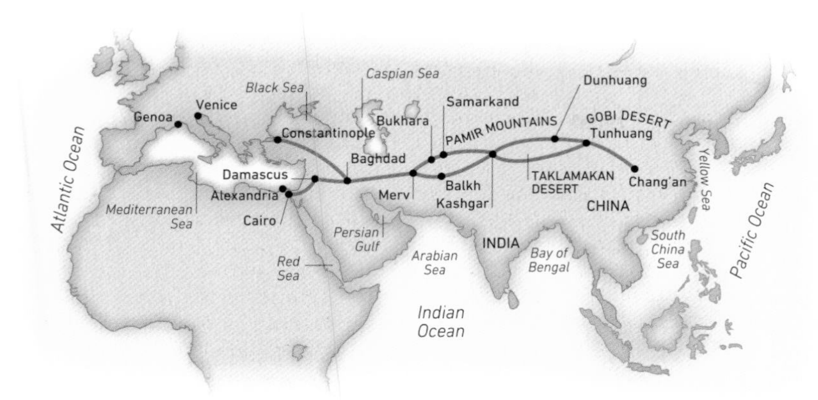

△ **Silk Road trade routes**
The Silk Road was not a single highway but a network of routes converging and diverging across Asia. Its total length has been estimated at around 7,000 km (4,300 miles).

formed caravans of hundreds or even thousands of merchants, porters, guides, and translators, and had long trains of pack animals. Along parts of the route, caravanserais (see pp.92–93) provided lodging and sustenance for men and beasts alike, but in spite of these, the journey could only be arduous. The road from the Black Sea coast around the south of the Caspian Sea, followed by many European merchants, was plagued by bandits. Paying protection money to bodies of armed men was a normal procedure for any Silk Road traveller. Much of the land was arid, and on particularly barren sections, all food and drink had to be carried for many days of travel. Crossing the Pamir Mountains and the Taklamakan Desert posed especially gruelling challenges to travellers.

Heyday and decline

By the 14th century, travel on the Silk Road had become common enough for guidebooks to be written. Italian merchant Francesco Pegoletti had success with *The Practice of Commerce*, a book of advice for travellers to China that itemized travel times on various stretches of the route, the likely costs to be incurred, what precautions should be taken, and the most profitable trade goods to carry. Pegoletti ended with a description of Cathay (China), where paper money had to be used, and the city of Khanbaligh (Beijing) which "has a circuit of a hundred miles and is all full of people and houses".

The golden age of the Silk Road came to an end with the Chinese revolt against Mongol rule, which led to the founding of the Ming dynasty in 1368. As the Mongol Empire declined, the rise of the militantly Islamic Ottoman Empire blocked Christian European access to the trade routes via Constantinople. But it was the development of oceanic trade in the 16th century, opening up sea routes from Europe to the Spice Islands and East Asia, that delivered the final blow to the Silk Road – it was one from which it never recovered.

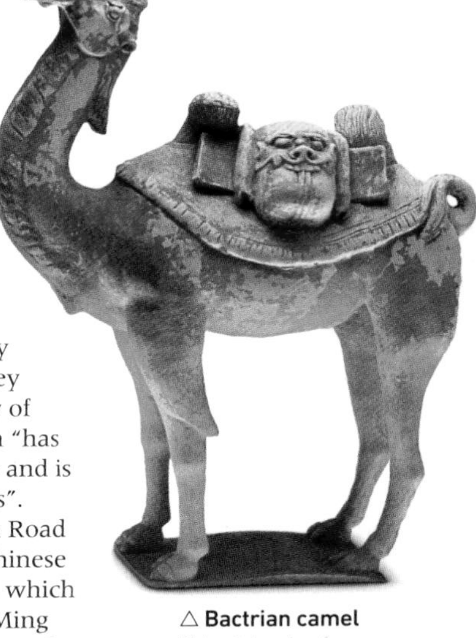

△ **Bactrian camel**
This statue is of a sturdy Bactrian camel, native to the Central Asian steppe, which was the standard pack animal of the Silk Road caravans.

IN PROFILE
The Black Death

The Silk Road carried not only trade goods, but also disease. Plague was endemic in Central Asia, and in the 1340s it spread along the Silk Road to trading posts on the Black Sea. From there, it was distributed by ship – possibly carried by fleas on rats infesting the ships' holds – to ports in western Europe. Known as the Black Death, between 1348 and 1350 the plague epidemic killed at least a third of the entire European population, which had no immunity to this exotic disease. The Black Death also struck Syria and Egypt, with equally disastrous effect.

A MEDIEVAL PAINTING OF A DOCTOR DOING HIS BEST TO TREAT PLAGUE VICTIMS

> "And from Utrar to Almalik it is **45 days** by **pack asses**. And you meet Mongolians **every day**."
>
> FRANCESCO PEGOLETTI, *GUIDE TO THE SILK ROUTE*, 1335

◁ **Desert tower**
A Chinese frontier town on the edge of the Taklamakan Desert, Dunhuang was a major stopping point on the Silk Road. Its towers and walls, now ruined, were a welcome sight to travellers arriving from the west.

The travels of Marco Polo

Marco Polo was a Venetian merchant who journeyed to China in the 13th century and lived at the court of the emperor Kublai Khan. His published account of his experiences made him the most famous traveller of his time.

△ **Venetian traveller**
Marco Polo was born into a wealthy family of Venetian merchants. He was not the only European to travel to China and back, but his compelling account of his travels by land and sea had a unique appeal.

When Marco Polo was born in 1254, the maritime city of Venice was a major player in the burgeoning trade between Europe and the East. Venetian merchants made fortunes purchasing luxury Asian goods such as silk and spices at trading posts in the eastern Mediterranean and the Black Sea, and reselling them in the European market. Marco's father and uncle, Niccolò and Maffeo Polo, were both engaged in this trade. Travelling to the Black Sea around the time of Marco's birth, they found themselves drawn even deeper into Asia. After many vicissitudes, they arrived at the court of the Mongol leader Kublai Khan, who found these European visitors an amusing novelty. He made them his emissaries, and charged them with establishing communication between him and the Pope.

By the time Niccolò and Maffeo returned to Venice in 1269, Marco was an adolescent. In 1271, he accompanied his father and uncle when they set off on a second journey to Kublai's court. This time, they sailed to the port of Acre in the Crusader-controlled Levant. From there, they followed a rambling course eastward, changing direction to avoid regions rendered hazardous by war or banditry. Travelling via Baghdad and the Persian port of Hormuz, they eventually picked

▽ **Travels to the East and back**
The Polos' 24,000-km (15,000-mile) journey took them through numerous dangerous territories, many of which would later become impassable. However, they returned rich, reputedly with gems sewed into their coats.

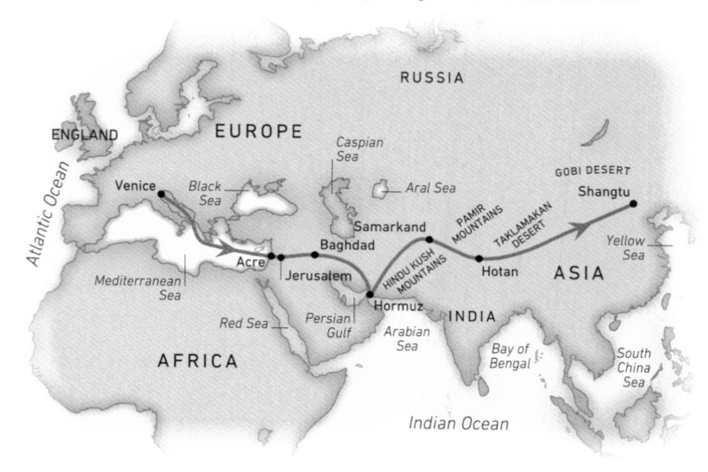

> "I did not tell **half of what I saw**, for I knew that **I would not be believed**."

MARCO POLO ON HIS DEATHBED, AS REPORTED BY JACOPO D'ACQUI, 1330

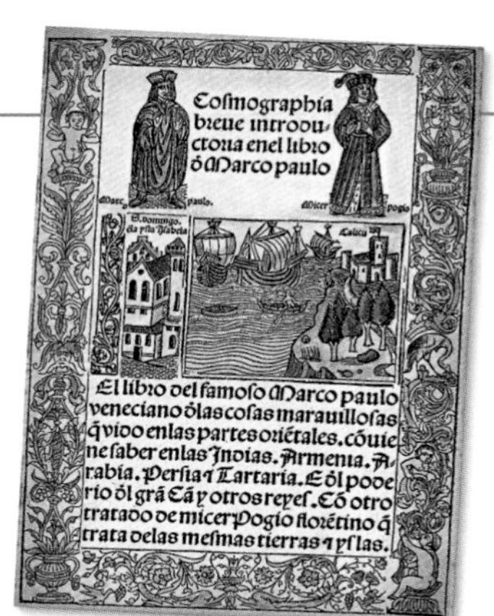

Cosmographia breue introductoria enel libro d Marco paulo

Marco paulo. Micer pogio

El libro del famoso Marco paulo veneciano d las cosas marauillosas q vido enlas partes orietales. conuiene saber enlas Indias. Armenia. Arabia. Persia y Tartaria. E dl poderio dl gra Ca y otros reyes. Co otro tratado de micer Pogio floretino q trata delas mesmas tierras y yslas.

◁ ***Il Milione***
This version of Marco Polo's *Travels* was published in 1503. The book is usually known in Italian as *Il Milione*, perhaps because it was maliciously said to contain a million lies.

up the main trail of the Silk Road (see pp.86–87) across the formidable obstacles of the Pamir Mountains and the Taklamakan Desert. All along the route, they found cities in ruin, still not recovered from the effects of the Mongol conquest that had completely devastated the region earlier in the century.

Agent of the emperor

The Polos' journey from Venice to China took three and a half years. They arrived at Kublai Khan's summer residence at Shangtu, north of Beijing, in 1275, and were taken into the emperor's service. The Mongol ruler of a restive Chinese population, Kublai welcomed the chance to employ foreigners who posed no threat to his rule. As the emperor's agent, Marco was sent on many journeys around Kublai's domains, giving him the opportunity to observe local customs.

Marco and his relatives stayed in Kublai's service for 17 years, until in 1292 they found an opportunity to leave. A Mongol princess in China was pledged in marriage to a Mongol prince in Persia, and the Polos were entrusted with escorting the bride to her promised husband. They made the journey by sea, following the maritime trade routes that carried spices across the Indian Ocean from Southeast Asia to the Muslim Middle East. Landing at Hormuz, they delivered the princess to Tabriz, the Mongol capital of Persia, and continued overland to the Black Sea, where they took a ship for Venice. They arrived home in 1295.

◁ **Marco Polo leaves Venice**
A medieval illustration shows Marco and his relatives departing from their home city, bound for the court of Kublai Khan. It would be almost a quarter of a century before they saw Venice again.

The Travels of Marco Polo

The world would never have heard of Marco's many adventures but for a chance encounter with an author. In 1298, Marco was taken prisoner by the Genoese, who were at war with Venice. In prison, he met the Italian romance writer Rustichello da Pisa, and told him the stories of his travels. Adding fantastical elements of his own, Rustichello wrote *The Travels of Marco Polo*, which was published around 1300. Although many people were sceptical of the facts narrated in *The Travels* (doubting, for example, that the Chinese could use paper as money), it became one of the most widely read books of its day. Marco Polo died in his bed in Venice, in January 1324.

A CHINESE PORTRAIT OF EMPEROR KUBLAI, WHO EMPLOYED MARCO IN 1275

IN PROFILE
Kublai Khan

A grandson of Genghis Khan, the founder of the Mongol Empire, Kublai Khan succeeded to reign Mongol-ruled northern China in 1251. He was proclaimed Great Khan, the overall leader of the Mongols, in 1260, but in effect mutated into an emperor of China, ruling from Beijing. Founding the Yuan dynasty, he completed the conquest of southern China in 1279, and attempted several expansionist ventures, including unsuccessful invasions of Japan in 1274 and 1281. His rule in China was characterized by religious tolerance, but despite adopting the traditions of Chinese government, he remained an alien ruler to his Chinese subjects.

CHARIOT, ROME, 200 BCE

LAUFMASCHINE, GERMANY, 1817

COVERED WAGON, US, 1850

HARLEY DAVIDSON, UK, 1916

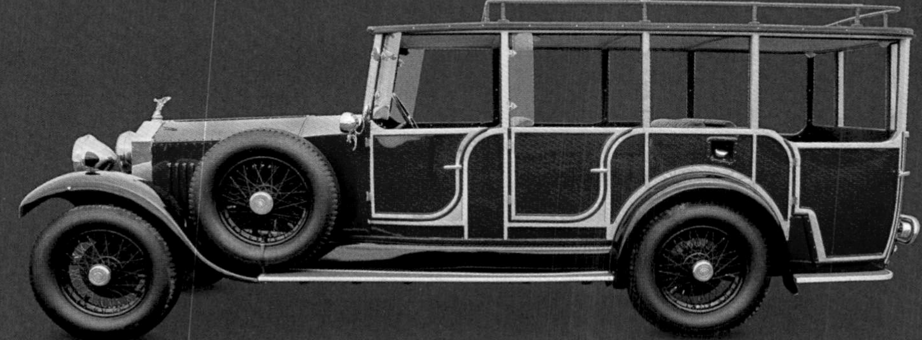
CHARABANC, UK, 1920

On wheels

For 5,500 years, people have been thinking up ever-more inventive ways of using and improving wheels to make transport easier.

The first potter turned a wheel on its side as an aid to transport in around 3,200 BCE, when the ancient Mesopotamians began to make horse-drawn chariots. It was another 1,600 years before spoked wheels, as strong as a solid wheel, but much lighter, appeared on Egyptian chariots. These early spokes were wooden – a far cry from the steel, aluminium, or carbon composites used to make lightweight bicycles today.

Wheels work best on a smooth surface, and rudimentary roads appeared soon after chariots. Improvements in wheels and roads went hand in hand. Around the same time that spokes were invented, iron bands were first used to reinforce wheel rims on wagons – these were the earliest tyres. Macadamized roads, made from compacted broken stone aggregate, were built in the 1820s, followed in 1846 by the first pneumatic tyres. Thanks to the internal combustion engine, wheels no longer needed horses to propel them, and transport hit the road fast in the 20th century. Tarmac was patented in 1901, boyhood friends William S. Harley and Arthur Davidson produced their first motorcycle in 1903, and Henry Ford brought out his "Tin Lizzie" in 1908. From its first hesitant revolutions, the wheel is still essential to transport, and the future seems inconceivable without it.

SCHOOL BUS, US, 1940

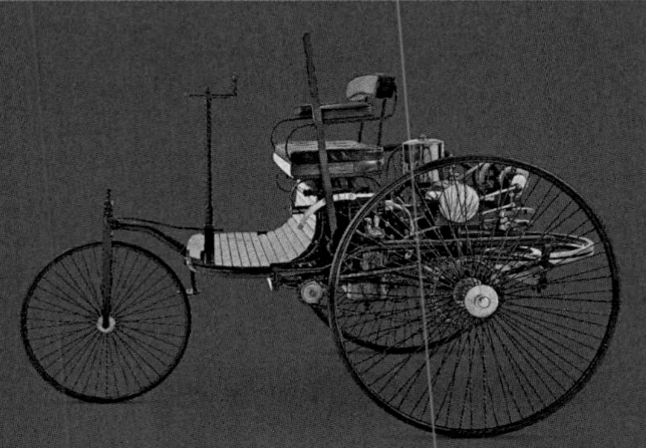
BENZ MOTORWAGEN, GERMANY, 1886

SPIDER PHAETON, UK, 1890

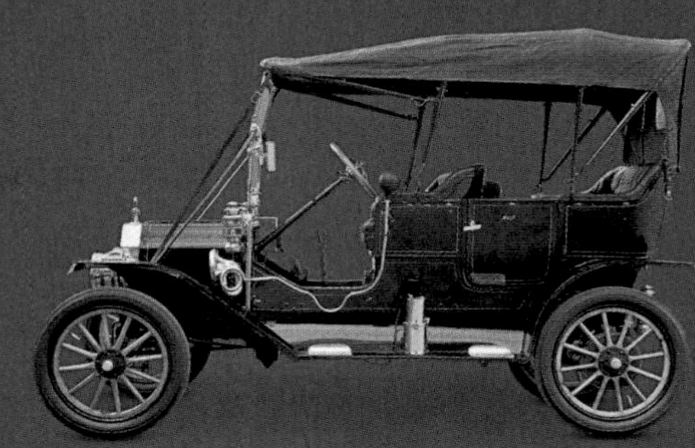
FORD MODEL T, US, 1908

AEC REGENT III RT BUS, UK, 1938

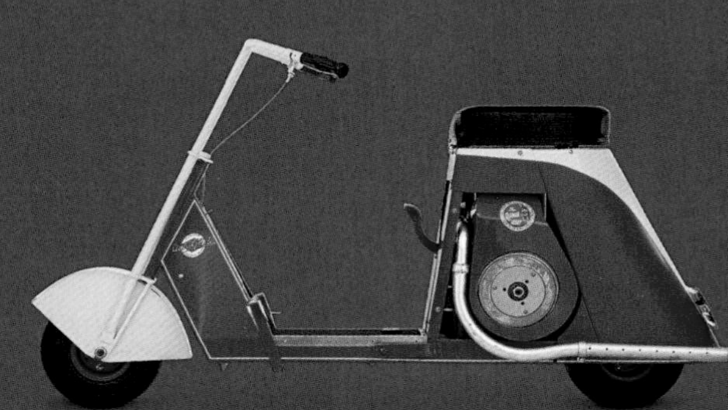
CUSHMAN AUTO-GLIDE MODEL 1, US, 1938

BIANCHI PARIS-ROUBAIX, ITALY, 1951

VOLKSWAGEN KOMBI, GERMANY, 1950-67

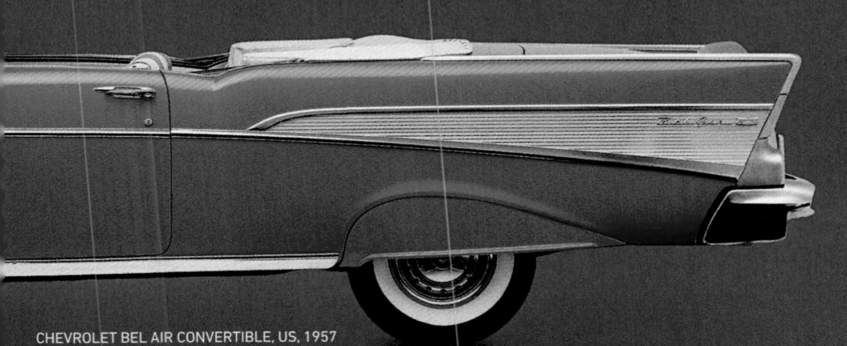
CHEVROLET BEL AIR CONVERTIBLE, US, 1957

NISSAN LEAF, JAPAN/US, 2010

Caravanserai

Caravanserais were way stations along medieval Asian trade and pilgrimage routes, offering secure lodging for travellers and their animals.

For any pilgrim or merchant, the sight of a caravanserai promised welcome refreshment and relief from the hardships and dangers of travel. Wayfarers entered these establishments through an arched gateway high enough to admit a heavily burdened camel, and passed into an open roofless courtyard. Around this courtyard were the sleeping rooms – bare chambers with no bed, table, or chair. The traveller was free to occupy any such room he found empty. The caravanserai also provided secure storerooms for trade goods and stables for the horses, donkeys, and camels, as well as a prayer room and a bathhouse.

From the outside, most caravanserais looked like forts, a high outer wall giving protection against bandits or wolves. Caravanserais were erected and maintained by local rulers obeying the religious injunction to facilitate the pilgrimage to Mecca and the commercial imperative to encourage trade. Basic lodging, food, and animal fodder were provided free. A foreign merchant might have found himself obliged to pay stiff local dues levied on trade, however.

Along the most frequented routes, there would be a caravanserai every 30 or 40 km (20 or 25 miles) – the distance of a day's travel for a caravan. As well as those situated in open country, there were numerous similar establishments in towns and cities. Filled with a wide mix of travellers, the caravanserais became busy marketplaces in which goods were traded with local people and between the travellers themselves. They were also places of cultural exchange, where people encountered different ideas and beliefs. The caravanserais played an essential part in Asian life for about a thousand years, into the 20th century. Now, many survive only as spectacular ruins, although some still function as marketplaces or have been renovated as tourist hotels.

◁ **Wikalat Bazar'a**
Built by a Yemeni called Bazar'a in the 17th century, Wikalat Bazar'a is one of 20 caravanserais left in Cairo. Like most caravanserais, it is built around a central arcaded courtyard.

Remote ruins
Many caravanserais provided the only shelter in remote, barren terrain, like this long-abandoned building in a mountain valley in Afghanistan. Caravanserais were found all across Asia, from Turkey to northern India and Kazakhstan.

Trans-Saharan salt caravans

Seven hundred years ago, one of the world's richest trade routes led north across the Sahara desert from Timbuktu, capital of the Mali Empire. Camel caravans transported enslaved people and cargoes of gold, salt, and ivory.

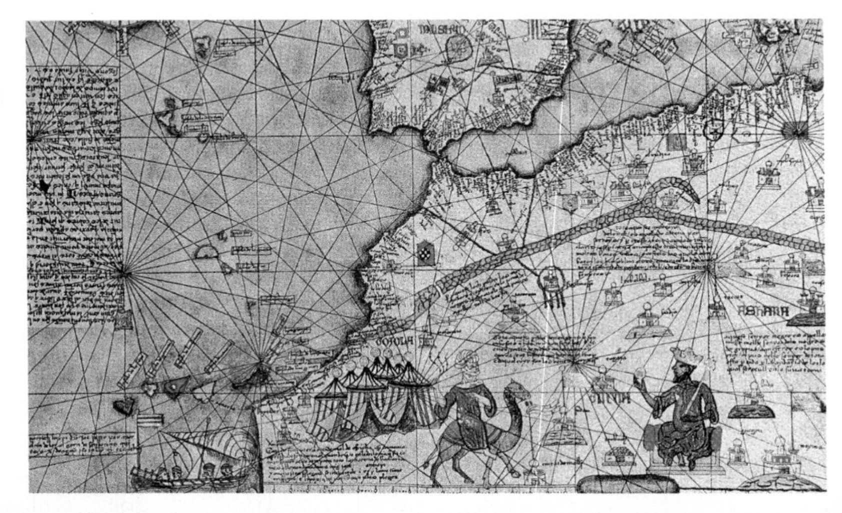

▷ ***Catalan Atlas***
In the *Catalan Atlas*, drawn in 1375, Mansa Musa, Malian ruler and "most noble lord of all this region", is shown seated on his throne south of the Sahara.

A landlocked Muslim state at the southern edge of the Sahara, the Mali Empire was mysterious even in its own day. The wider world only became conscious of its existence, and of its enormous wealth, in 1324, when its ruler, Mansa Musa, embarked upon a pilgrimage to Mecca. His arrival in Cairo, where he emerged unexpectedly from the desert with thousands of followers, caused a sensation. Mansa Musa's train of camels was burdened with such a quantity of gold that it had a major impact upon the Egyptian economy – at that time the most prosperous in the Islamic world.

Trans-Saharan trade

Founded in the 13th century, the Mali Empire drew its wealth from its control of the trans-Saharan trade. In the markets at Timbuktu and other Malian towns, enslaved people and commodities such as gold, copper, kola nuts, and ivory, carried up the River Niger from the forest country south of the Sahara were exchanged for salt and Berber horses brought by Tuareg nomads from

" There is **complete security** in their country. Neither **traveller** nor **inhabitant** has **anything to fear**… "

MOROCCAN EXPLORER IBN BATTUTA DESCRIBING THE MALI EMPIRE, 1354

▽ **Camel train**
Capable of surviving for long periods without water, dromedaries were essential beasts of burden in trans-Saharan trade.

the desert. The Tuareg then carried the valuable goods from the south right across the desert to Sijilmasa or other Moroccan towns at the northern edge of the Sahara. From there, other traders eventually passed them on to the Islamic Middle East and Europe – the gold coinage of late medieval European states was minted from African gold.

Fiercely independent, the Berber Tuareg had survived in their harsh desert environment since ancient times, living in characteristic blue tents and herding camels, goats, and sheep. They also exploited the salt deposits found in the heart of the Sahara at Taghaza, almost midway along the 1,600-km (1,000-mile) route from the Mali Empire to Sijilmasa. Great slabs of rock salt were strapped on to the backs of camels for the desert journey, fetching a high price in the Malian markets. Salt was so highly valued that it was sometimes known as "white gold".

◁ **Tuareg veil**
This Tuareg man wears the traditional indigo-dyed tagelmust, a combined turban and veil. It offers maximum protection from windblown sand.

Crossing the desert

The Tuareg trade caravans were impressively large. The Arab traveller Ibn Battuta, who crossed the desert southward from Sijilmassa in 1352, says a caravan typically numbered a thousand camels and could be much larger. Caravans travelled mostly by night to avoid the scorching desert heat, breaking their journey at rare oases. It took around two months to cross the desert, a hard and hazardous journey justified by the high profits to be made.

The rulers of Mali levied duties on the Saharan trade and used the revenue to make their empire a place of learning, religious piety, and architectural splendour. Timbuktu, in particular, flourished around the hub of its marketplace. In the early 15th century, its population may have reached 100,000. Scholars from across the Islamic world came to teach and study in the city, which boasted a university and one of the most extensive libraries in the world, containing tens of thousands of manuscripts.

In the course of the 15th century, the Mali Empire was supplanted by a local rival, the Songhai Empire, but the Saharan trade routes continued to flourish. In subsequent centuries, however, the vast expansion of oceanic trade, dominated by European maritime nations, reduced the international significance for trade of the interior of North and West Africa. The whole area eventually fell prey to French imperial ambitions in the 19th century.

△ **Sankore Mosque**
Timbuktu is famous for its mosques made of sun-dried mud and sand. The Sankore Mosque is part of the city's university, dating from the golden age of the Mali Empire.

TRAVELLER AND SCHOLAR, 1304–1369

Ibn Battuta

In medieval times, no one travelled further than Muhammad Ibn Battuta. Born in North Africa, he voyaged across deserts, steppes, and oceans, recording the splendours of Baghdad, Samarkand, Beijing, and Timbuktu.

Born into a family of judges and educated in Islamic law, Ibn Battuta left his home in Tangier, Morocco, at the age of 21 to make the pilgrimage to Mecca. Having embarked on the journey, he fell in love with travel for its own sake. He was not to return home to Tangier for 14 years.

In the early 14th century, great opportunities were open to a Muslim traveller. The world of Islam, united by religious custom, trade, and the language of the Qur'an, extended from Spain to Indonesia, and from the Central Asian steppe to Zanzibar. After completing his pilgrimage, Ibn Battuta made ever longer journeys across this interconnected world. His first foray took him overland around Iraq and Persia. He then travelled by sea down the east coast of Africa, where traders had spread Islam as far south as modern-day Tanzania. Between these journeys, he perfected his legal studies in Mecca, making himself readily employable in any Muslim state he might visit.

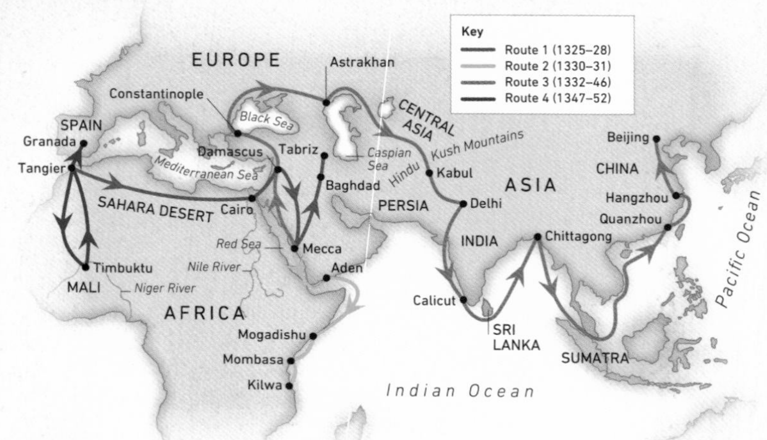

◁ **Routes through the East**
Ibn Battuta's travels took him across most of the world known to Muslims in the 14th century. Only Christian Western Europe was closed to him because of its religious intolerance.

▷ **Ibn Battuta in Egypt**
An Egyptian guide shows Ibn Battuta the ruins of ancient Egypt during his visit to Cairo in 1326. Cairo was then one of the world's greatest cities.

To the ends of the Earth

In 1332, he decided to seek a place at the wealthy court of the Muslim ruler of Delhi. Instead of going to India by sea routes from Arabia, he made a vast detour overland via Christian Constantinople, along the Silk Road into Mongol-ruled central Asia, and across the snow-capped Hindu Kush. The Sultan of Delhi, Muhammad bin Tughluq, was happy to be served by a visiting scholar, but Ibn Battuta found him a capricious employer. Eventually, sent on a mission to southern India, he decided not to return to Delhi. Instead, he followed the sea trade routes from Sri Lanka around southeast Asia to imperial China.

In the known world there was almost nowhere further to travel. In 1347, Ibn Battuta finally headed for home. His return coincided with the scourge of the Black Death, the effects of which he observed with horror. His parents had also died in his absence. Barely stopping at Tangier, he continued to travel, first into Muslim-ruled southern Spain, and then by camel caravan across the Sahara Desert to the fabled empire of Mali. Only then had he seen enough.

Having travelled some 120,000 km (75,000 miles) in 30 years, he settled in Morocco and began to dictate the story of his travels. Although the accuracy of some of the material has been queried, his *Rihla* (*Travels*) remains a key source of information about the world of the Middle Ages.

◁ **Hagia Sophia**
Ibn Battuta was impressed by the splendour of the cathedral of Hagia Sophia when he visited Constantinople, capital of the Byzantine Empire. He spent a month in the city in 1332.

> “ I set out **alone**, having **neither fellow traveller**... **nor caravan** whose **part** I might **join**. ”
>
> IBN BATTUTA, *RIHLA*, 1354

KEY DATES

- **1304** Born 25 February in Tangier, Morocco, to a family of Berber origin.
- **1325** Makes the pilgrimage to Mecca, via the coast of North Africa.
- **1326** Visits Cairo and Damascus en route to Medina and Mecca.
- **1327–28** Travels around Iraq and Iran.
- **1330–31** Sails down the Red Sea to Aden and south along the coast of Africa to Kilwa.
- **1332–34** Journeys to India via the Byzantine Empire, Central Asia, and Afghanistan.
- **1334–42** Serves at the court of the Sultan of Delhi.
- **1342–46** Travels to China via southern India, Sri Lanka, and Southeast Asia.
- **1352** Heads south across the Sahara to the Mali Empire and the River Niger.
- **1354** Returns to settle in Tangier and dictates the story of his travels.
- **1369** Dies in Tangier aged 64 or 65.

BYZANTINE EMPEROR ANDRONIKOS III PALAEOLOGOS, WHO MET IBN BATTUTA

THE PURPORTED TOMB OF IBN BATTUTA IN TANGIER, MOROCCO

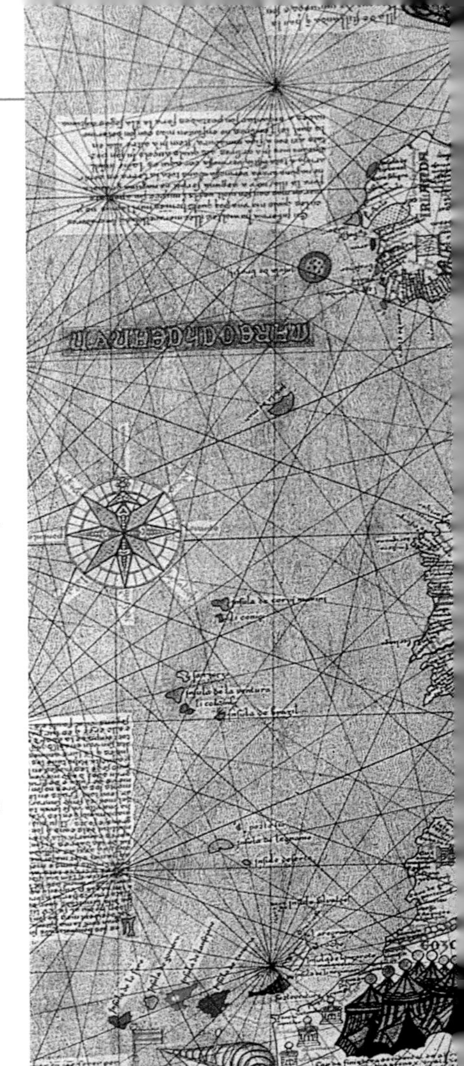

Medieval maps

Medieval maps were compendiums of myth, religious tradition, and inaccurate travellers' tales. Even as familiar regions became better mapped, much of the wider world was left in obscurity.

△ **Hereford Mappa Mundi**
The picture of the world presented in this mappa mundi is to a large degree mythical. Asia and Africa are peopled with monsters and figures from biblical stories and legends.

Medieval traders, pilgrims, and Crusaders setting out from Europe for the East did not consult maps. The maps which existed in Christian Europe were not intended to help travellers find their way to distant lands, but rather to represent the broad idea of the world as conceived by medieval Christians. A fine example is the mappa mundi (map of the world) in Hereford Cathedral, England, dating from around 1300. The world is shown as divided into three continents – Europe, Asia, and Africa – bordered by the Nile, the Don, and the Mediterranean, and surrounded by a circular outer ocean. The mapmakers have placed Jerusalem, because it is the site of Christ's crucifixion and burial, at the centre of the circle of the world. East is at the top. Superimposed on this general geographical scheme is a wealth of imagery relating to biblical stories and ancient legends. There are exotic animals and mythical beasts, the Garden of Eden, and the location of Noah's Ark, human monsters, and the Minotaur's labyrinth. Other medieval mappa mundi include the world map in the Béatus manuscript, made at Saint-Sever Abbey in France around 1050, and the Ebstorf map from northern Germany.

Medieval Muslim mapmakers produced more accurate world maps. They had two key sources not available to the makers of the Hereford Mappa Mundi: the works of the ancient geographer Ptolemy, long known to the Arabs but at that time still unavailable to Europeans, and information from Muslim traders and sailors who voyaged across the far-flung lands of Islam. As a result, Muhammad al-Idrisi, an Arab geographer working at the religiously tolerant court of King Roger II of Sicily in the 12th century, was able to make a map showing North Africa, Europe, and Asia with a fair degree of realism.

Ptolemy rediscovered

From the mid-13th century, a more practical tradition of mapmaking began to develop in Christian Europe. Made to assist with the navigational needs of mariners, portolan charts outlined coasts, islands, and reefs, with a scale marker to indicate distance and compass

> " The **globe**... is divided into **three parts**, which are called **Asia**, **Europa**, **Africa**. "
>
> ISIDORE, ARCHBISHOP OF SEVILLE, IN *ETYMOLOGIAE*, 633

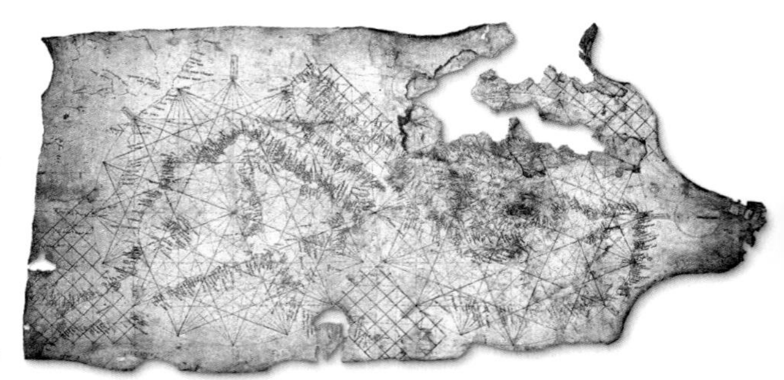

▷ ***Carta Pisana***
Dating from the late 13th century, the *Carta Pisana* is the oldest surviving portolan chart. Intended for use by Italian sailors, it accurately depicts the Mediterranean coastline and its ports.

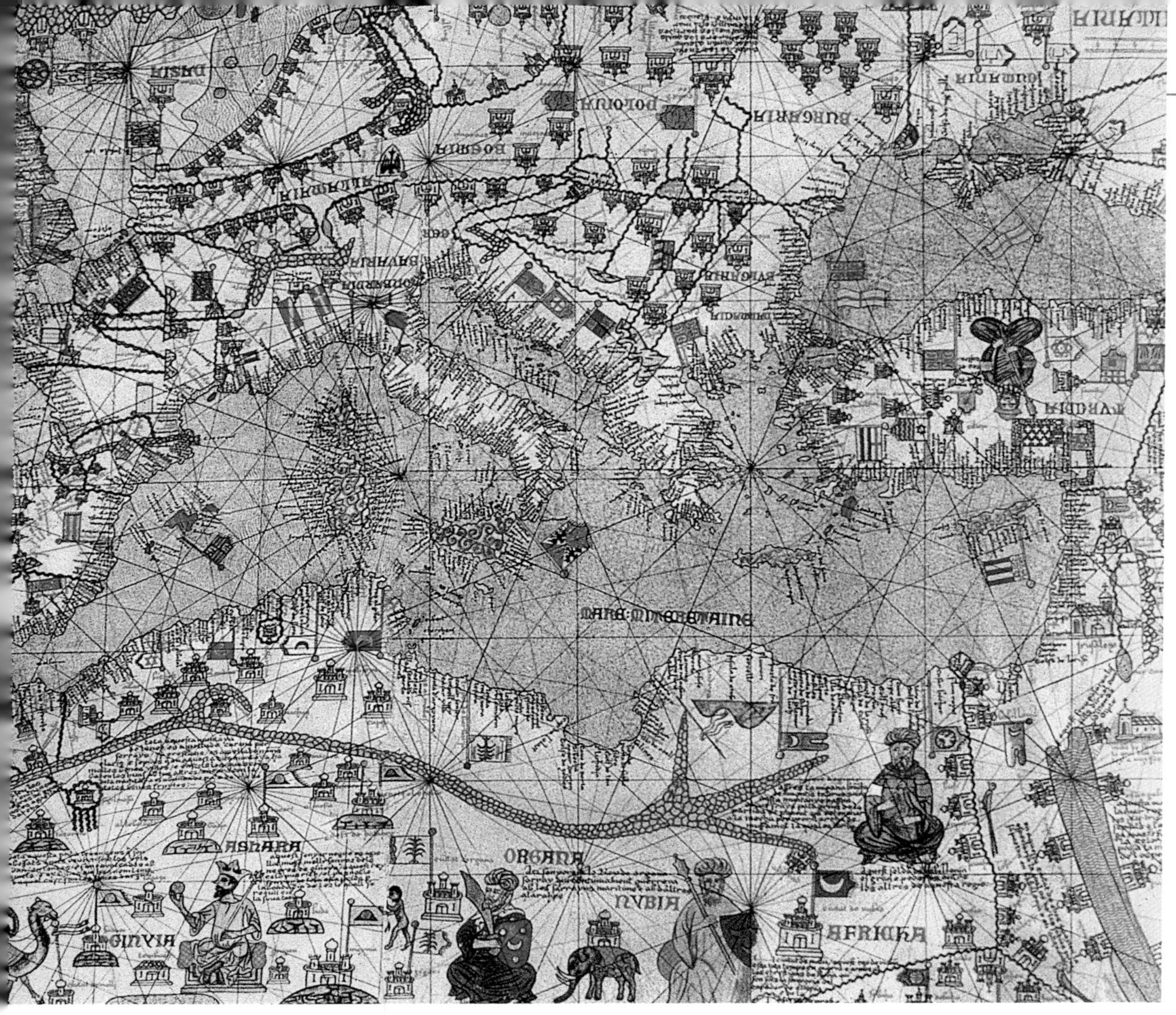

◁ ***Catalan Atlas***
Made in Majorca in 1375, the *Catalan Atlas* world map has many features of a portolan chart, with compass roses and rhumb lines. Superimposed on it are elements from medieval travellers' tales, both real and imagined.

roses to show direction. These charts were crisscrossed with networks of "rhumb lines" drawn from the points of the compass, to help pilots follow a precise course. Often extraordinarily accurate in tracing coastlines, given the technology available at the time, they were first produced in the maritime cities of Italy, later spreading to Catalonia and Portugal.

Portolans were practical regional maps, mostly covering parts of the Black Sea and Mediterranean coasts. Over time, however, the techniques acquired through making portolan charts fed into attempts to map the wider world. Pietro Vesconte, a Genoese cartographer working in Venice, made a reputation for himself with his portolans in the early 14th century. Around 1320, he combined his chart-making skills with the tradition of the mappa mundi to create a world map of much greater accuracy than previous efforts in Europe.

△ **Portolan charts**
This chart was made in 1559. It shows how much progress was made in cartography once the age of oceanic exploration had begun.

Another map that combined the portolan and mappa mundi traditions was the *Catalan Atlas* (1375). Produced by Jewish cartographer Abraham Cresques in Majorca, it incorporated information from the travel accounts of Marco Polo and the tales of Sir John Mandeville – including their fantastical elements.

The key event that led to the further development of European cartography was the rediscovery of the works of Ptolemy via Islamic sources in the 15th century. Ptolemy established the importance of defining places by coordinates of longitude and latitude. Although European mapmakers had always known the world was a sphere, Ptolemy gave them a precise estimate of its size. However, his estimation was wrong, as he had judged Earth to be only three-quarters of its true size. It was this miscalculation that would send Christopher Columbus westwards in search of a passage to China.

THE AGE OF DISCOVERY
1400–1600

THE AGE OF DISCOVERY, 1400–1600

Introduction

The period between 1400 and 1600 is known as the Age of Discovery because it was the time when many navigators from Western Europe made voyages of exploration and discovered places that were previously unknown to Europeans. As a result, countries such as Spain, Portugal, France, and England formed worldwide trading networks, and founded settlements in Africa, America, and Asia. This gave them a lasting influence, not only in the countries they settled, but also globally.

Maritime pioneers

The main reason for embarking on these often dangerous journeys into the unknown was trade. European merchants already imported costly goods, such as spices and silk cloth, from the Far East, but these items came via the long overland route across Asia. A sea route could provide a more reliable, and possibly faster, way of importing goods from the east, and, it was hoped, access to valuable new markets. Members of the ruling European families, such as Prince Henry of Portugal (also known as Henry the Navigator), Ferdinand and Isabella of Spain, and Elizabeth I of England, all encouraged explorers to search for new routes to Asia.

Among the most important of these early voyagers were Vasco da Gama, who was the first European to reach India via the Cape of Good Hope, and Christopher Columbus, who crossed the Atlantic to become the first European since the Vikings to land on American soil. These pioneers and their followers established sea routes, set up staging posts on the African and Caribbean coasts, and inspired like-minded adventurers to follow their example.

IN 1492, CHRISTOPHER COLUMBUS BECAME THE FIRST EUROPEAN SINCE THE VIKINGS TO REACH THE AMERICAS

IN 1501, AMERIGO VESPUCCI DISCOVERED THAT THE AMERICAS WERE NOT ATTACHED TO ASIA

SCHOLAR ANTONIO PIGAFETTA CIRCUMNAVIGATED THE GLOBE WITH FERDINAND MAGELLAN'S CREW

> " By **prevailing over all obstacles** and distractions, one may unfailingly **arrive at his chosen goal**. "
>
> CHRISTOPHER COLUMBUS

Further waves of explorers included Ferdinand Magellan, whose project to circumnavigate the globe was completed by his crew after his death. At the same time, conquistadors defeated the Aztec and Inca empires, beginning the long period of European rule in Mexico and South America. These later adventurers were motivated less by trade than by the quest for resources – they expected South America to be rich in gold and silver. However, trading never ceased, and when the French began to explore Canada, they profited from the fur trade as well as from the territory they claimed for France.

Exchange of goods

Most of these expeditions were small in human numbers. Columbus made his first voyage with just three ships; Pizarro set out to conquer Peru with 180 men; Cortés had 500 soldiers. However, they had the advantage of superior technology: navigational aids, such as the backstaff and compass, helped them find their way, and plate armour and firearms ensured victory over ill-equipped natives. As a consequence, the impact of these voyages was enormous.

The transatlantic explorers in particular created an entirely new network of exchange, through which crops, livestock, and technologies crossed the world, albeit in tandem with infectious diseases. Maize, tomatoes, and potatoes came to Europe for the first time, while chickens, pigs, and horses crossed from Europe to America. Such networks ensured the eventual industrialization of America, but they also paved the way for the slave trade. The Age of Discovery profoundly changed both the old and new worlds alike.

IN 1519, HERNÁN CORTÉS BEGAN THE PROCESS OF CONQUERING MEXICO FOR SPAIN

IN THE 1530S, FRANCISCO PIZARRO DEFEATED THE INCAS AND CLAIMED THEIR LANDS FOR THE SPANISH THRONE

A MAP OF THE EAST COAST OF NORTH AMERICA BASED ON THE FRENCH DISCOVERIES OF 1534–41

Fleet commander
Admiral Zheng He, who led seven Chinese naval expeditions between 1405 and 1432, was born a Muslim – his father had made the pilgrimage to Mecca. Like all officials at the Ming imperial court, he was a eunuch.

FLEET ADMIRAL, 1371–1433

Zheng He

In the early 15th century, before the beginning of the European Age of Exploration, Chinese admiral Zheng He commanded what was then the world's largest fleet on a series of epic voyages from China to India, Arabia, and Africa.

Zheng He was a court eunuch in the service of the Yongle Emperor, Zhu Di. At the time, the newly founded Ming dynasty was full of energy, determined to assert China's international status. Zhu Di ordered a great "treasure fleet" to be built to project China's power and prestige far beyond its shores. As the emperor's most trusted official, Zheng He was selected to command this fleet.

The fleet consisted of more than 1,600 of the largest wooden sailing ships ever built. Zheng He's task was to lead expeditions around Southeast Asia and into the Indian Ocean, to impel local rulers to acknowledge the Chinese emperor as their overlord and send tribute to the Chinese court. The fleet carried some 27,000 soldiers, in case the required homage was not willingly given.

Between 1405 and 1422, Zheng He made six voyages, each taking about two years. Setting off from Nanjing on the Yangtze River, the first three voyages sailed to Sri Lanka and southern India via the coasts of Vietnam and the islands of Java and Sumatra. The fourth voyage carried on beyond India to the Persian port of Hormuz, and the fifth and sixth voyages extended to Arabia, the Red Sea, and the east coast of Africa. Each voyage returned with ambassadors from Indian Ocean states and a wealth of tribute goods – including exotic animals such as giraffes and zebras for the imperial menagerie.

When voyages were suspended in 1424, Zheng He continued to hold high office. He resumed the role of admiral for a final voyage of the treasure fleet in 1432, but died the following year. Subsequently, imperial China turned inward, banning oceanic voyages and largely erasing the memory of Zheng He's missions for five centuries. In parts of Southeast Asia, however, he was venerated by many Chinese overseas, who built temples in his honour.

Key
First three voyages
Fourth voyage
Fifth and sixth voyages

Black Sea, Caspian Sea, Mediterranean Sea, TURKMENISTAN, CENTRAL ASIA, PERSIA, ASIA, CHINA, Nanjing, Yangtze River, Persian Gulf, Hormuz, Guangzhou, Red Sea, ARABIA, Medina, INDIA, Pacific Ocean, VIETNAM, Bay of Bengal, Arabian Sea, AFRICA, Calicut, SRI LANKA, PHILIPPINES, Mogadishu, Indian Ocean, JAVA AND SUMATRA

△ **Map**
Zheng He's voyages followed established coastal trade routes around the Indian Ocean from Malaysia to East Africa.

◁ **Mighty treasure ship**
This modern representation of the treasure fleet gives an impression of the size of the junks Zheng He commanded: each was almost 100 m (330 ft) long.

KEY DATES

- **1371** Born into a Muslim Hui family in the remote Yunnan region of China.
- **1381** His father is killed resisting the Ming conquest of Yunnan; he is castrated and made a servant of Zhu Di, Prince of Yan.
- **1402** Leads an army during Zhu Di's seizure of the imperial throne and is rewarded with high office at court.
- **1405** Commands the first voyage of the Chinese fleet built on Zhu Di's orders, sailing as far as southern India.

A GIRAFFE BROUGHT TO CHINA AS TRIBUTE

- **1407–19** Carries out four more voyages, travelling as far as Arabia and East Africa.
- **1422** Returns from his sixth voyage. Further sailings are suspended on the emperor's orders.
- **1422–31** Serves as military governor of the city of Nanjing.
- **1432** Embarks on a final voyage into the Indian Ocean. He dies in 1433 and is buried at sea.

TEMPLE DEDICATED TO THE VENERATION OF ZHENG HE IN PENANG, MALAYSIA

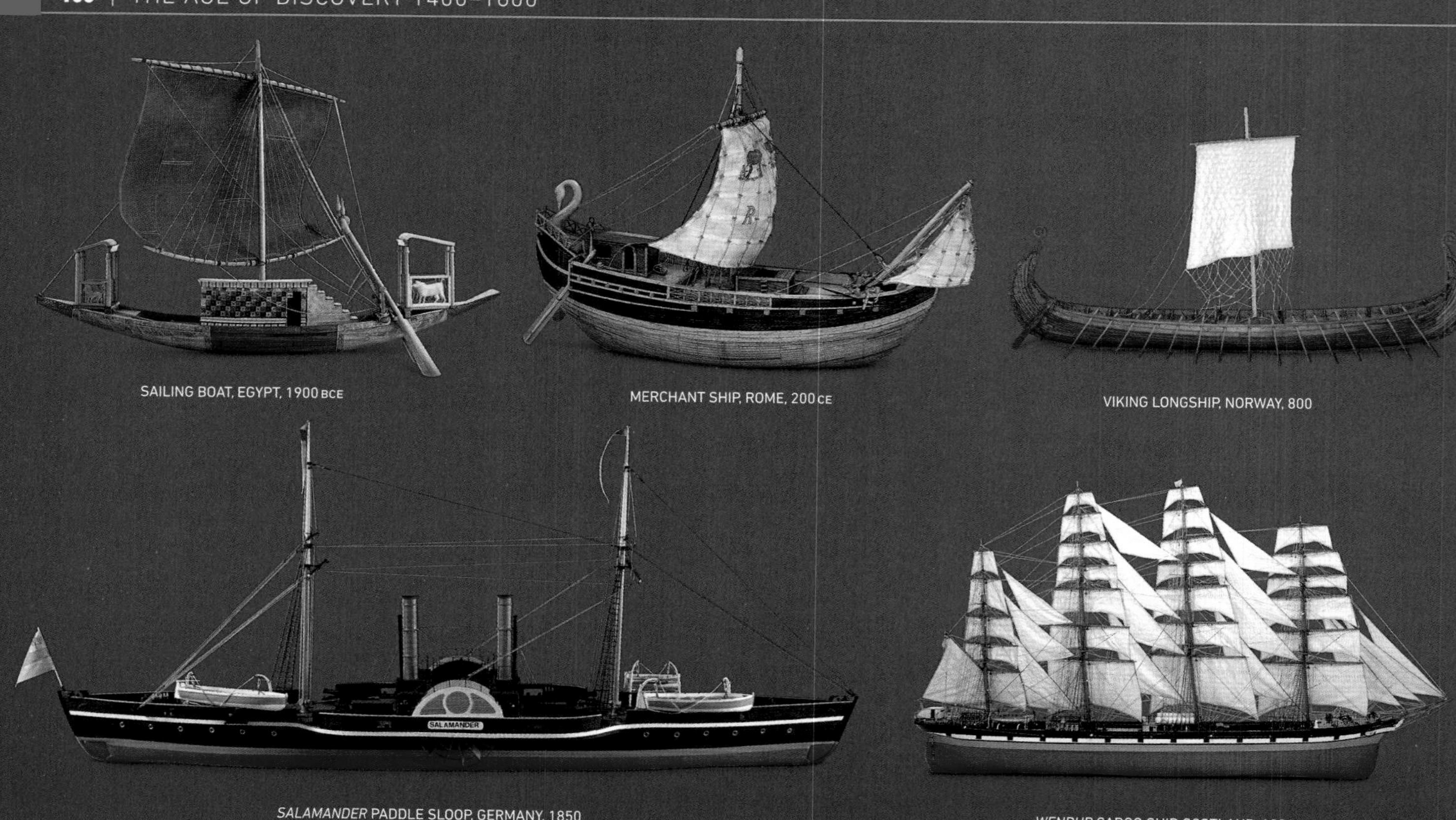

SAILING BOAT, EGYPT, 1900 BCE

MERCHANT SHIP, ROME, 200 CE

VIKING LONGSHIP, NORWAY, 800

SALAMANDER PADDLE SLOOP, GERMANY, 1850

WENDUR CARGO SHIP, SCOTLAND, 1884

Ships

For millennia, ships have been the primary mode of transport for carrying people and goods over long distances.

Many early civilizations depended on rivers for their survival: the ancient Egyptians on the Nile; the Mesopotamians on the Tigris and Euphrates; the earliest societies in what are now India and Pakistan on the Indus. As well as providing water for people and their crops, the rivers created an opportunity for water-borne transport.

The earliest known boats were made from animal skins and carved-out tree trunks. The Egyptians made sailing boats from papyrus reeds and the Greeks raised the art of rowing to a level that has never been surpassed. Later, the Vikings built sturdier vessels from trees and took to the oceans, propelled by wind and manpower.

In the 15th century, shipbuilders in Europe blended the best of what had gone before to construct massive, three- and four-masted vessels for the high seas, and added guns so that they could conquer other nations. Trade came in the wake of conquest, and huge merchant ships – the East Indiamen – were built to carry goods to trade with Asia.

The Age of Sail ended with the Industrial Revolution, when steam engines replaced sails to power transoceanic vessels. Steam engines have now mostly been replaced by diesel and gas-turbine engines, but many warships still use steam for propulsion, produced by onboard nuclear reactors.

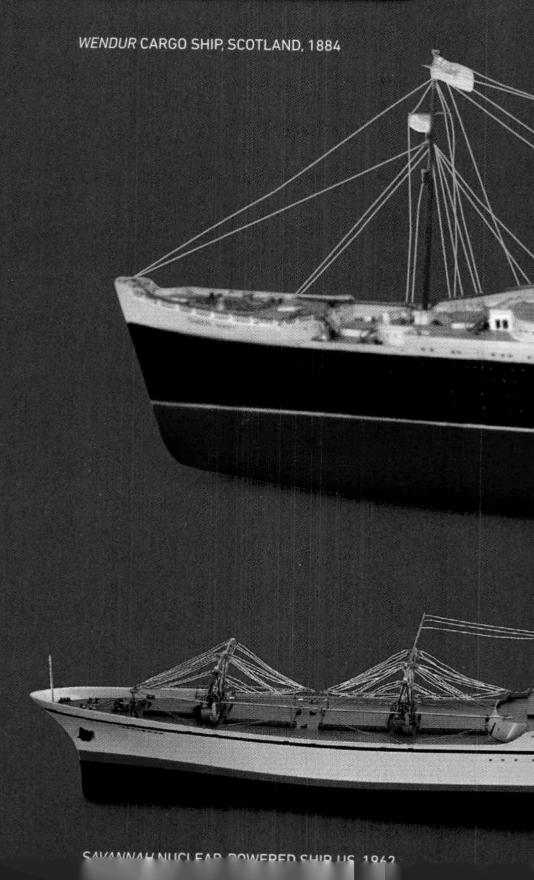

SAVANNAH NUCLEAR-POWERED SHIP, US, 1962

SANTA MARÍA CARRACK, SPAIN, 1492

SAVANNAH SAIL/STEAMSHIP, US, 1819

JUNK, CHINA, 1840

ANGUILLA FISHING STEAMSHIP, ITALY, 1923–24

RMS *QUEEN MARY* OCEAN LINER, UK, 1936

Around Africa to India

Seeking the source of one of the most valuable items of world trade – spice – the Portuguese monarchy sent a naval expedition on a pioneering ocean voyage to India around the Cape of Good Hope.

△ **Vasco da Gama**
After his famous voyage in 1497–99, da Gama led a second fleet to India in 1502. He was appointed governor of all Portuguese territories in the east in 1524.

Portuguese exploration of the African coast began in the early 15th century, sponsored by Prince Henry the Navigator. Seeking Christian allies to fight Muslims in North Africa, and eager for the gold he knew could be found somewhere south of the Sahara, the Portuguese prince promoted voyages as far south as Sierra Leone. By the time Henry died in 1460, Portugal had begun a profitable trade in West African slaves and gold.

◁ **Portuguese caravel**
The key ship used in the Portuguese exploration of the African coast was the caravel. Small and nimble, it could sail in shallow coastal waters and up rivers.

First attempts

After a lull of two decades, the accession of King John II in 1481 triggered new voyages of exploration. Although information on Africa was scant, contemporary knowledge suggested it was possible to sail around the south of the continent into the Indian Ocean, gaining direct access to the Asian spices whose valuable trade was monopolized by Muslim powers and their Venetian trading partners. The king assembled astronomers and mathematicians to produce a manual of navigation, and sent a secret mission to Egypt to scout the lands around the Indian Ocean.

At first, the new project floundered. In 1482 and 1484, Portuguese sailor Diogo Cão travelled as far south as modern-day Namibia without finding the expected eastward turn towards India. In 1487, the king sent another maritime expedition down the African coast, led by Bartolomeu Dias. Early in 1488, lost in the ocean, Dias was carried by wind and currents around the Cape of Good Hope without seeing it, before making landfall in southern Africa.

At this point, Dias's sailors insisted on turning for home, but he returned to Lisbon with the assurance that the route to India was open.

Da Gama's triumph

A decade later, under John II's successor, Manuel I, the Portuguese were finally ready to embark for India. Commanded by Vasco da Gama, a fleet of four well-armed ships set sail from Lisbon in July 1497. Da Gama was not a sailor but a nobleman and diplomat. His mission was to act as a Portuguese ambassador, establishing trade links with Indian states. The fleet travelled to the Cape Verde Islands, then followed an oceanic route to southern Africa, spending 13 weeks out of sight of land – an impressive feat of navigation. After rounding the Cape of Good Hope, the Portuguese sailed up the coast of East Africa to Malindi. There, they met Ahmad ibn Mājid, a Gujarati sailor who was prepared to guide them across the Indian Ocean to Calicut in southern India.

Da Gama was not warmly received by the local ruler, but he succeeded in loading up with pepper and cinnamon before embarking for home.

◁ **Bartolomeu Dias**
Leading an expedition of two caravels and a supply ship, Dias rounded the Cape of Good Hope in 1488.

“ God alone is **the master and pilot** who had to deliver them by **his mercy**... ”

VASCO DA GAMA, ACCOUNT OF HIS VOYAGE, c.1500

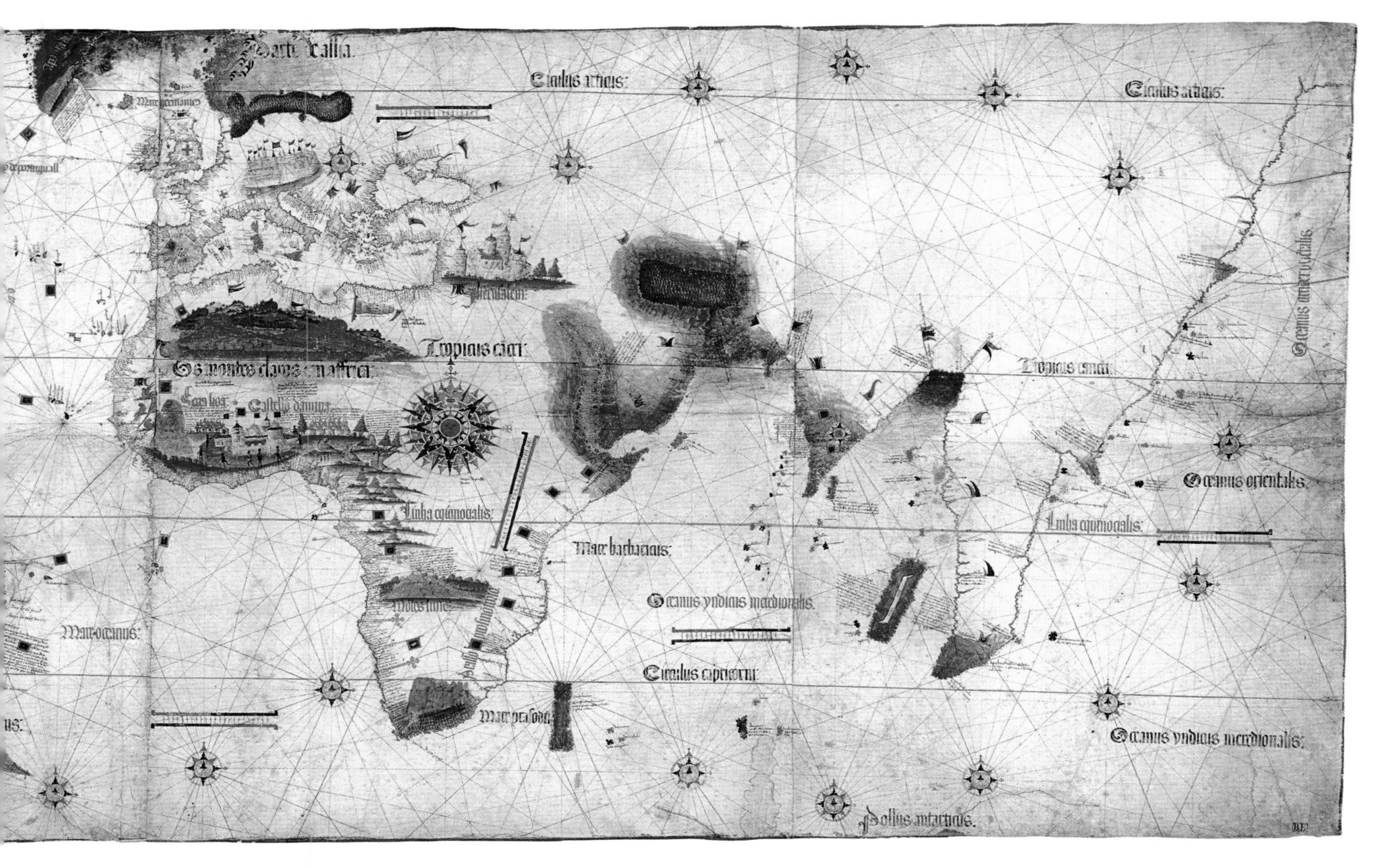

△ **The Cantino Planisphere**
Completed by an unknown cartographer in 1502, the Cantino Planisphere depicts the world as it was known to the Portuguese after da Gama's voyages of exploration.

Recrossing the Indian Ocean, the expedition was slowed by headwinds, battered by storms, and decimated by disease. Da Gama had to suppress a mutiny, and one of the ships was abandoned. When he reached Lisbon in September 1499, he had lost more than half his men. But da Gama's return was celebrated as a triumph. Subsequent expeditions cemented the Portuguese presence in India, and discovered Brazil. In the course of the 16th century, Portuguese mariners sailed as far as China and Japan, and the country grew rich on the trade in spices and other Asian goods.

◁ **Jerónimos Monastery**
This monastery at Belém, Lisbon, was built to celebrate da Gama's voyage to India, and was paid for with a tax on Portugal's trade with Africa and Asia.

Arrival on Hispaniola
Columbus's landing on Hispaniola is the subject of this 16th-century engraving by Theodore de Bry. Some of the explorer's men erect a cross by the coast, indicating their intention to convert the locals to Christianity.

A new world

From 1492 to 1502, explorer Christopher Columbus made four transatlantic voyages. These marked the beginning of the European colonization of America, and the forging of permanent links between the two continents.

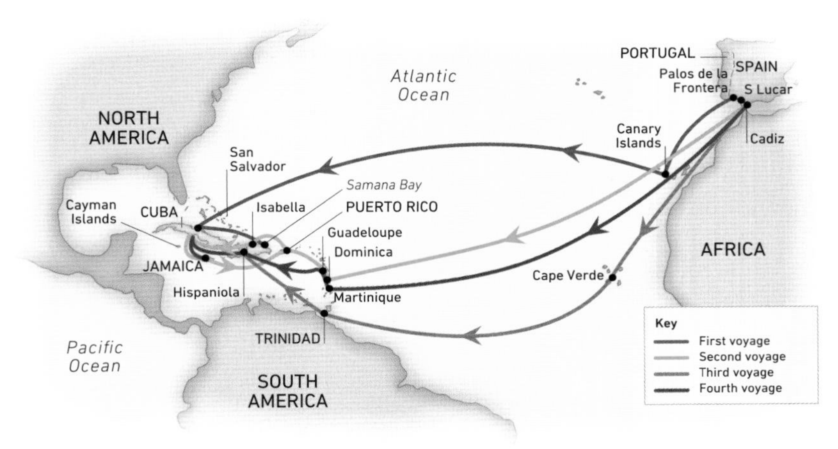

◁ **The four voyages of Columbus**
Columbus followed various routes across the Atlantic as he discovered more about the prevailing winds. The usual Spanish route westwards was to go south to the Canaries and then catch the trade winds blowing towards the Caribbean.

Christopher Columbus was born in Genoa, an Italian city-state with a strong tradition of maritime trade. An experienced navigator, he became convinced that the best route from Europe to the Spice Islands of Southeast Asia was westwards across the Atlantic.

During the 1480s, Columbus tried to find sponsors for a transatlantic expedition, but he was turned down by the rulers of Portugal, Venice, Genoa, and England, whose advisors knew that he had greatly underestimated the distance involved. However, Ferdinand and Isabella, the monarchs of Spain, decided to back Columbus, in order to prevent others from benefitting from the lucrative trade he might open up. He was able to crew and equip three ships – the *Santa Maria*, a large carrack, and two smaller caravels, the *Pinta* and the *Niña* – for the journey.

The first voyage

Columbus's small fleet set off from Palos de la Frontera, on Spain's southwestern coast, on 3 August 1492. First, they stopped for supplies and repairs in the Canary Islands, a Spanish possession, then Columbus and his men then sailed westwards into unknown waters for what turned out to be a five-week crossing of the Atlantic.

On 12 October, the lookout on the *Pinta* reported that he had spotted land – an island in the Bahamas that they called San Salvador, although the native Lucayan people called it Guanahani. Columbus noted that the weapons of the locals would be no match for his ships' weaponry if it came to a battle. However, nearly all the people Columbus met on his first voyage were friendly. As he sailed on, the winds took him to the island of Ayti (present-day Haiti), which he renamed Hispaniola. He found the inhabitants, the Taíno, to be handsome, gentle, and friendly. Admiring their gold ear ornaments, he kidnapped a number of them, hoping they would lead him to the source of the gold. The lure of treasure led the captain of the *Pinta*, Martín Alonso Pinzón, to break away in search of an island rumoured to have lots of gold. The *Santa Maria* and *Niña*, meanwhile, carried on exploring the coast of Hispaniola until the *Santa Maria* ran aground, on Christmas Day 1492. Columbus was forced to abandon his flagship, continuing with just one vessel until the *Pinta* rejoined them on 6 January 1493.

△ **Christopher Columbus**
This portrait of the explorer was engraved about 100 years after his famous voyages to America.

Viceroy of the Indies

Leaving 39 men to start a colony, La Navidad, Columbus continued along the Hispaniola coast. At the Bay of Rincón, he met the local Cigüayos people and hoped to trade with

IN CONTEXT
The carrack

Columbus's largest ship, the *Santa Maria*, was a carrack. A spacious vessel with several masts, and larger and more sturdy than his smaller caravels, a carrack was big and strong enough to survive rough seas, and was large enough to carry plenty of trade goods or supplies for a long voyage. Carracks were developed in the 14th century, and were used by Mediterranean seamen for trading voyages in the rough seas of the North Atlantic, as well as for journeys down the coast of Africa. They carried a mix of square and lateen (triangular) sails, the latter enabling them to tack against the wind, which was useful on long trips through variable winds.

MODEL OF THE THREE-MASTED *SANTA MARIA*

△ **Columbus with King Ferdinand and Queen Isabella**
After two years of negotiations, the Spanish monarchs agreed to back Columbus's 1492 voyage. They promised him 10 per cent of any income from newly discovered lands.

» them, but the Cigüayos refused and a clash broke out. Some of his men were injured, and Columbus headed for Spain with a group of captured Cigüayos, although some died en route. A severe storm forced them to take shelter in the Azores, where the local governor arrested some of the crew on suspicion of piracy. He released them after two days, but further storms delayed their return to Spain until 15 March.

Because he had miscalculated the distance to the Spice Islands, and because no-one knew America existed, Columbus was sure he had reached the East Indies. The Spanish rulers agreed and made him Admiral of the Seven Seas and Viceroy of the Indies. A series of return journeys was planned.

A new colony

Columbus's second expedition (1493–96) was the biggest, with 17 ships carrying some 1,200 men, including farmers, soldiers, and priests, who planned to found a colony. Columbus explored the coasts of Dominica and Guadeloupe, sailed along the Lesser and Greater Antilles, and landed in Puerto Rico. He also found that La Navidad had been destroyed in disputes with locals; many colonists had been killed. He therefore founded a new settlement at La Isabela on Hispaniola. Keeping the colony going proved difficult. The precontact Taíno culture was already well-organized and resisted the colonists' attempts to exploit them, and the colonists themselves were disgruntled not to find the many riches Columbus had promised them.

◁ **Coat of arms**
The coat of arms granted to Columbus in 1493 featured the lion and castle of Léon and Castile together with a group of islands.

Columbus humiliated

On his third voyage (1498–1500), Columbus sailed with six ships, three of which went directly to Hispaniola laden with supplies for the colonists. He took the other three to explore Trinidad and the South American coast. He became ill, however, and returned to Hispaniola, only to face a rebellion among the Spanish colonists. Their disillusion with their difficult life in the Caribbean, together with Columbus's authoritarian rule as governor, had finally boiled over. They put Columbus in chains and sent him back to Spain, where he was sacked from the governorship.

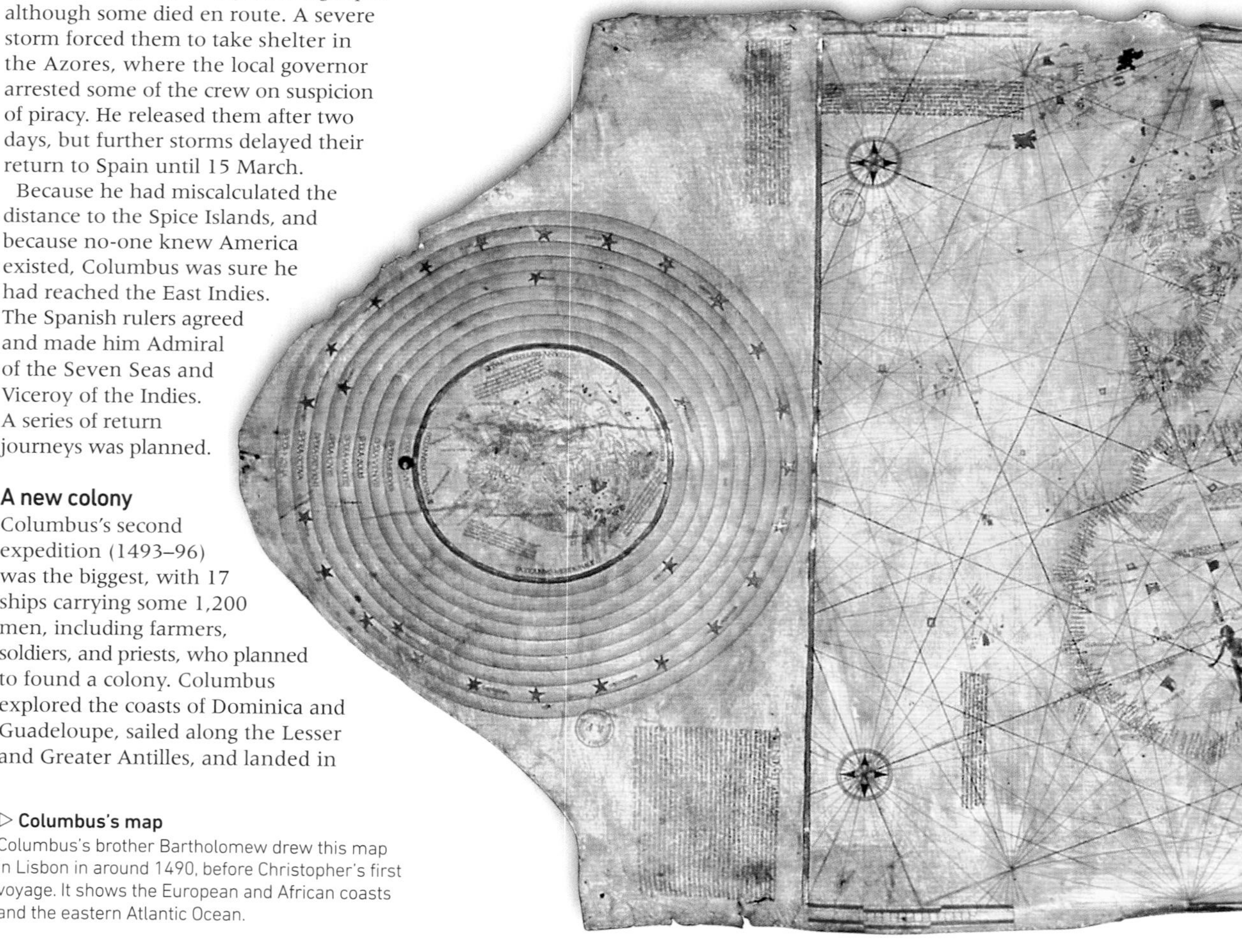

▷ **Columbus's map**
Columbus's brother Bartholomew drew this map in Lisbon in around 1490, before Christopher's first voyage. It shows the European and African coasts and the eastern Atlantic Ocean.

AQVILO
ASIA
AFRICA
Mare glaciale
SEPTENTRIO
90
80
MARE GLACIALE
SERICA REGIO
SCITHIA EXTRA IMAV
ASIA
SINVS GANGETICVS
EQVINOCTIALIS
MARE INDICVM
OCCEANVS INDICVS MERIDIONALIS
TROPICVS CAPRICORNI
MARE PRASSODV
MADAGASCAR
SEYLAM
NOTVS
AVSTER
130
140
150
160
170
180
190

△ **The *Victoria***
The sole ship in Magellan's fleet that made the full journey, the *Victoria* was a carrack of 85 tons, with a crew of 42. Magellan named it after the Santa Maria de la Victoria de Triana Church in Seville.

Circumnavigating the globe

Portuguese navigator Ferdinand Magellan launched the first full round-the-world sea expedition. He did not survive the journey, but a few of his seamen made it home, achieving a complete circumnavigation of the globe.

The transatlantic voyages of Christopher Columbus, Amerigo Vespucci, and others proved that there were lands on the western side of the Atlantic. Columbus believed that these lands were the Spice Islands of Southeast Asia, but Vespucci realised that they were in fact part of a new continent, which geographers named America (see pp.116–17). However, some navigators were still convinced that if they could find a strait through America, they would reach Asia and the riches of the Spice Islands. One of these was the Portuguese seaman Ferdinand Magellan.

Magellan changes sides

Magellan was a highly experienced sailor. From 1505 to 1514, he spent several years in India fighting in battles on the Portuguese side, sailed with the first Portuguese embassy to Malacca, and served his country in Morocco, before falling out with the Portuguese government when he was accused of doing illegal trade deals with the Moors. He then tried to interest the Portuguese Crown in a westward journey in search of the Spice Islands. If the route was an easier one than the eastward route around Africa and India, this would potentially give Portugal greater profits in the spice trade. However, Magellan was out of favour, and King Manuel I rejected the plan.

◁ **Ferdinand Magellan**
Having served his native Portugal over many years, Ferdinand Magellan lost favour with the Portuguese court and decided to sail under the flag of its rivals, Spain.

Still keen to get his voyage underway, Magellan turned to Portugal's rival, Spain. The Spanish were eager to find a western passage to Asia because, under the Treaty of Tordesillas (1494), the eastern route (and therefore, in effect, the spice trade) was allotted to Portugal. So, in 1518, Magellan was authorized by Spain's ruler Charles I (later on the Holy Roman Emperor Charles V) to journey westwards. Magellan was given royal funds and a number of benefits, including a 10-year monopoly on the resulting trade and the governorship of any lands he discovered.

Crossing the Atlantic

Magellan set sail with five ships and a crew of around 237. Key people on the voyage included Spanish merchant ship captain Juan Sebastián Elcano; Venetian scholar Antonio Pigafetta, who kept a record of the journey that is an invaluable source for historians; and Juan de Cartagena, who was in charge of the trading side of the expedition. The mix of nationalities on board brought Magellan trouble, because many of the Spanish resented being led by a Portuguese captain.

Magellan's five ships left Spain on 20 September 1519. Immediately, the Portuguese King Manual I sent ships in pursuit – Magellan was seen as a traitor for having mounted an expedition with the backing of Spain. However, the Portuguese did not catch Magellan's ships and, after they stopped at the Canary Islands for provisions, they sailed towards Cape Verde. Bad storms delayed them off the coast of Africa, but eventually they set a course southwest towards Brazil. They sighted Brazil on 6 December 1519, but this was a Portuguese possession, so they sailed south, putting down their anchors when they reached what is now Argentina, near modern-day Rio de Janeiro.

Trouble in the South Atlantic

Continuing southwards, they anchored for the winter in a natural harbour in Patagonia that Magellan called Puerto San Julián. Here, over Easter 1520, Magellan hit his first major problem: three of the five captains staged a mutiny. Their revolt was partly the result of continuing tension between the Spanish and Portuguese, and partly an unwillingness to sail further into the unknown, icy South Atlantic waters. Magellan suppressed the rebellion rapidly. Some of the rebel leaders were executed, and one of the »

IN CONTEXT
Wildlife

Magellan and his crew were the first Europeans to sail down the eastern coast of South America into the Pacific. On their way, they noticed a number of animals and birds that were unknown to European naturalists. Antonio Pigafetta noted them in his journal, which is the main source for information about the voyage. One was described as a camel without humps, which was probably a guanaco, although it may have been a llama, vicuna, or alpaca. Perhaps the most notable bird was the penguin, which they called a "goose" because it had to be skinned rather than plucked. Later, scientists named one species the Magellanic penguin, in honour of their discoverer.

MAGELLANIC PENGUIN

◁ **Juan Sebastián Elcano**
Like Magellan, Elcano had been out of favour with his sovereign, having surrendered a Spanish ship to the Genoese to pay off a debt. He joined Magellan's dangerous expedition to gain a pardon from Charles I.

> "...**unlike** the mediocre, **intrepid spirits** seek **victory** over those **things** that **seem impossible**."
>
> FERDINAND MAGELLAN

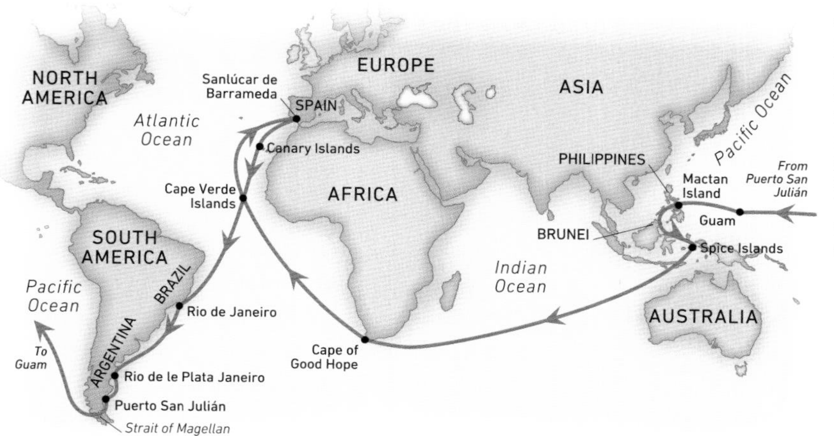

◁ **Magellan's route**
By starting with an Atlantic crossing and locating the strait into the Pacific, Magellan proved that Southeast Asia could be reached by a westerly route. The survivors of the trip made it home via the Indian Ocean.

◁ **Battle of Mactan**
This illustration from a 1626 book by Lenvinus Hulsius shows the Spanish fighting native peoples on the Philippine island of Mactan, where Magellan met his death.

△ **Antonio Pigafetta**
The Venetian scholar Pigafetta took notes on plants, animals, people, languages, and geographical features during the voyage. His *Report* was of great value to future navigators.

» rebel captains, together with the expedition's chaplain, was marooned. Their followers were pardoned and they continued southwards. Finally, on 21 October 1520, they found the strait that would take them west towards Asia. By the end of November, all the ships were through the 600-km (370-mile) passage and into the comparative calm of the waters beyond. Grateful for this respite from storms, Magellan named the sea *Mar Pacifico* (Pacific Ocean).

The Philippines

Sailing northwest, they crossed the Pacific, visiting Guam (where locals stole one of their ships' boats) and arriving in the Philippines on 16 March 1521. Magellan had a Malay indentured servant who helped him speak to the local people. At first, relations were peaceful, gifts were exchanged, and a local leader, Humabon of Cebu, was converted to Christianity. However, another leader, Lapu-Lapu, who was the enemy of Humabon, refused to convert. Humabon persuaded Magellan to launch an armed attack on him. The resulting battle took place on 27 April 1521, when a group of 49 Europeans landed on the island of Mactan and faced a much larger force of locals. Magellan was recognized as the European leader, injured with a bamboo spear, and set upon by islanders, who killed him. The rest of the Europeans escaped in their boats.

The voyage home

The expedition was now much reduced. Casualties on the way and in the Battle of Mactan left them with only enough men to sail two ships, so they burned one and the rest of the team set sail in the *Trinidad* and *Victoria*. They sailed to Brunei, where they saw the splendid court of the ruler and such wonders as tame elephants. On 6 November, they finally reached the Spice Islands and traded with a local sultan, who had not already made a trading alliance with the Portuguese.

Set to return to Spain loaded with valuable spices, they had a further mishap – the *Trinidad* sprang a leak. It looked as if it would take a long time

> "... **all** at once **rushed upon him**... so that they slew our **mirror**, our **light**, our **comfort**, and our **true guide**."
>
> ANTONIO PIGAFETTA, ON MAGELLAN'S DEATH, IN *REPORT*, 1525

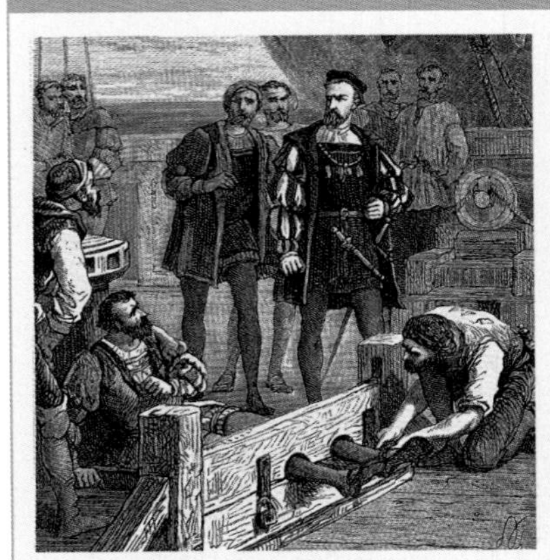

CARTAGENA IS PUT IN THE STOCKS BY MAGELLAN AS PUNISHMENT FOR MUTINY

IN PROFILE
Juan de Cartagena

Juan de Cartagena travelled on Magellan's expedition as Inspector General. This role gave Cartagena the task of overseeing any trading activities and reporting directly to King Charles. Although Cartagena was an accountant and not an experienced seaman, Magellan appointed him captain of one of the ships, the *San Antonio*, as a mark of his importance. However, the pair clashed, Cartagena criticizing Magellan for food rationing when they were delayed and had to ration supplies, and he was removed from his command. Cartagena resented this and organized the mutiny of April 1520. His punishment was to be marooned on an island off Patagonia, along with the priest Pedro Sanchez de la Reina.

to repair, so some of the men stayed with the *Trinidad* (which was later lost to the Portuguese) and the *Victoria*, under the command of captain Elcano, returned across the Indian Ocean towards the Cape of Good Hope and Europe. The *Victoria* and a small crew completed their journey on 6 September 1522, so it was Elcano who was the first to complete a successful circumnavigation of the globe, bringing with him not only a cargo of spices but also information about the Pacific Ocean, proof that the Spice Islands could be reached by a westward route, and more precise knowledge about the size of the Earth. However, the cost of this knowledge was heavy: of the original contingent of around 237 sailors on the expedition, some 219 died en route.

◁ **Strait of Magellan**
Magellan called the waterway through which he passed from the Atlantic to the Pacific, the *Estrecho de Todos los Santos* (The Channel of All Saints). Seven years later, it was renamed the Strait of Magellan.

▽ **Magellan's death**
This 19th-century engraving illustrates how Magellan was killed on Mactan by native people. The locals recognized Magellan as the leader and began to attack him with spears, before finally dispatching him with cutlasses.

Meeting with Moctezuma
Cortés meets Aztec Emperor Moctezuma at Tenochtitlan on 8 November 1519. The Aztecs offer gifts, believing Cortés to be a god, while Moctezuma's feet are kept from touching the ground by a servant.

Cortés and the conquest of the Aztecs

The soldier and colonizer Hernán Cortés led the Spanish conquest of the Aztecs, an event that proved to be a turning point in the history of Mexico and the surrounding regions.

After the voyages of Christopher Columbus and the first Spanish colonies in America and the Caribbean (see pp.118–21), the Spanish Crown authorized several individuals – invader-soldiers known as "conquistadors" – to explore parts of the New World. One of these was Diego Velázquez, who conquered Cuba for Spain, becoming its governor, and whom Spain granted the right to control further exploration of the American mainland. Another was Hernán Cortés, who moved to the Caribbean in 1504 and took part in the Cuban conquest. In 1518, Velázquez gave Cortés approval to explore Mexico, but he then had second thoughts, questioning Cortés's motives for exploring this reputedly rich area, and cancelled the commission. However, Cortés left for Mexico anyway, beginning a series of events that transformed the Mexican mainland forever.

△ **Hernán Cortés**
Cortés, a child of lesser nobility, was a Spanish conquistador who overthrew the Aztec Empire in the 1520s and helped to inaugurate the Spanish colonization of the Americas.

Ally and turncoat

Landing on the Yucatán Peninsula, Cortés defeated the locals' attempts to defend themselves. One woman, called Malintzin, or "La Malinche", who spoke both Maya and the Aztec language of

The leader of the 1541 Spanish expedition to Ecuador was Gonzala Pizarro, the younger brother of Francisco, the conqueror of Peru (see pp.124–25). Francisco de Orellana joined him as a captain and they set out from Quito with a large party, including native people from South America. Their main aim was to find spices, but some were keen to find gold and the mythical El Dorado.

The expedition had a disastrous beginning. While working their way through unknown forests, they swiftly ran low on supplies, and by the time they had covered 400 km (250 miles), they had used up most of their food and lost thousands of their company to disease, starvation, and desertion. Having consumed all of the pigs they had taken with them, the survivors began to eat their dogs. Finding the Rio Coca, they built a boat, and Orellana and 60 men were sent downstream to find food.

The journey downriver

Initially, they fared as badly as before, and were soon eating their shoes and the roots of forest plants, some of which were poisonous. Meanwhile, the current became so fast that they were unable to turn back, and were soon beyond hope of rejoining Pizarro and their comrades. Luckily, when they met a local tribe preparing to defend themselves from these intruders, Orellana was able to speak to them (he was an accomplished linguist) and they managed instead to exchange European goods for food. Pressing on, they reached the Rio Negro, near modern-day Manaus, where they found human heads displayed on posts – a clear warning to keep away from the local people.

Eventually, the river on which they were travelling emptied into a much larger one, which they named the Orellana. They survived by raiding pens of turtles kept by local people, and clashed with a tribe that included fair-skinned warrior-women, who were apparently naked to the waist. They named these warriors "Amazons", after the female warriors of Greek myth (indeed, European geographers later named the entire river after them), although the fighters may well have been men dressed in skirts. There were casualties, but most of the expedition survived the fight. The river proved to be vast, flowing past islands on which they stopped to make further repairs to their boats. Then the wind picked up, a hint that the ocean might be near.

When they reached the Atlantic, Orellana and his companions sailed northwards up the coast towards Guyana. Their boats were separated, but eventually they reunited on the island of Cubagua, off Venezuela. Orellana decided to return to Spain from here, hoping to lead a further expedition and eventually become governor of the Amazon region.

◁ **Orellana's boat on the Amazon**
This 19th-century American wood engraving shows Orellana's boat, under a combination of oars and sail, making its way down the Amazon. The rocky, tree-covered banks made landing difficult.

The aftermath

Orellana did indeed lead an expedition, which embarked in May 1545. However, it ended in disaster, with the deaths of most of its participants, including Orellana himself. Orellana is, nevertheless, remembered for his Amazon voyage, which brought the world's longest river to European notice for the first time. A Dominican friar called Gaspar de Carvajal had travelled with him, and his record of events includes descriptions of the native peoples they encountered, detailing their customs, rituals, and methods of warfare – a treasure-trove of information about the Amazon region at the time of the first European contact.

△ **Francisco de Orellana**
Based on a 16th-century woodcut, this portrait shows Orellana after he had lost an eye in Peru with Pizarro.

◁ **16th-century South America**
Joan Martines, cosmographer to Philip II of Spain, used information from Orellana's Amazonian expedition to produce this map of South America for his atlas of 1587.

IN PROFILE
El Dorado

The story of El Dorado (literally "the golden one") grew up among Europeans in the 16th century. It concerned a South American king who covered himself in gold dust as part of a ritual that was centred on a sacred lake. Eventually, the name El Dorado was applied not just to a person but also to the country in which he lived. Many travellers hoped to find El Dorado and come home rich with golden treasure, and the search for gold motivated the conquistadors and many early colonizers. A possible place of origin for the story is Guatavita in the Colombian Andes, where a number of gold artefacts have been found in a lake.

GOLD FUNERARY MASK FROM PERU, c.9TH–11TH CENTURIES

The Columbian Exchange

The arrival of Europeans in the Americas meant that animals, plants, goods, and diseases also travelled between the two worlds. This series of transfers is known as the Columbian Exchange.

The Columbian Exchange affected life on both sides of the Atlantic. The transfer of plants and animals brought important new food crops to Europe and led to changes in agriculture in the New World. Among the tools that the Europeans brought with them was the plough, which enabled farmers to cultivate on a much wider scale than before. Firearms, unknown in the Americas before Christopher Columbus landed in 1492, transformed not only warfare, but also hunting, enabling Native Americans to kill larger animals easily.

Columbus himself brought horses, pigs, cattle, sheep, and goats across the Atlantic. These too transformed the lives of many, especially the horse, which gave Native North Americans a beast of burden for the first time. Writing was also introduced, the Europeans being keen to propagate Christianity, although this impacted heavily on indigenous cultures and led to the spread of the English, Spanish, and Portuguese languages that are so widely used today. Unfortunately, slavery was also imported on a huge scale, especially in the Caribbean and the southern parts of North America where sugar cane plantations were established. On returning home, the Europeans took a variety of new food plants back from America, including beans, tomatoes, avocados, pineapples, maize, and potatoes.

A number of infectious diseases also travelled to America with the settlers, including smallpox, chickenpox, measles, and influenza. These spread rapidly and, as the local people had no immunity against them, completely decimated the indigenous populations of America. There were also diseases that travelled the other way, including syphilis (especially prevalent at the time of the Columbian Exchange) and polio.

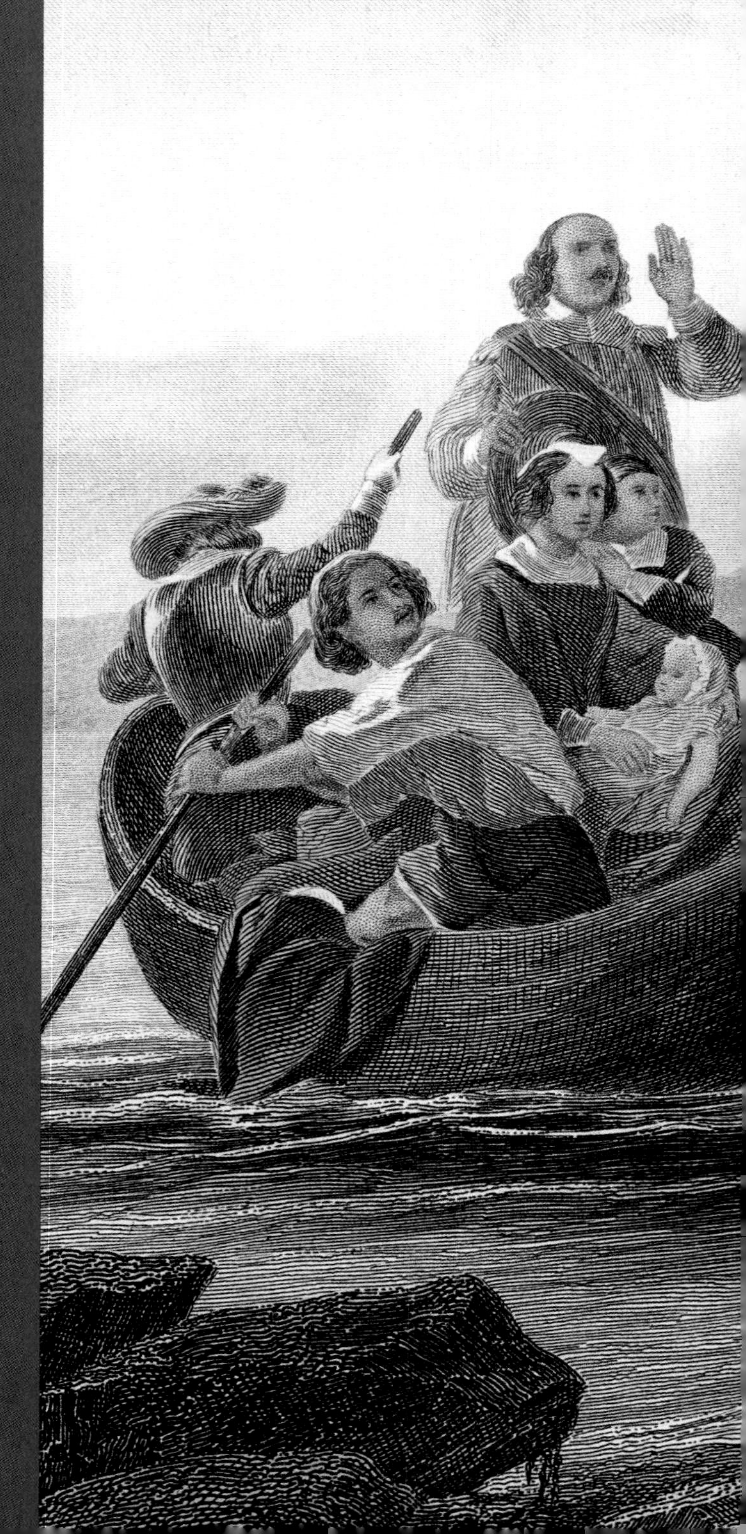

▷ **Roger Williams lands in the New World**
Based on a painting by Alonzo Chappel, this engraving shows a group of Native Americans offering the pipe of peace to British theologian and future founder of Rhode Island, Roger Williams, as his boat comes ashore.

" My sons, here you see how I fared because I was **too trusting** toward these **Spanish** people... "

MANCO INCA, PERUVIAN RULER, IN HIS DYING SPEECH TO HIS CAPTAINS

Cartier's North America

This map of North America's east coast was made in the early 1540s, using information provided by Jacques Cartier. Its coverage of the mouth of the St Lawrence River is especially detailed. Technically, the map is upside-down, with north pointing towards the bottom of the map.

> " It is this **art** that **drew** me to **love the sea** at a **very young age** and that **compelled** me to **challenge** its **treacherous waters**... "

SAMUEL DE CHAMPLAIN, ON THE ART OF NAVIGATION

◁ **Ruler of New France**
This image of Champlain shows him towards the end of his life, when his exploring days were almost over and he was Lieutenant-General of New France.

KEY DATES

- **1574** Born into a seafaring family in the province of Aunis, southwestern France.
- **1603** Makes his first transatlantic voyage.
- **1604** Joins a voyage led by Pierre Dugua de Mons, on which he is able to record the coast of New England.
- **1608** Founds Quebec City on the site of a fort previously built by his predecessor, Jacques Cartier.
- **1609** Leads an expedition up the Richlieu River and discovers Lake Champlain.
- **1615–16** Spends a winter exploring the Lake Huron area, becoming the first European to gain first-hand knowledge of the Great Lakes.
- **1632** The treaty of St Germain-en-Laye restores Quebec City to the French from the British. Champlain returns to the settlement as Lieutenant-General.

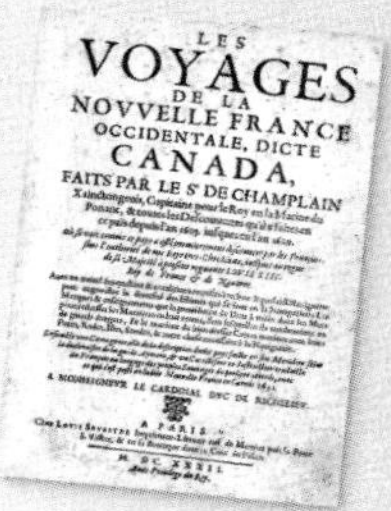

CHAMPLAIN'S *VOYAGES DE LA NOUVELLE FRANCE*, 1632

CHAMPLAIN JUDGING HIS LATITUDE WITH AN ASTROLABE

Early missionaries

When trade routes opened between the Americas and Europe, Christian missionaries, mostly Catholic clergy, accompanied the naval expeditions to spread the Gospel throughout the Americas, and later, parts of Asia.

△ **Jesuits in India**
This lithograph by Théophile Fragonard comes from the 19th-century book *The Dramatic History of the Jesuits*. It shows a Jesuit priest, in distinctive black robe, preaching to Hindu Brahmins.

The first Catholic missionaries were mainly Dominican and Franciscan friars, who set up "missions", or communities, in the places where European traders had settled in the Caribbean and North America. In 1493, Spanish Pope Alexander VI instructed the friars to convert the native people in these areas, and the following year he divided the New World between Spain and Portugal. Under the Treaty of Tordesillas, signed on 7 June 1494, Spain received all the lands to the west of a meridian that cut vertically through South America, while Portugal received everything to the east. As a result, Brazilians speak Portuguese, while the rest of Latin America speaks Spanish.

◁ **Missionary church in Samoa**
The early missionaries built churches wherever they could. These were generally modelled on European buildings, but incorporated elements of local architecture.

Catholics and Protestants

The principal missions to the north were established along the coast of California – San Francisco, Los Angeles, and Santa Cruz were all missionary settlements. However, during the 16th century, other Catholic orders arrived to spread the faith throughout South America. Together, these missionaries taught Spanish, Portuguese, Bible stories, and prayers to the native people, and learned the local languages in return. By the mid-16th century, parts of the Bible had been translated into 28 South American languages. Protestant missionaries also started to arrive, most notably the French Huguenots, who had settled in Brazil by 1550.

Although the missionaries adapted to local customs, many retained their European bias. Few, for example, tried to put an end to slavery, although there were notable exceptions, such as the Dominican friar Bartolomeo de las Casas. The invasion of the conquistadores disrupted civilizations that had flourished undisturbed for centuries, sowing lasting resentment among native people (see pp.122–25). In 1562, Diego de Landa, the Spanish-born Archbishop of Yucatán, destroyed the libraries of the Maya civilization, obliterating centuries of Mayan thought and tradition. For such reasons, missionaries were not always welcomed. Indeed, in 1583, five Jesuit missionaries were executed in Goa, a Portuguese colony on the west coast of India, and several were killed by Mohawks in North America in 1649.

Lasting legacy

These early missionaries did more than spread the Gospel. They were among the first Europeans to cross-fertilize European, Native American, and Asian cultures. In around 1600, the first Portuguese-Chinese dictionary was created, and in 1603, the Jesuits published the first Japanese-Portuguese dictionary. However, in 1638, the Japanese Shogunate enforced a total ban on Christianity, imposing the death penalty on anyone who converted. All missions were then curtailed until the 19th century, when freeing slaves became a focus of missionary activity.

18TH-CENTURY PORTRAIT OF FRANCIS XAVIER

IN PROFILE
Francis Xavier

Spanish nobleman Francis Xavier was at Paris University when he met Ignatius of Loyola in 1529. Ten years later, with five other friends, they established the Jesuit Order, which was approved by Pope Paul III in 1540. In 1542, at the behest of John III of Portugal, Francis embarked on a missionary journey to India and spent three years preaching the Gospel in Tamil along the coast of Goa. Although he struggled with the language, he was convinced of the need to adapt to local customs. After travelling through Ceylon, the Moluccas, and the Malay Peninsula, he spent several years in Japan, but died at the age of 46, as he prepared to enter China.

Jesuit missionary in China
Matteo Ricci was an Italian Jesuit who, from 1582, spent almost 30 years in China, fostering relations between China and the West. He learned Mandarin and was greatly respected by the emperor.

The Northwest Passage

In the 16th century, several explorers tried to find a northern passage from Europe to the spice markets of Asia. Although this route promised to be faster than the journey around Africa, it was fraught with danger.

Among the first Europeans to sail the Atlantic Ocean in search of the Northwest Passage was British navigator Martin Frobisher (c.1535–94). Frobisher crossed the ocean in 1576, losing two of his three ships in a storm before reaching the coast of Labrador. He explored the waters around Baffin Island, becoming the first European to visit the large inlet now called Frobisher Bay. However, he clashed with the local Inuits, who took five of his men prisoner. Frobisher failed to get them released, and returned to England with just one Inuit prisoner. He also took back a sample of black rock that he thought was coal.

Frobisher led two further expeditions, in 1577 and 1578, both with several ships and a large number of men, including miners. They had been ordered to investigate the potentially valuable ores and minerals that Frobisher thought could be found on Baffin Island. As well as setting the miners to work, Frobisher continued his navigation of the waters around Greenland and Baffin Island, adding to European knowledge of the area that might lead to the discovery of the Northwest Passage. He took back to England a large amount of rock, including some that he thought was gold, but which turned out to be worthless iron pyrites. After this disappointment, Frobisher made no more attempts to find the Northwest Passage, but sailed as a privateer with English sea captain Sir Francis Drake.

▽ **Map of routes**
Martin Frobisher and Henry Hudson explored the eastern fringes of the Arctic archipelago. The map below shows the routes taken by the two explorers.

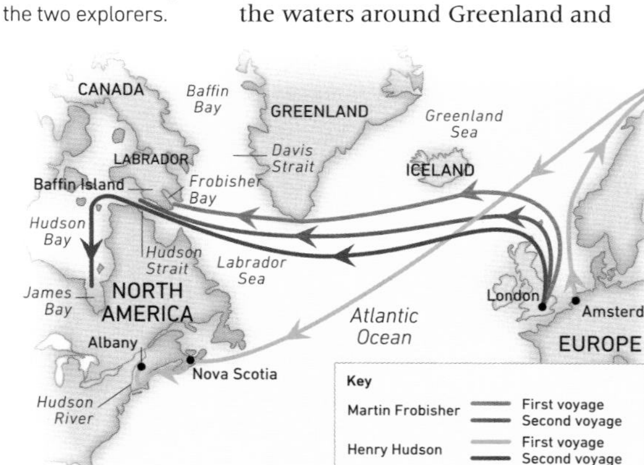

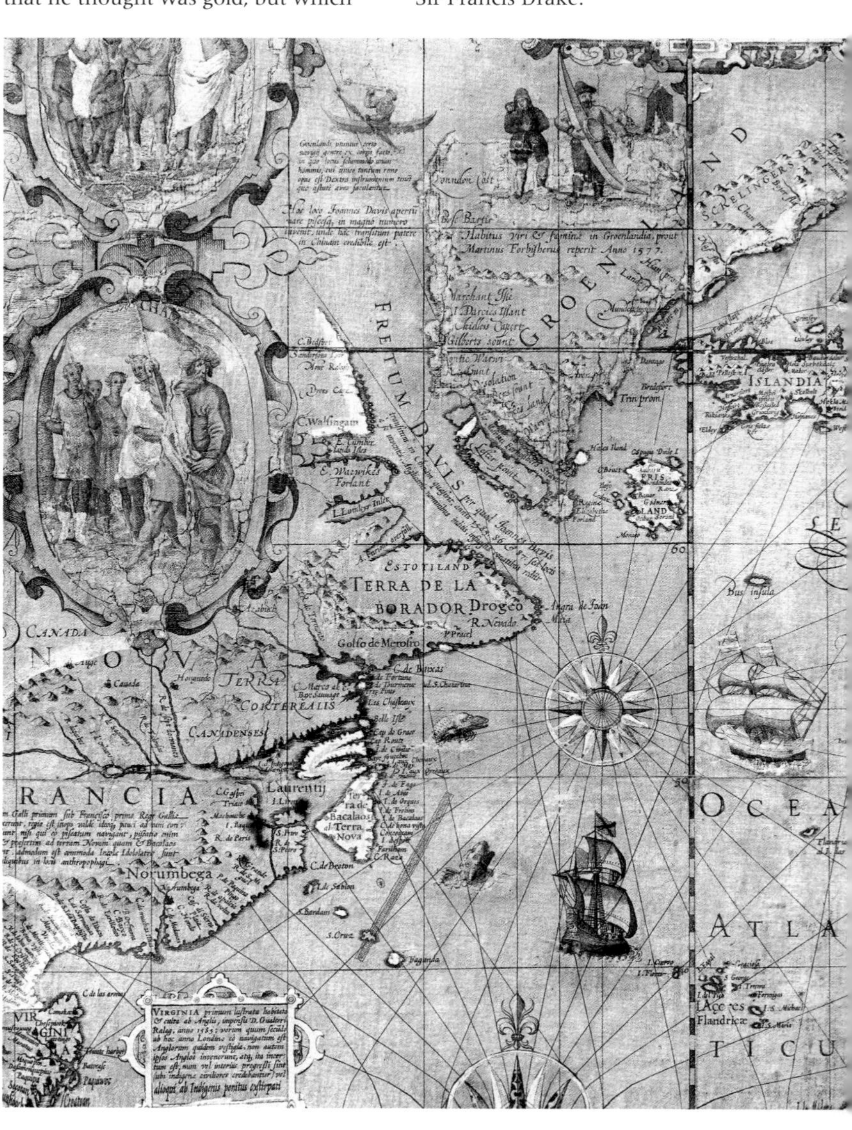

▷ **World map on Mercator's projection**
A detail from a map by Dutch cartographer Jodocus Hondius, made in 1608. Only the southeastern coast of Greenland is visible, but there is some detail around Nova Scotia and Baffin Bay.

Henry Hudson

Among the most intrepid of all northern navigators was British explorer Henry Hudson (c.1565–1611). In 1609, he was hired by the Dutch East India Company to find a Northeastern Passage by sailing north of Russia. However, his route was blocked by thick ice and so he turned back. Instead of returning home, however, he crossed the Atlantic and looked for a Northwest Passage across North America to the Pacific. Arriving in Nova Scotia, Hudson sailed south, finding the river now named after him, and then travelled up it as far as the site of the modern city of Albany. He traded with local people, purchasing furs, and established the Dutch claim on this part of North America.

▷ **Skirmish with Inuits**
Relations between European explorers and native peoples were sometimes difficult, as is shown in this watercolour by the artist John White, who accompanied Frobisher on his journeys. Some explorers, such as Hudson, however, managed to establish trading relations with them.

Hudson's last voyage

The following year, Hudson undertook a further transatlantic voyage, this time with funding from the British East India and Virginia companies. He crossed the North Atlantic and found the northern tip of Labrador. From there, he followed the strait to the northwest – now the Hudson Strait, the bay that today bears his name. It took Hudson some months to explore the waters and coasts of this vast bay, together with its southeastern extension, James Bay. By the time he had mapped the eastern coast of Hudson Bay, winter had arrived and his ship was stuck in James Bay, trapped in the ice. He overwintered there, but in the spring lost his life as the result of a mutiny.

Although they failed to find the Northwest Passage, Frobisher and Hudson hugely expanded European knowledge of the seas north of Canada. Hudson also opened up a valuable trade in furs with native peoples in North America. The stories of these men live on in the names of various places, from Frobisher Bay to the Hudson River.

IN PROFILE
Henry Hudson

British explorer Henry Hudson made a series of voyages in search of northern passages to Asia, including two attempts to find a northeastern route north of Russia, and two northwestern voyages into the waters of Canada. Hudson's final voyage in search of the Northwest Passage, however, ended in tragedy. Trapped by thick ice, Hudson and his crew were forced to spend the winter of 1610–11 in Hudson Bay.

When the ice melted in the spring, Hudson planned to further explore Hudson Bay, hoping to find the Northwest Passage. However, many of his crew wanted to return home, and when Hudson refused, they mutinied and forced their leader off his ship. Hudson, his teenage son, John, and a few loyal crew members were left to perish in an open boat.

PAINTING SHOWING HUDSON AND HIS SON ABANDONED IN AN OPEN BOAT

حصارلر
آياصوفيه
طوپقاپو

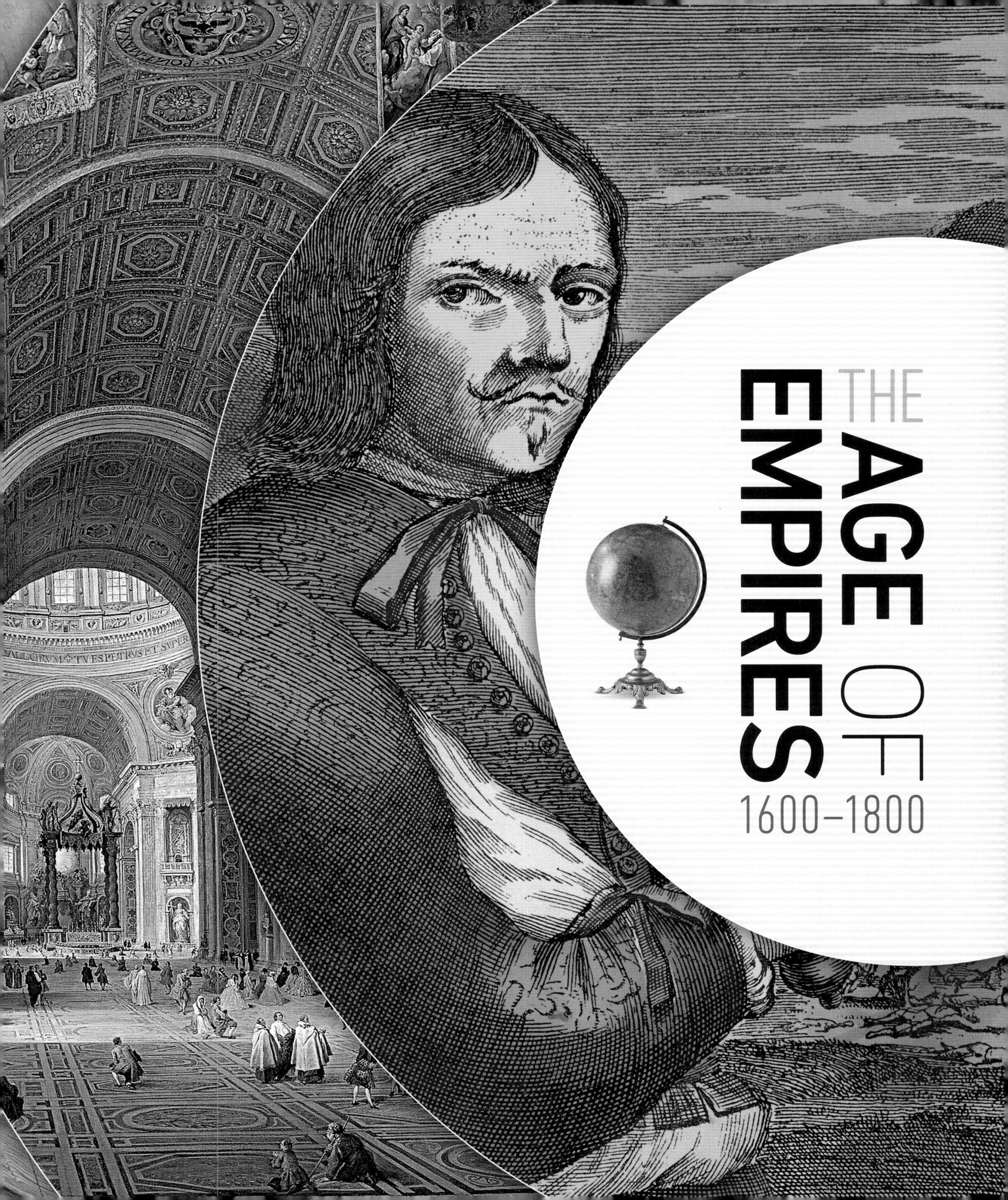
THE
AGE
OF
EMPIRES
1600–1800

THE AGE OF EMPIRES, 1600–1800

Introduction

As the world was explored, the challenge became how to profit from it. In the mid-17th century, the maritime dominance of Spain and Portugal was superseded by the skill and enterprise of Dutch and British seafarers. With the founding of the Dutch and British East India Companies – which became vast trading corporations – these two nations established outposts and colonies all over the world, from the East Indies and India to North America. They divided up the globe between them, little imagining the consequences. The Dutch secured parts of Southeast Asia, while handing over to Britain control of a fledgling outpost on an island called Manhattan. They initiated a global age of shipping trade, which introduced tea, coffee, tobacco, spices, and sugar to Europe, and approximately 12 million African slaves to the Americas.

Changing tastes and fashions

With the new goods came new places to meet, such as coffeehouses and intellectual salons. People wanted to know about the far-off lands from which these new goods originated, and it became fashionable among the wealthy to collect mementoes of distant lands. These small, private collections, known as *Wunderkammern* (cabinets of wonder), were symbols of status and identity, but also signs of a growing curiosity about the natural order of the world and man's place in it.

Science increasingly became a motive for exploration. Russian tsar Peter the Great sent sailor and cartographer Vitus Bering to the northern Pacific to chart the Siberian coastline and discover if there was a land bridge to Alaska. Catherine the Great sent him back to Siberia as part of the 3,000-strong Great Northern

BRITISH OFFICERS OF THE EAST INDIA COMPANY SECURED LUCRATIVE RESOURCES IN INDIA

COFFEE HOUSES BECAME PLACES WHERE CURRENT AFFAIRS WERE HEARD AND DISCUSSED

CAPTAIN COOK OPENED UP SEA ROUTES ACROSS THE WORLD, FROM ANTARCTICA TO POLYNESIA

> “ Do just once **what others say you can't do**, and you will **never pay attention** to their limitations again. ”
>
> CAPTAIN JAMES COOK

Expedition. At the time, this was the largest scientific venture ever undertaken, and it marked the beginning of Russian expansion into the east.

Voyages of learning

Smaller in size but even wider in scope were the expeditions headed by Captain James Cook on behalf of Britain's Royal Society. Over the course of three lengthy voyages, Cook criss-crossed the Pacific, circumnavigated Antarctica, and charted previously unknown parts of Australia and New Zealand. However, it was the on-board scientists, such as Joseph Banks, who really made the voyages so momentous. They discovered thousands of new species, all of which were carefully collected, catalogued, and taken back to Britain for examination. Cook's expeditions also produced a staggering collection of drawings and paintings of places, flora, and fauna, which together made a massive contribution to the study of the natural history of the world.

Generally speaking, however, it was not necessary to journey as far as Captain Cook to improve one's knowledge; a trip to Rome would do. The Italian capital was the ultimate goal for the multitudes who embarked upon what came to be known as the Grand Tour. In the 18th century, it was thought that there was no better finishing school for a young gentleman than to send him off with a chaperone to visit the major cultural capitals of Europe, in the hope that he would find enlightenment along the way. War in Europe brought the Grand Tour to a halt at the end of the 18th century. However, the notion of travelling for no other reason than self-fulfillment had taken root.

VITUS BERING TWICE EXPLORED THE NORTHERN PACIFIC FOR RUSSIA, AND MADE LANDFALL IN ALASKA

FOR MANY, THE GRAND TOUR ENDED IN NAPLES, WITH ITS VIEW OF MOUNT VESUVIUS

THE FIRST BALLOON FLIGHT WAS MADE IN PARIS IN 1783, PRESAGING A NEW MODE OF TRANSPORTATION

The spice trade

There have been many reasons for mankind's constant voyaging, but in the exploration of the Southeast Asian seas there was one goal above all, the quest to find the source of the treasure of the age – spice.

When the sole surviving ship of the Magellan expedition finally limped into port after completing the first circumnavigation of the world, among the meagre treasures contained in its hold were 381 bags of cloves, acquired in the Far East. Taking into account the four ships that had been lost en route, the advances paid to the 237 sailors who had set out, the back pay for the 18 who survived, and all the other expenses, selling the bags of cloves enabled the three-year-long expedition to still turn a small profit.

Spices were immensely valuable commodities. It was not just the pep they brought to the table – cinnamon, cloves, ginger, mace, nutmeg, and pepper were also believed to help combat illness, guard against pestilence, dispel demons, and even to act as aphrodisiacs. Few knew exactly where the spices originated from, only that they came to Europe via a complicated route, which involved passing through the hands of a succession of middlemen in Asia and the Middle East, who each ratcheted up the price.

▷ **Trading post in Hooghly, Bengal**
This painting, dating from 1665, depicts one of the numerous Dutch trading posts in Asia. Two ships can be seen on the Ganges, ready to carry goods back to the Netherlands.

Portuguese takeover

By the end of the 15th century, Portugal had established maritime dominance along the west coast of Africa – the next step was to voyage eastwards in search of the source of the spices. Vasco da Gama successfully reached Calicut on the southeastern coast of India in 1498, and was delighted to find a thriving market where the Chinese delivered spices to Arab and Italian merchants to transport overland to the Mediterranean. Within a year of his return to Lisbon, a fleet commanded by Pedro Álvares Cabral was dispatched with the express purpose of taking the spice trade out of their hands.

Cabral essentially turned the Indian Ocean into a "Portuguese lake." The pace of expansion accelerated, with successive expeditions pressing deeper into the heart of maritime Asia. In 1505, the Portuguese extracted tribute from Sri Lanka. Six years later, their galleons crossed the Bay of Bengal and seized control of Malacca, the richest port in the Far East at the time. Using this as a base, the Portuguese were eventually able to locate the heart of the fabled Spice Islands – a small archipelago called the Moluccas, which lies off the coast of modern Indonesia.

Although the Portuguese lost little time exploiting the wealth of the islands, their empire was already in decline. When English and Dutch ships began to appear in Asian waters at the close of the century, Portuguese domination was over. The first Dutch ships called at the Moluccas in 1599, and returned to Amsterdam laden with cloves. Two years later, in 1601, they were joined by ships sailing under the auspices of the newly formed British East India Company.

▷ **Company shields**
On the left, the Dutch East India Company's coat of arms, flanked by Neptune and a mermaid. The other shield shows that of Batavia, flanked by Dutch lions.

" After the year **1500** there was no **pepper** to be had at **Calicut** that was not dyed **red with blood**. "

VOLTAIRE, COMMENTING ON THE PORTUGUESE CONQUEST OF INDIA, 1756

◁ **Chinese porcelain**
In addition to spices, the Europeans traded in Asia for tea, textiles (especially silks), and porcelain.

East India rivalries

In 1602, the Dutch formed their own East India Company, known as the Vereenigde Oost-Indische Compagnie, usually abbreviated to VOC. In Asia, the company established a fortified base at Batavia (now Jakarta, the capital of Indonesia), with hundreds of subsidiary posts around the region. The Dutch had a far larger merchant fleet and navy than the British, and the VOC managed to more or less edge the English East India Company out of the spice trade, thereafter enjoying huge profits from its monopoly.

Around the middle of the 17th century, the balance of power began to shift as England invested heavily in its navy to support its growing imperial ambitions. The result was a series of Anglo-Dutch wars. The second of these was brought to an end by the Treaty of Breda (1667), in which the English agreed to renounce claims to any of the Spice Islands. In return, the Dutch acknowledged English sovereignty of a territory called New Netherland that England had seized from them. Subsequently, the English renamed it New York.

The VOC went on to become the world's largest trading company. At the height of its wealth and power, it owned more than half of the world's sea-going shipping, but was finally declared bankrupt in 1799.

◁ **British officials in India** Outmanoeuvred by the Dutch in Southeast Asia, the British contented themselves with a commercial monopoly in India and the new territories of America.

Wonder cabinets

Stocked with exotic artefacts collected from distant lands, cabinets of curiosities were the predecessors of the modern museum.

Before there were museums, there were *Wunderkammern*, or cabinets of wonder. First created in the 16th century, these were private collections containing all manner of marvellous and exotic objects brought back from places unknown, and typically associated with the discovery of "new worlds". Ideally, the collections showcased three main themes: *naturalia* (products of nature, such as fossils, shells, and preserved exotic animals and marine life), *arteficialia* (the products of man, including native arts and crafts), and *scientifica* (mechanical or scientific objects, such as astrolabes, clocks, and scientific instruments). Some *Wunderkammern* had specific themes, such as the one created by Dutch anatomist Frederik Ruysch (1638–1731), which displayed body parts and preserved organs alongside exotic birds, butterflies, and plants.

These collections were displayed in multi-compartmented cabinets and vitrines. By displaying such items together, viewers could make connections between objects, leading to an understanding of how the world functioned, and what humanity's place in it was. As time passed, the small private cabinets were absorbed into larger ones. In turn, these were bought by aristocrats and even royalty, and merged into cabinets so large that they took over entire rooms. Frederik Ruysch's *Wunderkammer*, for example, was bought by Tsar Peter the Great and now forms part of the Kunstkammer Museum in St Petersburg, while the cabinet of the London apothecary James Petiver (c.1665–1718) became the basis of the British Museum. A *Wunderkammer* known as The Ark, assembled by naturalist John Tradescant Senior (c.1570–1638) and regarded as one of the wonders of 17th-century London, ended up as part of the Ashmolean Museum in Oxford.

" **Such a collection** of... many **outlandish and Indian curiosities**, and things of nature. "

ENGLISH DIARIST JOHN EVELYN IN MILAN, 1646

▷ ***Cabinet of Curiosities*** **by Andrea Domenico Remps**
Remps's painting shows the types of object typically found in a cabinet of curiosity, such as archaeological items, scientific objects, animal specimens, and religious artefacts.

New Holland

As the great European commercial companies established themselves in the East, investigations were made into Terra Australis Incognita – the Unknown Southern Land.

△ **Abel Tasman**
An experienced merchant skipper in the employ of the Dutch East India Company, Tasman was charged with exploring the uncharted tracts of the southern hemisphere.

In 1606, Captain Willem Janszoon of the Dutch East India Company was dispatched from Bantam in Java to explore the coast of New Guinea in search of economic opportunities. Bearing south, he missed the Torres Strait and continued down along what he thought was an extension of New Guinea's southern coast. It was, in fact, the western coast of the Cape York Peninsula in modern Queensland. He put ashore, but after a hostile reception from the native people that resulted in the deaths of several of his men, he turned around and headed back to Java. Without knowing it, he and his crew had been the first Europeans to step foot on the land that would later be called Australia.

Over the next 160 years, many more crews from a range of European nations sighted, and sometimes set foot on, this uncharted and unexplored landmass. Most of these were merchant ships from the Dutch East India Company, and the new territory first appeared on one of the company's maps in 1622, erroneously labelled "Nueva Guinea". One notable visitor was Jan Carstensz, commissioned by the Dutch East India Company to follow up on the reports made by Janszoon in his 1606 voyages to the south. Carstensz named the large shallow sea he sailed across in 1623 the Gulf of Carpentaria in honour of Pieter de Carpentier, governor-general of the Dutch East Indies. Four years later, another Dutchman sailed south and mapped the lower reaches of Australia, from what is now Albany on the very southwest tip, halfway along to what is now Cebuna.

◁ **Maori club**
The Maori that Tasman encountered were armed with stone and hardwood weapons, such as this elaborately carved club, shaped to look like a bird's head.

Abel Tasman

Nearly 20 years passed, then, in August 1642, the Dutch East India Company dispatched sea captain Abel Tasman on a voyage to explore the region that it now called the Great South Land. Sailing west with the prevailing winds from Batavia (present-day Jakarta), and then southeast, he actually missed Australia altogether, and alighted instead on a landmass he named Van Diemen's Land after another governor-general of the Dutch East Indies. It was not until well over a century later that explorers established that this was in fact an island, and it was renamed Tasmania in honour of the Dutchman.

Tasman sailed on eastwards, and duly became the first European to reach New Zealand. He anchored his ships off the northern end of South Island, but quickly departed again after they were attacked by the Maori tribesmen. He returned to Batavia via several Pacific Island groups (now known as Fiji, Tonga, and the Solomon Islands) and New Guinea.

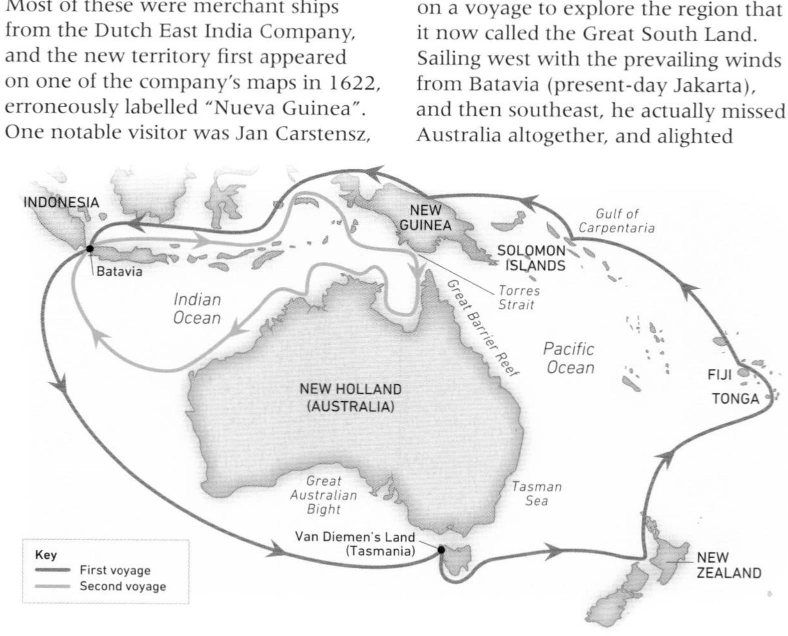

▷ **The voyages of Abel Tasman**
Tasman made two exploratory voyages. On the first, he discovered Van Diemen's Land and New Zealand; on the second, he mapped the northwest coast of Australia.

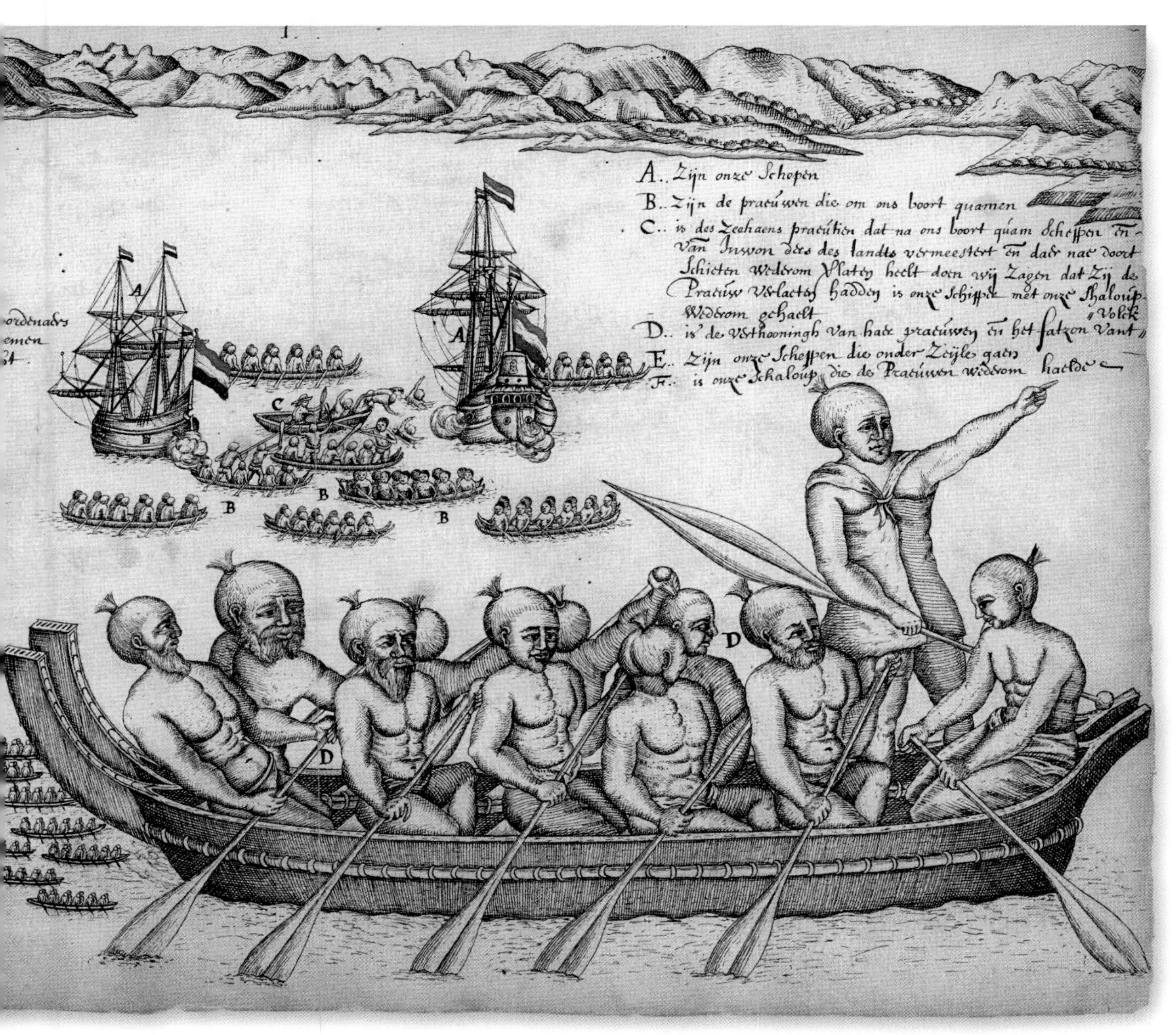

◁ **Murderers' Bay**
This drawing, made by Abel Tasman's artist, shows the Dutch skirmishing with Maori warriors at what became known as Murderers' Bay (present-day Golden Bay).

New Holland

In 1644, Tasman made a second voyage, this time following the south coast of New Guinea eastwards and then south. Again, he failed to find the Torres Strait, which would have provided the direct access east to South America the Dutch were hoping for. Instead, he sailed back west along the north coast of the Great South Land. He mapped the continent's coastline and made observations on its geography and inhabitants. He also named it New Holland.

From the point of view of the Dutch East India Company, however, Tasman's explorations had proved to be a big disappointment. He had not found a new shipping route, and New Holland offered nothing in the way of trading opportunities. The Dutch interest in the continent had dissipated, and Tasman's would be the last voyage to Australia until those of William Dampier and James Cook, 55 and 126 years later respectively (see pp.172–75).

▷ **Coronelli globe**
The globes made by Venetian Vincenzo Coronelli depicted the world of the late 17th century. New Holland already appears on it, but has no detail on its eastern coast, which was not explored until the arrival of James Cook in 1770.

△ **Here be monsters**
John White produced this map (complete with ships and sea-monsters) showing the American coast from Chesapeake Bay down to the Florida Keys during an expedition to Virginia with Sir Walter Raleigh in 1585.

Settling America

Once Portugal and Spain had laid claim to most of Central and South America, the powers of northern Europe were left to contest the colder and seemingly less promising coastline of North America.

By the end of the 16th century, North America had been well and truly "discovered". After Christopher Columbus (see pp.110–13), the Spaniard Ponce de Leon had landed in what is now Florida in 1513, and the Portuguese Estêvão Gómez had sailed along the Maine coast and may have entered New York Harbour in 1525. The conquistador Hernando de Soto had also explored deep inland, becoming the first European to cross the Mississippi River, which is where he died in 1542.

So far, however, the hardships involved in settling this New World had largely proved too daunting, and the rewards meagre. By 1600, the only permanent colony in America was that established by the Spanish at Saint Augustine in Florida in 1565.

Colonizing North America

The English had attempted to establish a colony on Roanoke Island off North Carolina in 1585, but it vanished without trace, a mystery that has never been solved. Undaunted, in 1606, a group of merchants founded the Virginia Company of London and raised money for an expedition in the hope that the new colony would provide profits in the form of gold, silver, and gems. The company duly dispatched three ships carrying 104 settlers, which reached the east coast of America in May 1607, and they founded a settlement named Jamestown after the English king.

The colony struggled. There was no gold, and all attempts to farm potentially lucrative cash crops failed. Of the 500 men in the colony in October 1609, only 60 were left when spring arrived. For the next few years, the settlement was only able to survive thanks to fresh supplies sent from England. Its fortunes changed in 1612, however, when the colony managed to cultivate and export tobacco. By 1619, the settlement was considered permanent and sufficiently family-friendly for a ship to be sent over carrying "respectable maidens".

◁ **Jamestown**
This reconstruction of Jamestown shows the first permanent English settlement in America, as it was around 1615.

▽ **Secotan Indians**
When the first English settlers built a fort at Roanoke, John White painted the local Secotan warriors. No Englishman had ever painted anything in North America before.

The Pilgrims

Commercial gain was not the only motivation for settlement. Modern America's boast of being the land of freedom is rooted in its foundation. From the beginning, it was a place of refuge for those fleeing religious persecution in Europe. As early as 1564, an expedition of around 300 French Huguenots fled certain death at home to form a short-lived settlement called Fort Caroline, near what is now Jacksonville in Florida. The following year, they were brutally massacred by a Spanish force from Saint Augustine, keen to nip any French territorial ambitions in the bud.

The English Puritan evangelists met with more success. They viewed their nation's church as beyond redemption because of its Catholic past, and set out for a new and untainted country in which they could practise a "purer" form of worship. In November 1620, their ship, the *Mayflower*, landed on the shores of Cape Cod, in present-day Massachusetts. After scouting locations, the group, who came to be known as the Pilgrims, eventually put ashore at Plymouth Harbour in December. Despite suffering similar hardships to those experienced by the Jamestown settlers (by February 1621, half of the group were dead), the Pilgrims were able to raise a harvest in their first summer, which they supplemented with abundant fish.

In the next few years, as life for Puritans became increasingly difficult in England, more of them made the journey across the Atlantic. »

▷ **Jamestown Church**
A ruined brick tower is all that survives of Jamestown Church, built by the settlers in 1639. It replaced several earlier structures, but remains one of the oldest buildings in the US.

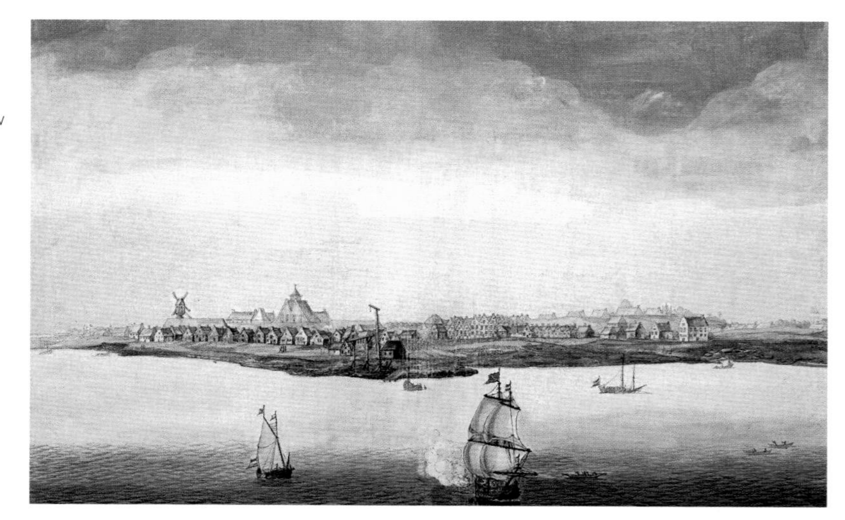

▷ ***Gezicht op Nieuw Amsterdam*, 1664**
This painting by Johannes Vingboons shows the port of New Amsterdam in the same year that the English would seize it from the Dutch and rename it New York.

» By 1630, some 10,000 Puritans had arrived to form new colonies in Massachusetts, a number swollen that year by the arrival of a fleet of 11 ships that carried around another 1,000 people and their livestock. This group formed a settlement that would soon become the thriving port of Boston.

New Netherland

In the meantime, other European groups were arriving to settle America: the French, in what is now Canada; scattered bands of Finns and Swedes along the mid-Atlantic coast; the Spanish in Florida and the far southwest (today's New Mexico); and the Dutch on the east coast in what they were calling New Netherland. In 1609, Henry Hudson, an English explorer in the employ of the Dutch East India Company, sailed along the coast of North America exploring inlets on a mission to find a northwest passage through to the Indies. He travelled up the river that now bears his name, as far as modern-day Albany, and claimed the territory for his employers. He did not find a Northwest Passage, but the territory proved to be an excellent source of fur – a valuable commodity.

The Dutch began arriving in North America soon after Hudson's voyage. In 1614, Dutch merchants formed the New Netherland Company, and the following year they established a post called Fort Orange, near Albany, to trade with Native Americans, exchanging cloth, alcohol, firearms, and trinkets for beaver and otter pelts. Six years later, the newly incorporated Dutch West India Company began organizing a more permanent Dutch settlement, and in 1624, the ship *Nieu Nederlandt* (New Netherland) departed with the first wave of settlers.

▽ ***The Landing of the Pilgrim Fathers***
This 19th-century engraving by John C. McRae depicts the Pilgrim Fathers arriving on the shores of North America on 21 November 1620. Their ship, the *Mayflower*, is seen in the background.

New York

Once in America, these new Dutch immigrants spread out over several settlements in the territory claimed by the company. Some of these outposts proved impossible to sustain, and others were too dangerous because of the local Native Americans. So, in the early summer of 1626, the Dutch purchased the island of Manhattan from the Native Americans for around 60 guilders' worth of trinkets. There, they started building Fort New Amsterdam, which became the hub of a province that by 1655 numbered 2,000 to 3,500 inhabitants. One significant way in which the

◁ **Pilgrim hat**
Beaver fur was a source of income for the settlers. It was processed into felt to make hats like this one, said to have belonged to Pilgrim Constance Hopkins.

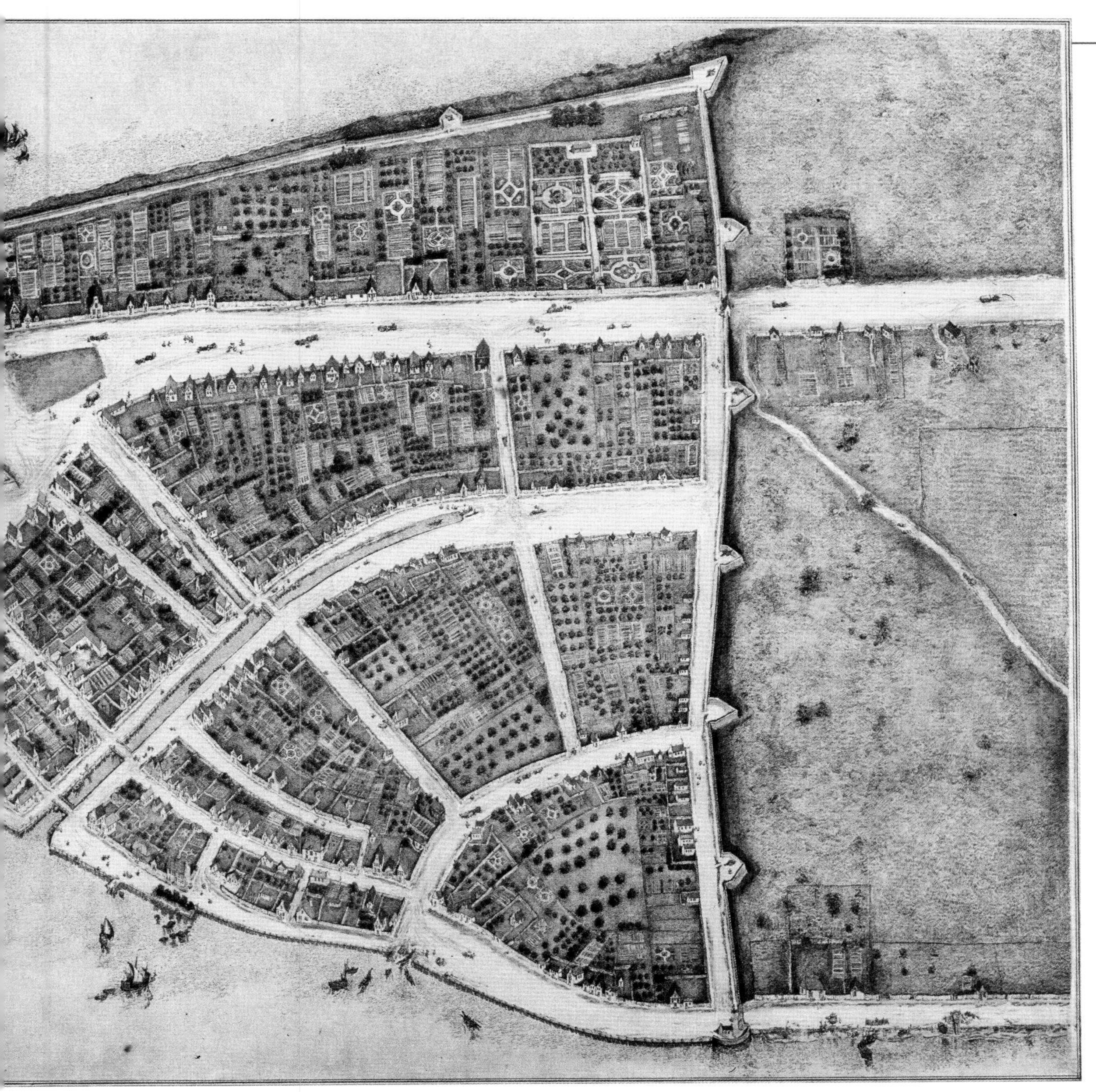

◁ **New Amsterdam**
This 1916 redrawing of a map of 1660 called the *Castello Plan* shows the Dutch settlement of New Amsterdam at the southern tip of Manhattan. The town was renamed New York in 1664.

New Netherland province differed from the British colonies to the north was demographics. Up to one half of the New Netherland population was not Dutch, and included Germans, Swedes, and Finns, as well as French, Scots, English, Irish, and Italians. All of these nationalities lived under Dutch rule.

As New Netherland prospered, the British set their sights on it. Back in 1498, the Genoese-born John Cabot, who was in the employ of Britain, had explored the coast of America from Newfoundland down to Delaware, and as this trip predated Hudson's by more than a century, the British felt they had prior claim to the land. In 1664, a fleet under the command of the King of England's brother, the Duke of York, was sent to seize the colony, and the settlement of New Amsterdam and the entire colony of New Netherland were both renamed New York.

Spain and France held far larger parts of North America, and the English settlements were confined to a narrow strip along the Atlantic coast, but as most immigrants arrived along the eastern seaboard, English speech and common law became dominant in American life.

> " On the **island** of **Manhate** there may well be **four** or **five hundred** men of **different sects** and **nations**. "
>
> FATHER ISAAC JOGUES, A VISITOR TO NEW AMSTERDAM IN 1643–44

سلیمانیه جامعی
حصارلر
بسم الله الرحمن الرحيم
ایا صوفیه
اولیا چلبی
اسکدار
مهرماه سلطان

EXPLORER, 1611–82

Evliya Çelebi

Although little known outside his native Turkey, the Ottoman explorer Evliya Çelebi spent most of his life journeying and left behind one of the greatest works of travel literature ever written.

Evliya Çelebi (*çelebi* means "gentleman" in Turkish) was a courtier to the sultans when the Ottoman Empire (or Sublime Porte) was at the height of its power. He was born in Istanbul in 1611, and from his first expedition in 1640 to his death around 1682, he spent most of his life travelling. By the time he died, he had visited the lands of 18 monarchies from Russia to Sudan, witnessed 22 battles, and heard 147 different languages. All of this was meticulously described in his *Seyahatname* (*Travelogue*), a sprawl of biography and journal that runs to 10 volumes.

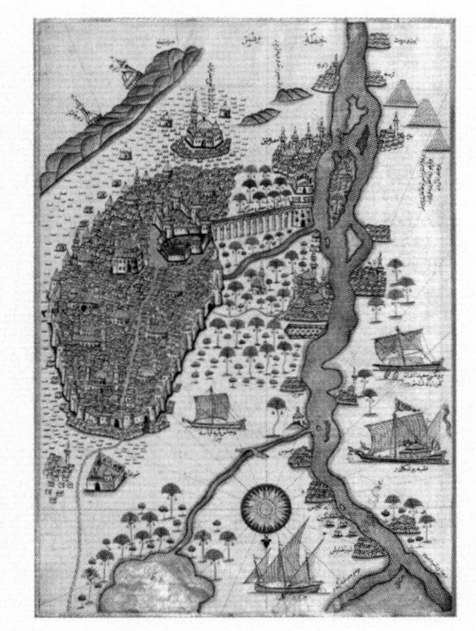

◁ **Piri Reis's map of Cairo**
Maps by Ottoman admiral Piri Reis, including this early 16th-century map of Cairo, were commonly used by 17th-century travellers, including Evliya.

Childhood dream

By his own account, it all started at the age of 20 with a dream in which the Prophet Muhammad blessed his intention to travel the world. However, the first volume of Evliya's account begins at home, with a depiction of Istanbul and the area around it. In the second volume, he travels to Anatolia, the Caucasus, Azerbaijan, and Crete.

Possessed of inherited wealth and powerful family connections (his father had been an imperial goldsmith), Evliya could well afford to travel, but he worked his passage, often accompanying diplomatic delegations as a minor functionary. In 1665, for example, he went to Vienna as part of the delegation sent to sign a peace treaty with the Habsburgs. Although a devout Muslim – he was able to recite the Qur'an from memory and did so every Friday – Evliya expressed great admiration for St Stephen's Cathedral in Vienna, where, he reports, the choir fills "one's eyes with tears." However, he also made scurrilous claims about other cultures, such as: "Hungarians are more honourable and cleaner infidels. They do not wash their faces every morning with their urine as the Austrians do."

Evliya's path crossed Buddhists, sorcerers, snake charmers, and tightrope walkers, but he was not above embellishing his work – he included, for example, a story about a Bulgarian witch who transformed herself into a hen and her children into chickens. Despite these flourishes, his book remains an important guide to life in the Ottoman Empire of the 17th century.

◁ **Writer at work**
This modern-day miniature shows Evliya at his desk writing an account of his travels.

▷ **Travel by sea**
This three-masted, Ottoman passenger ship is typical of the kind of vessel Evliya would have used on his voyages.

KEY DATES

- **1611** Born in Istanbul to a wealthy family from Kütahya.
- **1631** Dreams of a visit from the Prophet Muhammad in which he is told to travel.
- **1640** Makes his first journeys to Anatolia, the Caucasus, Azerbaijan, and Crete.
- **1648** Visits Syria, Palestine, Armenia, and the Balkans.
- **1655** Visits Iraq and Iran.
- **1663** While a guest in Rotterdam, he claims to have met Native Americans.
- **1671** Undertakes the hajj to Mecca.
- **1672** Visits Egypt and the Sudan, and settles in Cairo.
- **1682** Dies in either Cairo or Istanbul.
- **1742** Manuscript of *Seyahatname* discovered in a Cairo library and taken to Istanbul, where it achieves renown.

TRAVELLERS SPEND THE NIGHT AT A TYPICAL 17TH-CENTURY CARAVANSERAI

Coffee

Few commodities illustrate the growing global interconnections of the 17th century better than the coffee bean.

Many legends surround the discovery of coffee, but all agree that it took place in Ethiopia, or possibly Yemen. All also extol the invigorating effects of this bitter, brown brew, created by boiling up roasted berries. From its origin in the ports of the Horn of Africa, the drink is recorded as being enjoyed in Mecca as early as 1511, and it was the beverage of choice across much of the Middle East and Turkey by the end of the century – spread, no doubt, by the thousands of pilgrims visiting the holy city each year from all over the Muslim world.

Coffee was drunk in the home, but more significantly, it was also drunk in the new public coffeehouses. Here, coffee-drinking and conversation were complemented by itinerant musicians, professional storytellers, and games such as chess. The coffeehouses were sources of gossip and news, places people went to if they wanted to know what was going on. Not surprisingly, the authorities saw them as places of potential sedition, and coffee-drinking was frequently banned. Indeed, when the absolutist Murad IV claimed the Ottoman throne in 1623, he forbade coffee. Anyone found brewing or drinking it received a beating – and anyone caught a second time was sewn into a bag and thrown into the Bosphorus.

It did not take long for coffee to travel to Europe. Initially, it was viewed with suspicion because of its whiff of Islam, but this was quickly overcome, particularly when no less a personage than Pope Clement VIII gave the drink his blessing. From around 1650, coffeehouses sprang up all over Europe. Just as in the Arab and Turkish worlds, they became forums for lively discussion and outlets for streams of newsletters and pamphlets. So influential was their role in fostering intellectual debate that coffeehouses were nicknamed "Penny Universities", a penny being the cost of a coffee. By 1675, there were reckoned to be more than 3,000 coffeehouses in England, many of which, in a nod to coffee's origins, bore names such as the Turk's Head, The Saracen's Head, or The Sultan.

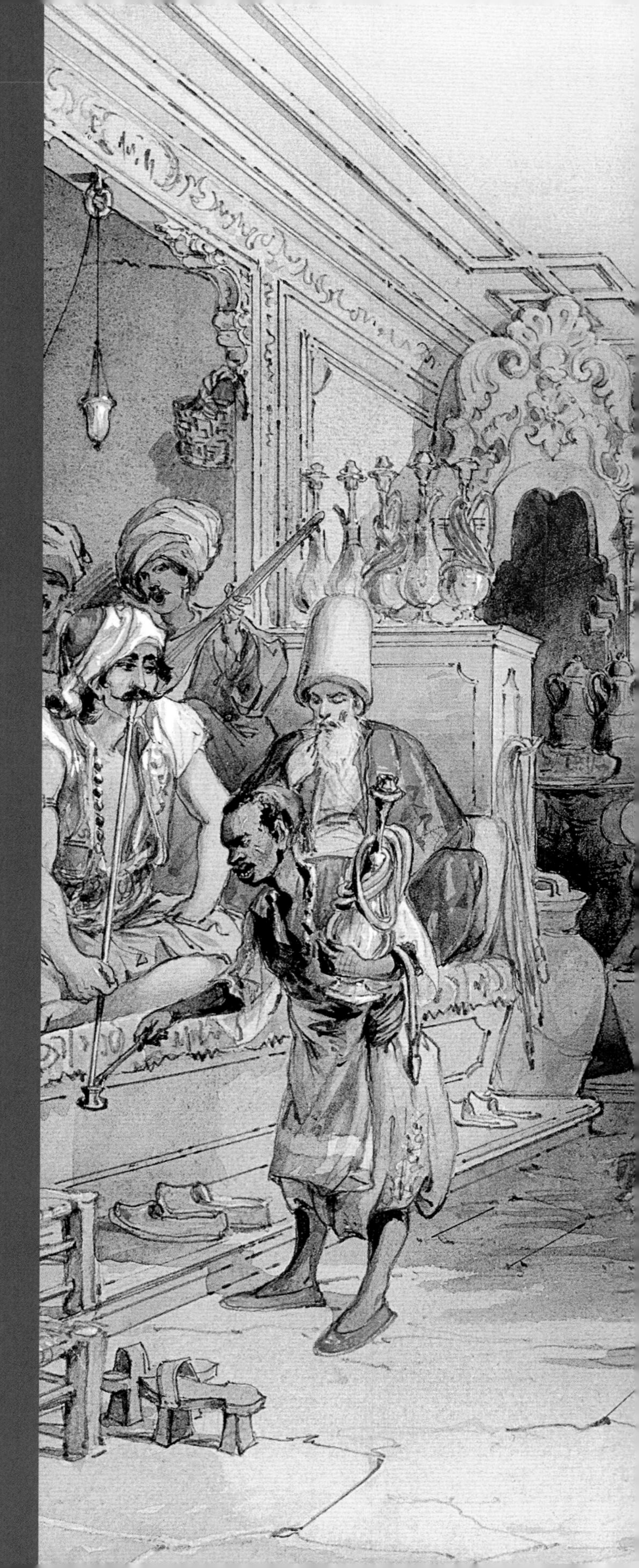

▷ **A coffeehouse in Constantinople**
This watercolour painting by Amadeo Preziosi, painted in 1854, shows all manner of characters including two Greeks in red caps, a whirling dervish in his conical hat, a Persian in purple robes, and an African boy attending to the pipes. In the corner, two large coffee pots stand on a stove.

A life of piracy

Although usually associated with tales of pillage and plunder, several of the figures involved in high-seas piracy made some extraordinary journeys in their quests for pieces of eight.

The second half of the 17th century was arguably a "golden age" of piracy for Europeans in terms of accruing wealth. It did not even necessarily mean criminality, as pirates (or buccaneers or freebooters, as they were also known) often had the backing of national governments, which gave them license to attack and loot the ships and ports of enemy states.

Francis Drake, who carried out the second circumnavigation of the world, raided Spanish possessions along the western coast of the Americas in the late 16th century, and was knighted for his efforts by Elizabeth I of England. Henry Morgan, who sacked Panama City, and whose life inspired the fictional tales of Captain Blood, was duly made Governor of Jamaica.

For an educated gentleman, piracy as a means of earning a living offered travel, adventure, and, of course, riches. In his quest to explore the world, for example, at the age of 16, Frenchman Raveneau de Lussan travelled to the French-controlled Caribbean island of Hispaniola, where he joined the crew of Dutch pirate Laurens de Graaf. By his early 20s, he was leading his own crew. After a few years raiding in the Pacific – sacking towns and ambushing the Spanish navy – he and his fellow buccaneers decided that they had seized enough treasure to return home, so set off on a journey back across Guatemala. It took 59 days, during which 84 of the 480 men perished of disease or were lost in the jungle. Eventually, they reached the Caribbean and sailed back to Hispaniola. There, they were given a warm welcome by the governor, who described their adventures as: "the greatest and finest voyage of any in our age".

The Barbary Coast

Another great region of piracy was the Barbary Coast of North Africa, where Muslim pirates terrorized the North Atlantic, the Mediterranean, and West Africa's Atlantic seaboard. Their main aim was to capture Christians for the Ottoman slave trade. One of the more colourful Barbary pirates was a Dutchman, Jan Janszoon van Haarlem, also known as Murat Reis. Janszoon captured the island of Lundy off western England and held it for five years. He raided Iceland, and in 1631, put ashore at the small town of Baltimore in western Ireland to abduct some 108 people, only two of whom ever returned home again. In honour of his endeavours, he was made Grand Admiral of the Corsair Republic of Salé, a pirating enclave in what is now Morocco.

△ **Gold doubloon**
This gold coin, minted in 1714, was one of thousands of pieces of gold stolen from the Spanish colonies by pirates.

△ **Sacking the Spanish**
To the English, Francis Drake was a hero – to the Spanish, he was merely a pirate. In 1585, he raided several Spanish bases in the Caribbean, including Santiago, shown besieged on this illustrated map.

The South China Seas

Arguably the most successful pirate of all was Zheng Yi Sao, also known as Ching Shih, who controlled a vast fleet of ships and crew in the early 1800s. Although she eventually negotiated a surrender and "retired", for a brief period the Pirate Confederation she led after her husband's death caused terror up and down the Guangdong coast.

PIRATES ANNE BONNY AND MARY READ DRESSED AS MEN, READY FOR WORK

IN CONTEXT
Pirate queens

Although women were not allowed aboard European pirate ships, Irish-born Anne Bonny countered this by dressing and acting as a man while first mate on Calico Jack's ship. She also happened to be his lover. Another cross-dresser was Englishwoman Mary Read, who joined the British military as Mark Read. Sailing to the West Indies, the ship she was on was captured by pirates and she joined their ranks. She was sailing with Anne Bonny and Calico Jack on their ship *Revenge* in 1720, when the ship was captured by pirate hunter Captain Jonathan Barnet. Bonny's last words to Jack were allegedly: "Sorry to see you there, but if you'd fought like a man, you would not have been hang'd like a Dog".

Travels in the Mughal Empire

As India became more accessible to Europeans in the 16th and 17th centuries, a succession of travellers reported back on the glories, sophistication, and architectural magnificence of the Mughal Empire.

Before Marco Polo's travels in the 13th century (see pp.88–89), the European view of India consisted largely of myths and fables. This began to change during the next 200 years, as more European travellers arrived in India to see the place Polo had described as "the richest and most splendid province in the world".

First impressions

One of the first Europeans to write a detailed account of India was Niccolò de' Conti, an Italian merchant, who from 1419 travelled in the footsteps of Marco Polo, including a spell in India. Half a century later, a Russian merchant named Afanasy Nikitin travelled via Azerbaijan and Persia to India. He spent three years there, and recorded his observations in a book called *The Journey Beyond Three Seas*, a rich source of information on the people and customs of India at the time.

De' Conti's travels provided a source for the 1450 Fra Mauro map, which suggested that there was a sea route from Europe around Africa to India. When the Portuguese proved this route existed, trade and cultural exchanges between Europe and India grew.

△ **Elephant fight at Lucknow**
This engraving from François Bernier's *Travels in the Mogul Empire* shows an elephant fight in Mughal India. Elephants were used for hunting, military campaigns, and sport.

Inevitably, conquest and exploitation were high on the European agenda, and soon the Portuguese had staked a claim to large swathes of southern India. Trading rivalries among the seafaring powers spurred the British, the Dutch, and the French to establish outposts in India over the following century. However, not everyone went there purely for profit.

Italian Cesare Federici travelled to India in 1563 to "see Eastern parts of the world". He spent 18 years in Asia and published an account of his travels in Venice in 1587. Jean-Baptiste Tavernier (1605–89) was a French gem merchant who combined business with a yearning for travel that took him to Persia and India. He documented his career in *The Six Voyages of Jean-Baptiste Tavernier* (1675). Perhaps most interesting of all was François Bernier (1625–88), a Frenchman who arrived in India during the reign of the Mughal emperor Shah Jahan, the builder of the Taj Mahal.

Court physician

Born the son of a farmer, and educated in Paris, Bernier managed to become a medical doctor on the strength of a three-month intensive course.

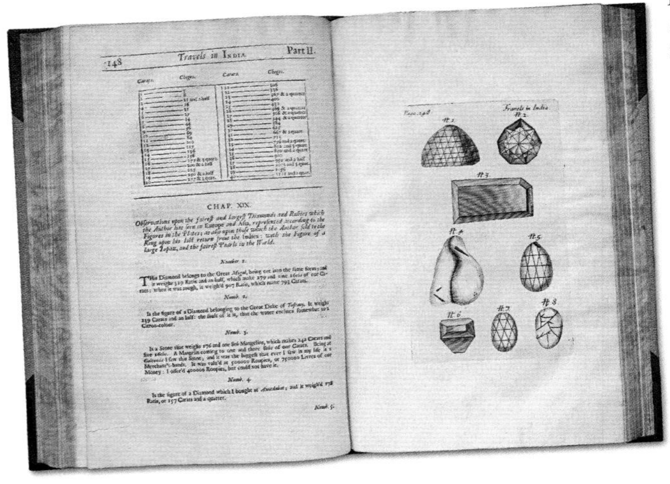

148 *Travels in* INDIA Part II.

CHAP. XIX.

Observations upon the fairest and largest Diamonds and Rubies which the Author has seen in Europe and Asia, represented according to the Figures in the Plates; as also upon those which the Author sold to the King upon his last return from the Indies: with the Figure of a large Topaz, and the fairest Pearls in the World.

Number 1.

THis Diamond belongs to the Great *Mogul*, being cut into the same form; and it weighs 319 Ratis and an half, which make 279 and nine 16ths of our Carats; when it was rough, it weigh'd 907 Ratis, which make 793 Carats.

Numb. 2.

Is the figure of a Diamond belonging to the Great Duke of *Tuscany*. It weighs 139 Carats and an half: the fault of it is, that the water enclines somewhat to a Citron-colour.

Numb. 3.

Is a Stone that weighs 176 and one 8th Mangelins, which makes 242 Carats and five 16ths. A Mangelin coming to one and three 8ths of our Carats. Being at *Golconda* I saw this Stone, and it was the biggest that ever I saw in my life in a Merchant's-hands. It was valu'd at 500000 Roupies, or 750000 Livres of our Money: I offer'd 400000 Roupies, but could not have it.

Numb. 4.

Is the figure of a Diamond which I bought at *Amedabat*, and it weigh'd 178 Ratis, or 157 Carats and a quarter.

Numb. 5.

▽ **Diaries of a gem merchant**
Jean-Baptiste Tavernier was a French gem merchant who became famous for his six voyages to the East. His account of his travels included diagrams of the diamonds he found in India.

> "It should not escape **notice** that **gold and silver**... come at **length** to be **swallowed up** in **Hindoustan**."
>
> FRANÇOIS BERNIER, COMMENTING ON THE WEALTH OF INDIA

IN CONTEXT

Indians in Europe

When European ships first reached Asia, they also provided a new route for Indians trading with Europe. From around 1600, many Indians headed west, but it was not until much later that any of them wrote about it. The first three to do so wrote in Arabic or Persian, but the fourth, Sake Dean Mahomed, wrote in English. Mahomed was born in northeast India, and became a trainee surgeon working for the British East India Company. In 1782, he travelled to England, where it is claimed he opened England's first Indian restaurant and introduced "shampoo" baths. In 1794, he published *The Travels of Dean Mahomed*, the first Indian travelogue in English.

SAKE DEAN MAHOMED

Impressively, he entered service as physician to the Mughal court, and when Shah Jahan died, he was kept on by his successor, the Emperor Aurangzeb. Bernier was a keen observer of the proceedings of the court, with its fusion of different cultures and religions, and of the bitter rivalry between Shah Jahan's sons, all of whom squabbled for their father's throne. As part of the court, Bernier travelled far and wide across northern India, the Punjab, and Kashmir, where he was the first, and for a long time the only, European to venture. He later recounted his experiences in *Travels in the Mogul Empire, AD 1656–1668*.

▷ **The young Aurangzeb**
This painting of 1635 shows the three younger sons of Shah Jahan. Aurangzeb (middle) was the last great ruler of the Mughal dynasty. He reigned for 49 years, from 1658 to 1707.

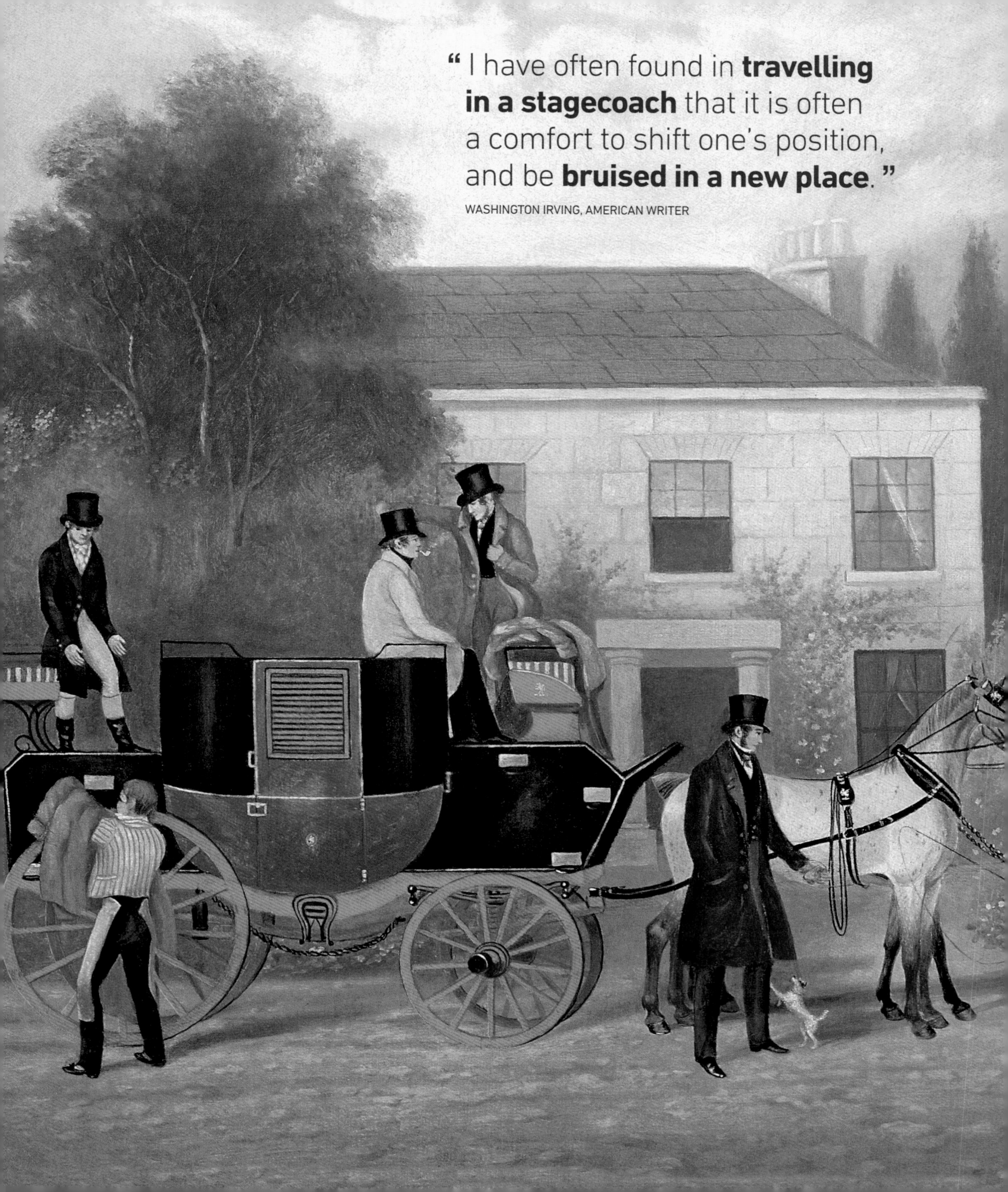

“ I have often found in **travelling in a stagecoach** that it is often a comfort to shift one’s position, and be **bruised in a new place**. ”

WASHINGTON IRVING, AMERICAN WRITER

The stagecoach

In the 16th century, coach services began running between towns, sparing people the discomforts of travelling on foot or on horseback.

Early travellers over land had no choice but to ride on horseback or to walk. Carriages or coaches were reserved for royalty, and generally female royalty at that. Such personages had the resources to send servants out beforehand to scout the most passable routes, and, where necessary, to clear and repair the rutted cross-country tracks that passed for roads. Elizabeth I of England is recorded to have spent a fortune on coaches during her reign (1558–1603), and it is perhaps for this reason that coaches subsequently became fashionable with the gentry – although it was generally considered effeminate for a man to prefer a coach to a horse.

The popularity of the coach was not confined to England. By the mid-17th century, horse-drawn stagecoaches had begun to travel between many European cities. It is around this time these vehicles became known as stagecoaches – a reference to the fact that they travelled in stages of 16–24km (10–15 miles). This was one of the reasons that journeys on the road were painfully slow. In England, in 1657, the 292km (182 miles) from London to Chester took six days. It was also uncomfortable. Carriages lacked suspension, up to eight people might be crammed inside, second-class travellers sat in a large open basket attached to the rear of the coach, and third-class passengers risked being bounced off perches on the roof.

Then, in the 18th century, turnpike trusts were established to provide better roads, charging a toll for usage, and a network of coaching inns emerged to provide travellers with overnight accommodation. These developments were spurred on by the innovation of carrying the mail by coach, so roads were now used by both stagecoaches and mail coaches. Stagecoaches themselves became more comfortable. A German design, known as the Berlin, featured curved-metal spring suspension and a coach body for four with a door on each side. Later improvements included glass windows to keep out the wind and rain, where previously there had been only blinds. In the US, the Concord coach had leather straps for suspension, prompting Mark Twain to describe it as "a cradle on wheels". The stagecoach remained the common form of long-distance transport until the arrival of the railways in the 1830s.

◁ ***The Stage Coach Preparing to Depart***
In its day, the stagecoach offered the most comfortable way of travelling long distances. This painting by Charles Cooper Henderson shows passengers embarking from a coaching inn sometime in the early 19th century.

The frozen east

Unable to expand further to the west, in the late 16th century Russia turned its gaze to the east. Intrepid explorers set out to explore and colonize the vast frozen expanse of Siberia.

During the reign of Ivan the Terrible (r. 1547–84), Russia's first tsar, Russian imperial ambitions had been thwarted to the west by Poland and Sweden, and to the south by the Crimean Tatars. Seeking new lands to conquer, Russia turned to the east, to the Ural Mountains and the vast expanse of Siberia beyond. While in the West the colonization of North America from coast to coast would take around 250 years, Russian explorers went from the Urals in the west to the Pacific Ocean in the east in around 65 years.

The movement into Siberia was spearheaded by Russian Cossacks hunting for ermines, foxes, and sables. As in America, early explorers tended to follow river routes, and one of the first Russian settlements, Tobolsk (created in 1585), was founded at the confluence of the Irtysh and Tobol rivers. Also as in America, the colonizers encountered many tribes of indigenous peoples as they advanced. Some of these offered assistance, including sharing geographical knowledge, while others were more hostile. Whether friendly or not, the Russians viewed it as their colonial right to claim all territory as their own and to civilize these "savages". These people would be decimated by slaughter, disease, and alcoholism.

Reaching the Pacific Ocean

In 1627, the Russians, led by Cossack explorer Pyotr Beketov, reached the region of Buryatia in central Siberia. Five years later, they founded Yakutsk as a far and lonely base camp, over 4,880 km (3,032 miles) east of Moscow. At this point,

Lake Baikal
It was not until 1643 that a Russian, the Cossack Kurbat Ivanov, first set eyes on Lake Baikal. The lake freezes in the winter months between January and May, and the temperature on the land around the lake can fall to as low as -19°C (-2°F).

▷ **Siberian Route**
A road was built connecting European Russia to China, known as the Siberian Route or *Sibirsky trakt*. Its surface was often frozen in the winter, so dog-sleds were used, as shown here.

they were only around 800 km (500 miles) from the eastern coast of Siberia, but it would not be until seven years later, in August 1639, that explorer Ivan Moskvitin and his party became the first Russians to reach the Pacific Ocean. Their reports were used to prepare the first map of the Russian Far East, which was drawn in 1642 by Kurbat Ivanov.

Ivanov was an explorer himself, and was keen to fill more of the empty spaces on his charts. He undertook his own expedition the following year, leading a party of 74 men up the River Lena in search of a rumoured vast body of water. This was Lake Baikal, the world's deepest lake, and the largest freshwater lake by volume. Ivanov became the first Russian to chart and describe it.

In a very short time, Russia had added new territory of around 13 million sq km (5 million sq miles), or about a twelfth of the total land surface of the globe, stretching from the Arctic to Central Asia, and from the Urals to the Sea of Japan. However, there was plenty more exploring to do.

Beyond the Stanovoy Range

Little of the territory found so far was suitable for agriculture, so in 1643, an administrator from Yakutsk, Vassili Poyarkov, was sent south to explore the lands bordering China. He set off from Yakutsk with 133 men, following various rivers south up into the Stanovoy Range. Beyond these mountains, he discovered Siberia's great southern river, the Amur, and a fertile plain fit for farming. However, as they crossed over the Stanovoys, Poyarkov and his men were driven back by the Chinese, and it was not until 1859 that the area was annexed by Russia.

By the mid-17th century, Russian conquests in the frozen east had ensured that its borders were roughly similar to those of modern-day Russia. What remained uncharted was the Arctic coastline and the northeastern peninsula of Kamchatka. These would not be completed until the Great Northern Expedition in the following century.

IN CONTEXT

A place of exile

Almost as soon as Siberia was colonized, it became a place of banishment. It was both a convenient way to cleanse European Russia of those deemed undesirable, and a way to populate the eastern wilds. "In the same way that we have to remove harmful agents from the body so that the body does not expire, so it is in the community of citizens," declared the bishop of Tobolsk in 1708. "That which is harmful must be cut out."

Exiles had to walk to Siberia, a journey that could take up to two years, spent shuffling in chains along the *Sibirsky trakt*. At Tobolsk, 1,770 km (1,100 miles) from Moscow, the prisoners' leg irons were removed – at this point they were so far into the wilderness that there was nowhere for them to run. Their sentences began only once they had arrived at their designated place of exile.

CHAINING PRISONERS, SAKHALIN, RUSSIA, 1890s

The Great Northern Expedition

With the colonization of Siberia underway, the Russian tsars commissioned two major expeditions to probe the far northern and eastern limits of their expanding empire. The second became known as the Great Northern Expedition.

△ **Vitus Bering**
Although Danish, Bering served the Russian tsar in his explorations of what became known as the Bering Strait. He prepared the way for a Russian foothold on the North American continent.

Peter the Great (r. 1682–1725) was the tsar who transformed Russia into a major European empire. He made modest territorial gains during his reign, mainly at the expense of the Ottomans and the Swedes, but arguably his greatest military achievement was the development of a modern Russian navy. For the navy to be effective, it needed detailed maps of the country's coastline, but at the time, there were still a considerable number of blank areas on the charts.

The man selected to rectify this was Vitus Bering, a Danish cartographer and seaman who had joined the rapidly-expanding Russian navy in 1704 and then served for 20 years. On the orders of Peter the Great, in 1725 he travelled to the northern Pacific to see if there was land connecting Siberia with Alaska.

The first two years of the project were spent moving men and materials from the new capital of St Petersburg across Siberia. It was not until July 1727 that Bering reached the shanty settlement of Okhotsk on the Pacific. Two small ships were built from materials transported for the purpose – ropes, sails, and iron parts, including the anchor (everything except wood). These ships carried the expedition to the remote Kamchatka Peninsula in northeastern Russia, where the men constructed another ship, the *Archangel Gabriel*, in which they took to the open sea. Bering kept so close to the Russian coast that he did not notice Alaska, which was a mere 110km (68 miles) away. Nevertheless,

▷ **Shipwreck**
Bering's ship was wrecked on one of the barren Commander Islands, 175km (109 miles) off the coast of Kamchatka. Bering and 30 of his crew died there, and the island was later named after Bering.

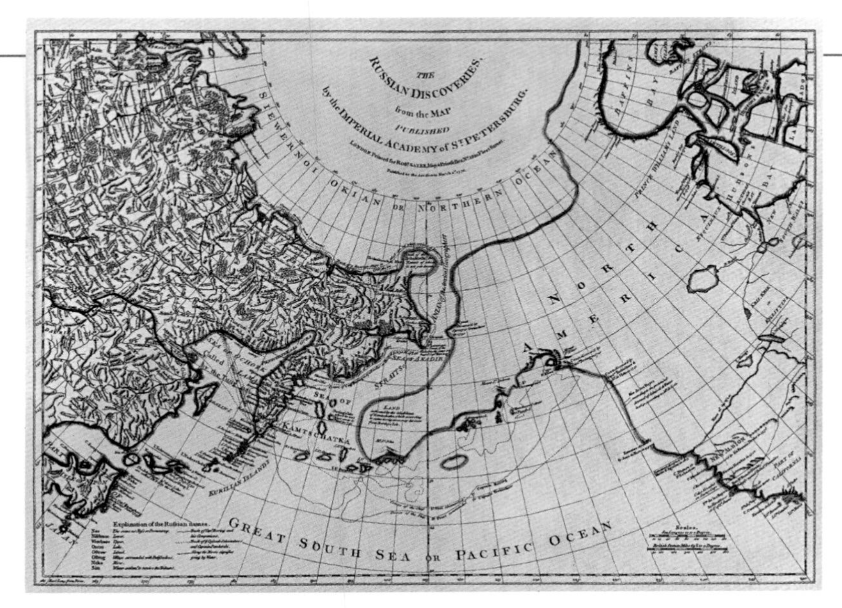

◁ **The Russian discoveries**
This English map is based on a 1754 map published by the St Petersburg Academy of Science, showing the sea routes taken by Bering and Chirikov during the expedition.

he had discovered that there was no land bridge connecting Russia and Alaska, but just open sea.

In the summer of 1730, Bering returned to St Petersburg, where he was criticized by admiralty officials for failing to see the American coast. Three years later, however, Peter's successor, Catherine I, commissioned Bering to make a second voyage.

A tragic adventure

This second expedition involved some 3,000 men, making it perhaps the largest scientific venture to date. It was organized into three groups: one to chart the northern coast of Siberia; one to make a scientific survey of Siberia; and another (Bering's) to map the North American coast and establish Russia's interests in the wider Pacific.

After 10 years of preparation, the Pacific explorers left Kamchatka in June 1741 in two ships built at Okhotsk – the *St Peter*, commanded by Bering, and the *St Paul*, commanded by Alexei Chirikov, his lieutenant. Within days, the two ships had lost sight of each other in fog, but on 16 July, Bering sighted land. Two days later, a small landing party, including German naturalist Georg Wilhelm Steller, went ashore on Kayak Island, becoming the first Europeans to set foot on Alaskan soil.

Still separated from the *St Paul*, and with supplies running low, Bering decided to head back west. Battered by storms, and with a crew sick with scurvy, the *St Peter* finally sighted land on 4 November. Tragically, what they hoped was Kamchatka turned out to be an uninhabited island. The crew spent winter there, living in driftwood huts dug into the permafrost, but 31 men died, including Bering. When the weather improved, the 46 survivors built a boat from the wreckage and made it back to Kamchatka. The *St Paul* also returned, and discovered new islands en route, but lost half of its crew. Despite its losses, the Great Northern Expedition mapped much of the Arctic coast of Siberia and discovered Alaska, the Aleutian Islands, and the Commander Islands. Bering was commemorated by having not only the Bering Strait named after him, but also the Bering Sea and Bering Island. The reports, maps, and samples from the expedition paved the way for Russian expansion into Alaska, which it occupied until it was sold to the US in 1867.

△ **Steller's sea cow**
Named for the expedition naturalist Georg Wilhelm Steller, this large mammal was hunted to extinction just 27 years after its discovery.

▽ **Kamchatka**
This sketch by German naturalist Georg Wilhelm Steller, a member of Bering's expedition, depicts the remote volcanic peninsula of Kamchatka in northeastern Russia. It was the launching point for both of Bering's expeditions.

> " We have **only** come to **America** to take **water** to **Asia**. "
>
> GEORG WILHELM STELLER, ON BEING TOLD THAT THEY WERE ONLY STAYING ON KAYAK ISLAND LONG ENOUGH TO TAKE ON WATER

Calculating longitude

Due to difficulties in calculating longitude, many of the great voyages of discovery owed as much to luck as to skill. It took a humble carpenter to provide the means for sailors to know exactly where they were.

▽ **The Scilly Isles disaster**
When four ships and more than 1,300 men were lost in 1707 due to a navigational error, the British government paid for a competition to solve the problem of measuring longitude.

Considering the many great voyages of discovery that were made in the 15th, 16th, and 17th centuries by the likes of Christopher Columbus, Ferdinand Magellan, and Vasco da Gama, it seems extraordinary that once they were out of sight of land, sailors were quite literally "all at sea". Despite charts and compasses, their inability to determine longitude with any reliability left even the most experienced captains largely at the mercy of luck or the grace of God.

Dead reckoning

Determining position at sea requires knowing both latitude and longitude. Latitude gives the north-south position, and longitude the east-west. Any sailor could calculate the former by the height of the sun or known guide stars above the horizon. Christopher Columbus simply followed a straight line of latitude on his historic journey of 1492.

However, to gauge their distance east or west of a home port, sailors relied on a method called dead reckoning, by which a navigator threw a log overboard and observed how quickly it receded from the ship. This crude estimate of the ship's speed, combined with the direction of travel (taken from the stars or a compass), and the length of time the ship had been on a particular course, plus or minus ocean currents and winds, gave a sailor a rough estimate of his longitude – of how far west or east he had travelled. It was a very imprecise method, and all too often fatally so. On 22 October 1707, four homebound British warships miscalculated their position and ran aground on the Scilly Isles near the southwestern tip of England, and over 1,300 men lost their lives.

△ **John Harrison**
Harrison was a self-educated English carpenter turned clockmaker who invented the marine chronometer, which enabled sailors to calculate longitude at sea. Thanks to Harrison, mariners could know their location anywhere on Earth.

Artist in the rainforest

As explorers sallied forth on voyages of discovery, amateur artists also embarked on a grand engagement with the natural world.

Throughout the 17th and 18th centuries, a steady influx of wonderful plants and animals from overseas provided European artists with new subjects to paint. Some of them decided they needed to see these wonderful things in their native settings. One of the first of these adventurous travelling artists was, unusually for the time, a middle-aged divorced woman.

Maria Sibylla Merian was born in 1647 in Germany to a Swiss printmaker and draughtsman and his wife. From an early age, Merian was passionate about both insects and painting, and lovingly painted fruit, flowers, and the insects she captured and bred. Married at 18, she had two daughters, and combined looking after them with working as an artist and teacher, until she left her husband to live in a commune in Denmark. Later, she moved to Amsterdam, and then, at the age of 52, she sold everything she owned, wrote a will, and set out with her younger daughter for the Dutch colony of Surinam in South America.

There, Merian made her own way. With great difficulty, she ventured deep into the rainforest to collect plants and insects: "One could find a great many things in the forest if it were passable: but it so densely overgrown with thistles and thorn bushes that I had to send my slaves ahead with axe in hand to hack an opening for me." Eventually, malaria or yellow fever forced Merian to return to Amsterdam, where she raised the money to publish her work in a series of portfolios, including *The Metamorphoses of the Insects of Surinam*, published in 1705. The resulting, mostly life-sized, paintings are phenomenally beautiful, and every bit as colourful and astonishing as the life of the woman who painted them.

" The **heat** in this **country** is **overwhelming**. It nearly **cost me** my **life**. Everyone is **amazed** that I **survived** at **all**. "

MARIA SIBYLLA MERIAN ON HER TIME IN SURINAME

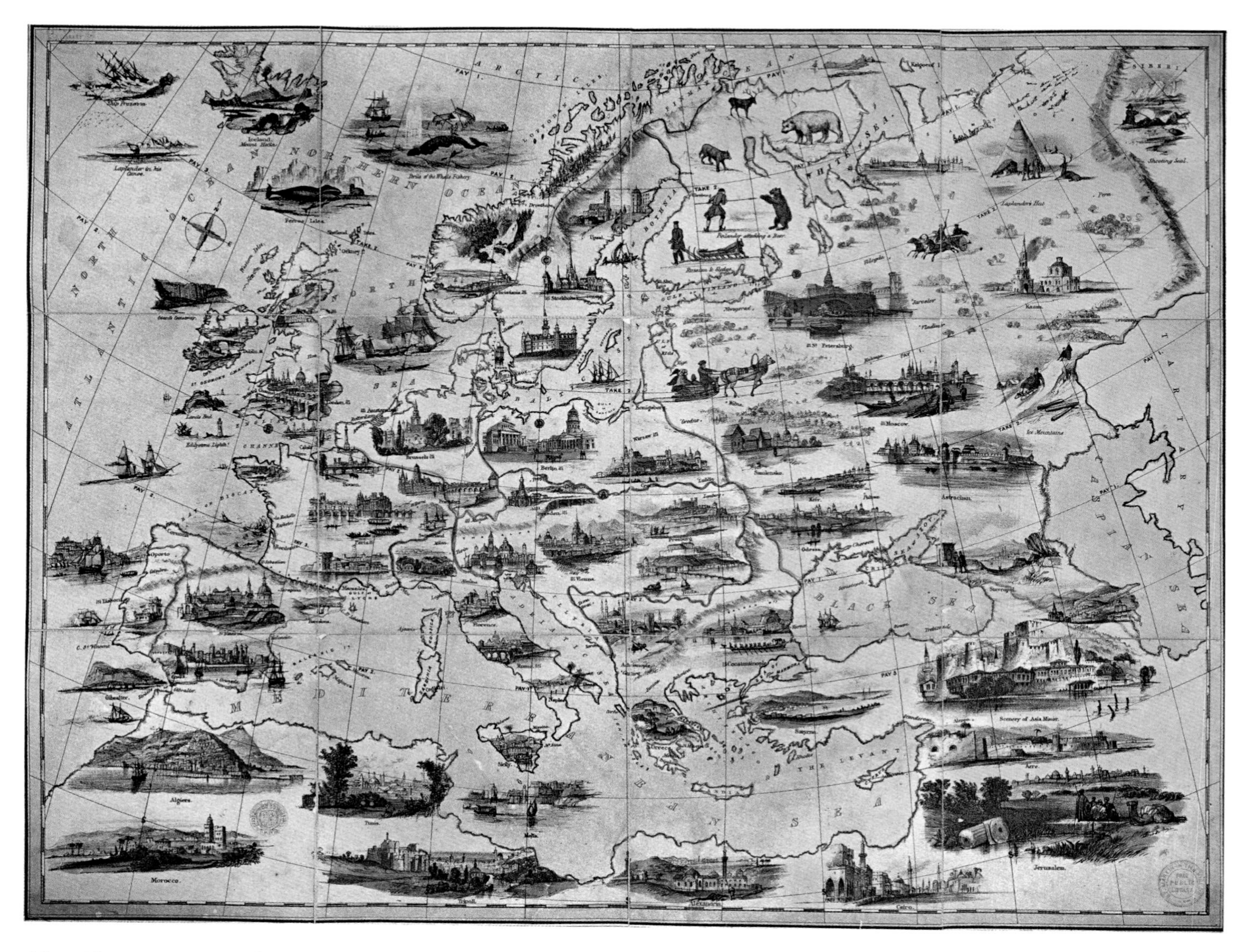

△ **A world of learning**
The Grand Tour of Europe began as a way for the sons of wealthy British families to gain a cultural education through acquaintance with the great art and architecture of the past.

The Grand Tour

From the 17th century onwards, wealthy young Englishmen began travelling across Europe to discover the roots of Western civilization – which, it was generally agreed, lay in Rome.

"A man who has not been in Italy is always conscious of an inferiority," proclaimed English man-of-letters Samuel Johnson. This was somewhat self-effacing considering Johnson had never visited the country. He was, however, acknowledging a belief widely held at the time, namely that an extended trip to Italy was fundamental to a man's education.

Italy, and Rome in particular, had been a popular destination for northern European artists, intellectuals, and diplomats since at least the 17th century, and a good education invariably meant training in the classics, with a working knowledge of Latin and Greek. Both ancient and Renaissance Rome were esteemed to be the source of all that was important in Western civilization, and a trip there was considered a sort of finishing school.

Cultural pilgrimage

In 1670, a book called *Voyage of Italy*, or *A Compleat Journey Through Italy*, was published, in which its author, Englishman Richard Lassels, called this phenomenon of cultural pilgrimage the "Grand Tour". The Grand Tourists were almost exclusively male and, at least in the beginning, mostly British. As the world's wealthiest nation, Britain had a substantial upper class with enough money and leisure time to travel. Usually accompanied by a tutor or guardian, called a Bear Leader, the young aristocrats might spend anything from a few months to several years journeying across Europe.

ANCIENT ROME, **GIOVANNI PAOLO PANINI (c.1691–1765)**

IN CONTEXT
By invitation only

There were no public museums in the 17th and 18th centuries, so as well as visiting churches, palaces, and other sites of artistic merit, travellers also made calls on the homes of nobles who were known to have collections of art. It was often possible to gain access with the right credentials, which generally meant a letter of introduction. In France, it was even possible to visit the Palace of Versailles, the permanent residence of the royal family. It was necessary to look like a gentleman, which meant being immaculately dressed and carrying a sword – although those lacking a sword could hire one at the palace.

▽ Inventing the Grand Tour
Richard Lassels earned a living acting as a guide for travellers in Europe. He first coined the phrase "Grand Tour" in his guide to Italy, published in 1670.

THE
VOYAGE
OF
ITALY;
OR
A Compleat Journey through
ITALY.
In Two Parts.
With the *Characters* of the *People*, and the Description of the Chief *Towns*, *Churches*, *Monasteries Tombs*, *Libraries Pallaces*, *Villa's*, *Gardens*, *Pictures*, *Statues*, and *Antiquities*.
AS
Also of the *Interest*, *Government*, *Riches*, *Force*, &c. of all the *Princes*.
With Instructions concerning *Travel*.

By *Richard Lassels*, Gent. who Travelled through *Italy* Five times, as Tutor to several of the *English* Nobility and Gentry.

Never before Extant.

Newly Printed at *Paris*, and are to be sold in *London*, by *John Starkey*, at the *Mitre* in *Fleet-street* near *Temple-Barr*, 1670.

The experience was intended to broaden their intellectual horizons and teach them about art, architecture, history, and politics in preparation for a career in public life back home. However, distance from home loosened social conventions and, particularly later in the 18th century, the tour often saw any pretence at education jettisoned in favour of sex, gambling, and drinking.

As with conventional pilgrimages, the route and destinations of the tour were fairly well-defined. After crossing the English Channel, the first stop was Paris, then straight on through provincial France to cross over the Alps into Italy. Here, the great cities of Rome, Florence, Naples, and then Venice were visited (in that order of importance) before the return home, typically via Switzerland, Germany, and the Netherlands.

Practicalities

By the mid-18th century, when Grand Tourism was at its peak, travel was surprisingly well organized. There were regular cross-Channel sailings, and travellers could, for example, arrange for transport across much »

> “ **Led by hand** he saunter'd **Europe round**, And gather'd every **vice** on **Christian ground**. ”
>
> ALEXANDER POPE, POET AND GRAND TOUR SCEPTIC (1688–1744)

▽ A grand souvenir
It was fashionable for Grand Tourists to have their portrait done. This painting is by Louis Gauffier, a French artist living in Italy, who made a living selling his work to British visitors.

» of the Continent from either London or Paris. There were even guidebooks, such as Thomas Nugent's *Grand Tour* (1749), offering descriptions of Europe's major cities and essays of moralistic advice in four volumes.

Once on the Continent, there were several options for onward travel. The wealthiest could purchase or hire a private carriage and horses. It was, however, more common to rent a carriage and travel "post", which meant hiring horses and a driver at designated points along main routes. This had the advantages of a private carriage without the expense of keeping horses. There were also the public coach services, which were cheap but slow.

▷ **Modest memento**
Not every tourist could afford to have their portrait painted or to purchase antiquities – some had to settle for more humble mementoes, such as this hand-painted goatskin fan.

The itinerary

Paris was still a medieval city at the time – smaller and more densely populated than London. Grand Tourist Horace Walpole, son of Robert Walpole, the British prime minister, who toured the Continent in the early 1740s, called it, "the ugliest beastliest town in the universe". Even so, tourists would stay for several weeks, visiting churches, palaces, and any homes of noblemen that contained collections of art.

Few bothered stopping anywhere else in France, making straight from Paris to the Alps, where carriages were dismantled and hauled by pack animals over the Mont Cenis Pass into Italy. The travellers would traverse the mountains in a sedan chair carried by porters.

Once on the other side, Turin was the first stop, although most tourists considered it too provincial to detain them for long. By contrast, the principle centre of the Renaissance, Florence, was one of the most popular cities, worthy of a stay of several weeks.

" No one who **has not been** here can have **any conception** of what an **education Rome is**. "

JOHANN WOLFGANG VON GOETHE, 1816

▽ **The eternal city of Rome**
Englishman Charles Thompson spoke for many Grand Tourists when, in 1744, he said he was "impatiently desirous of viewing a country so famous in history" – namely Italy. The treasures of Rome are depicted here by Bernardo Bellotto.

◁ **The end of the tour**
Naples, seen here in a painting from the end of the 17th century, was as far south as most travellers went. Here they visited the lava-trapped remains of Pompeii and Herculaneum, but also just luxuriated in the balmy climate.

Even in the mid-18th century, there was already a considerable English community here, so visitors could spend time socializing with their fellow countrymen.

Visiting Rome

A brief stop in Siena was usually all that then lay between the tourists and their ultimate goal of Rome. A popular guidebook recommended six weeks as a minimum to view all of the city's ancient ruins and more recent sights. William Beckford, a writer who visited Rome in 1782, thought even five years was not enough. For assistance, the visitor could call upon the services of a small army of tour guides, both Italian and foreigners who had settled in Rome. Generally speaking, the city lived up to expectations. In the opinion of German writer Goethe: "Here the most ordinary person becomes somebody, for his mind is enormously enlarged even if his character remains unchanged".

Eager to procure mementoes, many Grand Tourists had their portrait painted, often in a studio against the backdrop of some famous monument. Some bought art or antiquities, perhaps statuary and fragments of ancient buildings. Many private art collections in Britain were launched with artworks bought on a Grand Tour, and over time, many of these collections found their way into national museums.

A change in sensibility

The furthest south most tourists went was Naples. Apart from visits to the archaeological excavations at Herculaneum and Pompeii (begun in 1738 and 1755 respectively), most of those arriving in Naples simply enjoyed themselves. They revelled in the Mediterranean climate, and the exotic and colourful local coastal villages. Many also enjoyed the hospitality of William Hamilton, the British ambassador there from 1764 to 1800, who was known for his entertaining, as well as for his beautiful wife, Emma, who later became the mistress of Admiral Horatio Nelson.

The tradition of the Grand Tour came to an abrupt halt with the French Revolution and the Napoleonic Wars. Europe had become far too dangerous. By the time that peace returned in 1815, the Grand Tour was essentially a thing of the past. What remained was the idea of travel for enlightenment and pleasure, which, with the advances in technology in the 19th century, would lead to the expansion of "tourism".

IN CONTEXT

Building on the Grand Tour

As well as the tangible acquisitions made on the Grand Tour, there were also intangibles: new ways of thinking about history, civilization, aesthetics, and, especially, architecture. Time spent in Italy, in particular, engendered an appreciation of Classical architectural forms, especially the work of 16th-century Venetian architect Andrea Palladio. In his book *Italian Journey*, Johann Wolfgang von Goethe describes Palladio's unfinished Convent of Saint Maria della Carità in Bologna as the most perfect work of architecture. In England, aristocrats applied what they learned in Italy to their own country houses and gardens, inspiring a movement known as Palladianism, which evolved into Neoclassicism. The influence of Palladio even spread to America: Thomas Jefferson was an admirer, and the US Capitol building was inspired by Palladio's work.

THE NEOCLASSICAL CAPITOL BUILDING, WASHINGTON DC, US

First flight

In 1783, man's dream of flying was finally realized when the hot air and hydrogen balloons were invented. These liberated people from what Victor Hugo called "the ancient, universal tyranny of gravity".

◁ **The birth of modern ballooning**
The balloon designed by the Robert brothers, shown here taking off from the Tuileries, was filled with hydrogen and had a valve in the crown to release gas for the descent.

The year 1783 was a milestone for sheep, ducks, and roosters. That September, one of each of these animals was secured in a basket and sent aloft above the Palace of Versailles in a hot air balloon. The escapade lasted eight minutes – then the balloon came down a few miles away. The animals seemed unharmed by the experience, so two months later, on 21 November 1783, two French men, Jean-François Pilâtre de Rozier, a young Parisian doctor, and the Marquis d'Arlandes, a military officer, undertook the world's first untethered hot air balloon voyage, ascending from the outskirts of Paris and travelling nearly 9 km (6 miles).

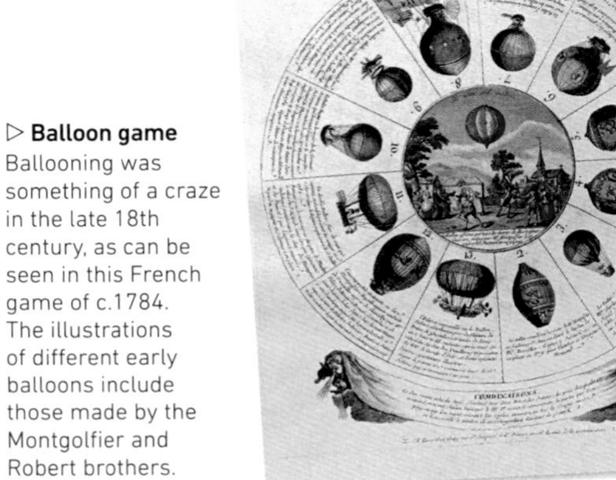

▷ **Balloon game**
Ballooning was something of a craze in the late 18th century, as can be seen in this French game of c.1784. The illustrations of different early balloons include those made by the Montgolfier and Robert brothers.

An unpredictable ride

Both of these balloons had been built by the Montgolfier brothers, world pioneers in the field of aviation. The two, Joseph-Michel and Jacques-Étienne, came from a family of papermakers. Not coincidentally, their prototype balloons were made of layers of paper, covered by buttoned-together sackcloth. The brothers described their creation as "putting a cloud in a paper bag". The balloons that carried the animals and humans aloft were grander affairs of silk-like taffeta, gloriously decorated with designs by wallpaper manufacturer Jean-Baptiste Réveillon.

Just 10 days after the Montgolfier's manned flight, on 1 December 1783, another pair of French brothers, *les frères* Robert, launched the world's first manned hydrogen balloon. A reported 400,000 spectators watched it rise from the Tuileries Gardens in Paris, including Benjamin Franklin, the great inventor and the diplomatic representative of the United States.

The following year saw the first flight by a woman, when Frenchwoman Elisabeth Thible went aloft dressed as Minerva, thrilling onlookers with her bravery. Also, the Robert brothers attempted to tackle a major deficiency of the balloon – the fact that it would only go where the wind blew it – by making one that was elliptical and, it was hoped, could be steered with oars and umbrellas. The experiment was not a success. Other flights achieved better results, however. In 1785, for example, when the wind was blowing in the right direction, a Franco-American team successfully ballooned across the English Channel. Unfortunately, Pilâtre de Rozier had died attempting the same trip just a few months earlier, making him one of the first air-crash fatalities. Nevertheless, "balloon-mania" captivated Europe until the death of Napoleon's "Aéronaute des Fêtes Officielles", Sophie Blanchard, who was famous for her daring flights, custom balloon, and airborne pyrotechnical displays.

▽ **The first manned flight**
This painting shows Jean-François Pilâtre de Rozier and the Marquis d'Arlandes becoming the first people to fly. Their hot air balloon was designed by the Montgolfier brothers.

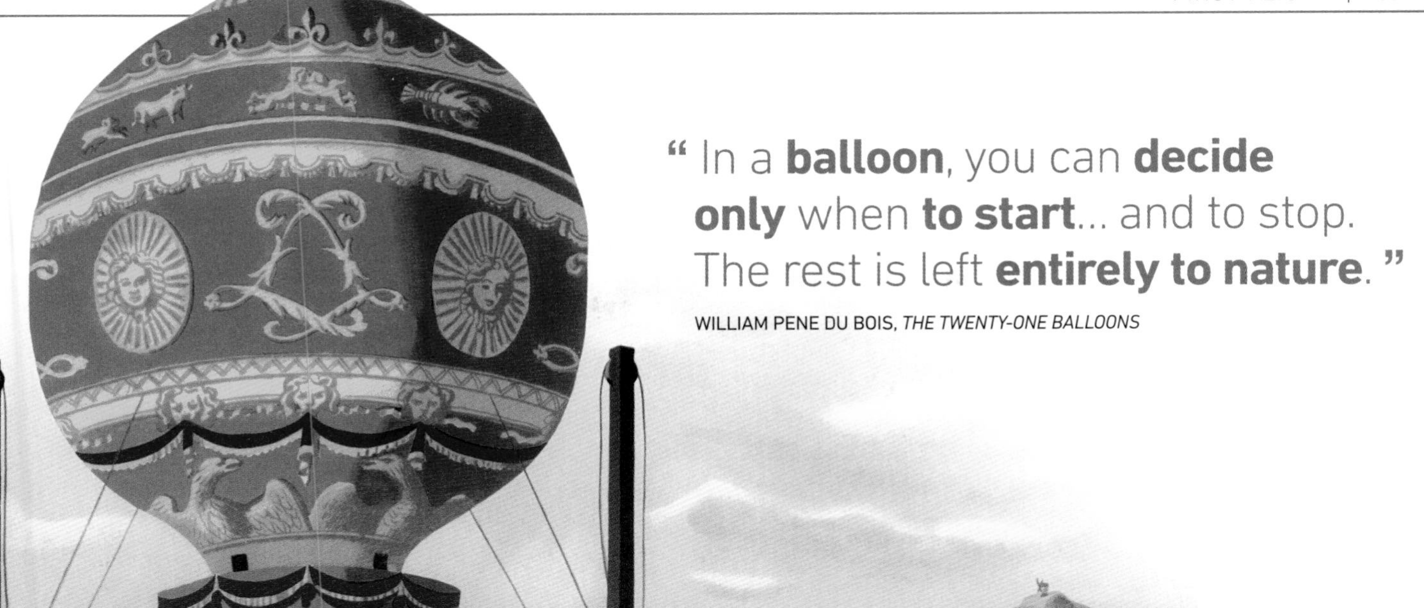

> “ In a **balloon**, you can **decide only** when **to start**... and to stop. The rest is left **entirely to nature**. ”
>
> WILLIAM PENE DU BOIS, *THE TWENTY-ONE BALLOONS*

▽ **Polar tragedy**
In 1897, Swedish explorer S.A. Andrée attempted to reach the North Pole by hydrogen balloon. The balloon came down, as seen here, and its passengers soon perished.

A specialist role

Its unpredictability meant that the balloon had little future as a means of public transport, but its value was recognized in other fields. During the American Civil War, the Union Army deployed a corps of balloons to spy on enemy positions, and in 1870–71, more than 60 balloons managed to fly away from Paris during the siege of the city by the Prussians.

Balloonists pioneered the science of meteorology, taking airborne readings and discovering from experience the limits of human endurance, including the effects of frostbite. This became pertinent when balloons were used in a tragic attempt to reach the North Pole in 1897. The novelist Victor Hugo claimed that the balloon had not only freed people from the effects of gravity but could go on to play a role in “the liberation of all mankind”.

Even if ultimately the balloon proved to be of limited use, it did provide the first ever aerial view of the world – one that showed the curvature of the Earth, the patterns of the landscape below, and the extent of man’s impact upon it.

Bound for Botany Bay

In 1787, the British Government attempted a new experiment – sending a fleet halfway around the world to turn a distant continent into a penal colony. In doing so, it caused great hardship for the indigenous populations.

On 13 December 1786, Francis Fowkes appeared before a judge at the Old Bailey, the central criminal court in London, accused of stealing a coat and a pair of men's boots from a tavern in Covent Garden. He was pronounced guilty, and the sentence was severe: he was to be "transported for seven years".

▽ **Arthur Phillip**
A Royal Navy admiral when he founded the penal colony that later became the city of Sydney, Arthur Phillip was appointed the first Governor of New South Wales.

Exile overseas

To be "transported" meant being sent into exile. In the later 18th century, crime in Britain was rampant, as there was no professional police force. Transportation was an attempt to purge the country of petty criminals by making them serve their sentences in a distant penal colony (perpetrators of more serious crimes were executed). Until the 1775–83 American War of Independence put paid to that option, the British had used North America for this purpose.

Of the alternatives, the east coast of New Holland, recently mapped by Cook's expedition (see pp.172–75), was deemed the most suitable, and a colonization party was duly sent forth in 11 ships under the command of Admiral Arthur Phillip. This "First Fleet" left Portsmouth on 13 May 1787 and sailed, via Rio de Janeiro and Cape Town, the 25,588 km (15,900 miles) to its destination of Botany Bay, arriving on 20 January 1788. Botany Bay was swiftly judged unsuitable for settlement. The

△ **Native inhabitants**
On landing, the naval officers were shown where to find water by local Aboriginals. Thereafter, relations swung between cooperation and armed conflict.

▽ **Port Macquarie Penal Station**
Established in 1822, this isolated island colony, in what is now Tasmania, took the worst convicts and those who had escaped from other settlements on the mainland.

> " ... From rectitude's **path** we did **stray**, So they **shipped us** across the **salt ocean**, To do time at **Botany Bay**. "

PART OF THE AUSTRALIAN FOLK SONG, *PINK 'UN*, 1886

fleet moved to Port Jackson, a more promising bay to the north, which had been charted by Cook. This became the site of the first permanent European colony on Australian soil.

Convict settlers

Among the 732 convict settlers, with 247 marines and their families, was the unfortunate Francis Fowkes. A skilled artist, he produced a hand-drawn map of the new colony that illustrates the very beginnings of the new state and what would become its capital city, Sydney.

Many of the new settlers died in the early days as they struggled to grow enough food to feed themselves, but their numbers were replenished by waves of new arrivals. Two more convict fleets arrived in 1790 and 1791, and the first free settlers arrived in 1793.

△ **Manacles from Port Arthur**
By the time Britain stopped penal shipments in 1868, around 164,000 convicts had been sent to Australia.

The last convicts

By the mid-1800s, convicts were also being sent to other newly founded colonies in Australia, including Port Macquarie and Moreton Bay further up the east coast, Van Diemen's Land (now known as Tasmania), western Australia, and tiny Norfolk Island in the South Pacific. Most of the convicts sent were English, Welsh, and Irish, but some were from British outposts such as India, Canada, and Hong Kong. Of those transported between 1788 and 1852, around one in seven were women. Good behaviour qualified a convict for a "ticket of leave", which was, in effect, the freedom to live independently in Australia, or to return to Britain if they so wished.

By the time of the last shipment of criminals in 1868, roughly 164,000 men and women had been transported. By this time, the total population of the colonies was around one million and they were completely self-sustaining. For the indigenous population who'd lived there for thousands of years, however, the colonies were disastrous: vast numbers were wiped out by European diseases, violent clashes with colonists, and hardship caused by disruption to their traditional way of life by settling farmers.

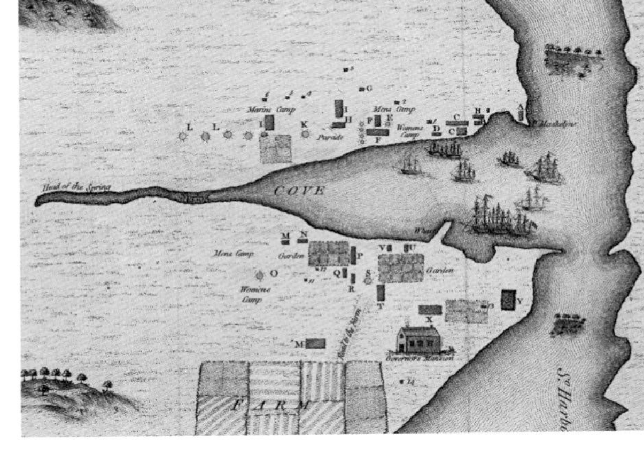

△ **Port Jackson**
A 1788 map drawn by Francis Fowkes, one of the first convict settlers, shows the origins of Sydney. The remains of the Governor's Mansion (the large red building) can still be visited today.

IN CONTEXT
Ten Pound Poms

After World War II, Australia needed workers for the country's booming industries. The government subsidized the cost of travelling for emigrants from Great Britain, charging them only £10 for the fare and promising employment. Many were only too happy to seize the opportunity of an escape from post-war austerity, and as a result, the scheme attracted over one million migrants from Great Britain to Australia from 1945 to 1972. The scheme was also extended to migrants from other countries, including Italy and Greece.

POSTER FOR AN AUSTRALIAN GOVERNMENT SCHEME PROMOTING IMMIGRATION, 1957

LONDON
TIME

THE
AGE OF
STEAM
1800–1900
18 201

THE AGE OF STEAM, 1800–1900

Introduction

The 19th century was an age of wonder. The achievements of mankind seemed boundless: iron towers reached to the heavens; vast factories thrummed with power; electricity lit up cities. But of all the inventions of the industrial age, the harnessing of steam had the most profound impact. Steam drove the engines that powered the factories, which created the wealth that enabled the colonial powers of Europe to extend their reach ever deeper into Africa, Asia, and Australia. In these far-flung lands, they harvested and mined raw materials that were then sent back to feed the factories and add to the wealth of the nations. Britain, France, and numerous other countries dispatched explorers to blaze trails through deserts and jungles, and sail up mighty rivers, such as the Mekong, the Niger, and the Nile, in search of further treasures.

By this time, America had emerged victorious from a war of independence in which it had shrugged off the rule of the British. It was now engaged in exploring, consolidating, and exploiting its own ever-expanding territories, which had various riches in abundance. In California, for example, a wealth of gold was discovered in 1849, prompting thousands of Americans to migrate to the West Coast.

Iron roads

Steam trains and steam ships revolutionized the economies of Europe. In America, the effect was even greater. River steamers opened up swathes of territory in the Midwest to settlement, while the completion of the first transcontinental railway line in 1869 literally bound the new nation together. Most importantly, steam travel was fast.

NAPOLEON'S EXPEDITION TO EGYPT IN 1798 PROMPTED MANY TO EXPLORE THE RIVER NILE

THE FIRST COMMERCIAL STEAMBOAT, *CLERMONT*, TOOK TO THE HUDSON IN 1807

THE FIRST COMMERCIAL STEAM TRAIN OPENED BETWEEN LIVERPOOL AND MANCHESTER IN 1830

> “ The prejudices which **ignorance has engendered** are **broken by the roar of the train** and the whistle of the engine awakens thousands from **the slumber of ages**. ”
>
> THOMAS COOK, 1846

A journey from the Atlantic to the Pacific had taken months; now it took days. In Europe, this opened up new travel possibilities for a growing class of people with the money and leisure time to indulge themselves. Few people could spare the months that were previously needed to tour the Continent, but with railways connecting nearly every country in Europe, the trip could be made in a matter of weeks. And as Napoleon's armies had occupied and exhaustively documented Egypt, many people took steamers from Italy and followed the general's footsteps up the Nile.

Inspiration for such journeys came from poets and novelists who rhapsodized over the romance of southern climes, ancient ruins, and untamed landscapes. Many early travellers also went with journal in hand, and returned home to add their own accounts to the rapidly proliferating library of 19th-century travel literature. These included Isabella Bird and Charles Dickens in Europe, and Washington Irving and Mark Twain in America.

The business of travel

By the mid-19th century, the desire to travel was such that several shrewd entrepreneurs spotted the potential for a profitable business. Grand new hotels and railway stations were built in major cities, and in 1841, Thomas Cook launched his eponymous tour company, initially in the hope that offering people day trips would broaden their horizons. Not long afterwards, Messrs Murray and Baedeker became wealthy pioneers in the guidebook business. By the end of the century, there were few places in the world that Thomas Cook & Son did not take people, or to which Karl Baedeker did not publish a guide.

ROMANTIC POET PERCY BYSSHE SHELLEY INSPIRED MANY TO MAKE A TOUR OF EUROPE

THE DISCOVERY OF GOLD IN SAN FRANCISCO DREW THOUSANDS OF IMMIGRANTS TO CALIFORNIA

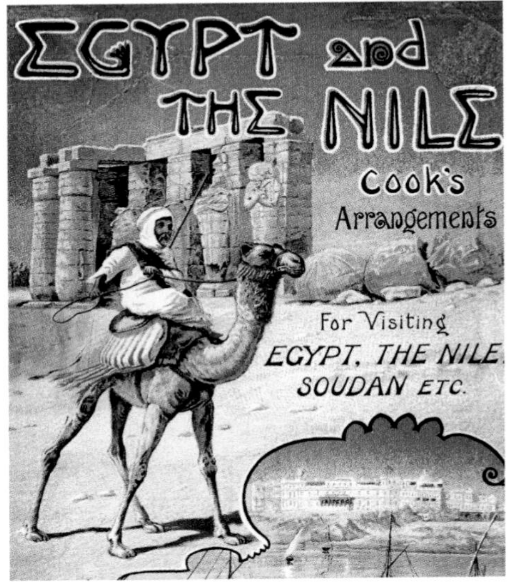

ENTREPRENEURS THOMAS COOK & SON BROUGHT HOLIDAYS TO THE MASSES

Humboldt at work
This portrait, made by Friedrich Georg Weitsch in 1806, shows the explorer at the age of 37. He had made his survey of South America, but was yet to make his epic crossing of Russia.

NATURALIST, 1769–1859

Alexander von Humboldt

Charles Darwin called him the "greatest scientific traveller who ever lived" and he has more plants, animals, minerals, and places named after him than anyone else, but Alexander von Humboldt's most significant contribution of all may be as the first prophet of climate change.

That there is a Humboldt lily, a Humboldt crater on the Moon, and a Humboldt squid that swims in the cold Humboldt Current off the coast of Peru, indicates the breadth of interests pursued by the polyglot genius who was born in Berlin in 1769. The second child of an aristocratic Prussian family, he lost his father when he was nine, and had a cold and distant mother. He sought solace in collecting plants, insects, and rocks.

Sent to university to study finance, Humboldt met Georg Forster, a naturalist who had accompanied Captain Cook on his second voyage (see pp.172–75). The two travelled to Europe where Humboldt met another of Cook's scientific compatriots, Sir Joseph Banks, with whom he developed a close friendship.

After university, Humboldt became an inspector in the Ministry of Mines, which allowed him to indulge his interest in geology. It was not until his mother's death in 1796, when he inherited a windfall, that he was properly able to pursue his wider scientific interests.

◁ **Mount Chimborazo** Humboldt's 1807 drawing of this volcano in Ecuador showed for the first time how different zones of vegetation are linked to altitude and temperature.

Latin American expedition

In 1799, Humboldt set off for South America with botanist Aimé Bonpland. Landing in what is now Venezuela, they canoed up rivers, trekked through the rainforest, and scaled some of the highest peaks in the Andes. In her 2015 biography *The Invention of Nature*, Andrea Wulf portrays Humboldt as insanely intrepid: trekking barefoot when his shoes disintegrated; swimming in crocodile-infested waters; and conducting experiments on electric eels bare-handed.

In the course of his adventures, Humboldt kept detailed journals and measured everything he could, from rainfall levels and soil composition to the blueness of the sky. He identified 2,000 new plant species and crossed the magnetic equator. While on Mount Chimborazo in today's Ecuador, he was struck by the idea that the Earth was one single great living organism in which everything is connected. He also reasoned that by disrupting this natural order, man might bring about catastrophe – a message so far ahead of its time that many still have trouble accepting it even today.

Returning to Europe, he wrote a monumental 30-volume account of his findings. It was this work that introduced Charles Darwin to the idea of scientific exploration. When he boarded the *Beagle* on his own voyage of discovery in 1831, Darwin took with him seven volumes of Humboldt's work.

Crossing Russia

Most scientists would be content to devote themselves to a lifetime studying the fruits of such an extended trip (Humboldt returned with 60,000 specimens), but the Prussian was not one to sit still.

In 1829, at the age of 59, he embarked on a six-month expedition to the Ural Mountains and Siberia. When he died, aged 90, his funeral in Berlin drew tens of thousands of mourners, and American newspapers lamented the end of the "Age of Humboldt".

◁ **Humboldt penguin** This South American penguin is named after the icy cold current in which it swims, itself named after Humboldt.

KEY DATES

- **1769** Born 14 September in Berlin.
- **1799** Sails from Spain to what is now Venezuela in the company of botanist Aimé Bonpland.
- **1800** Sails to Cuba with Bonpland, where they conduct scientific surveys for three months.
- **1804** Travels to the United States and has an audience with President Jefferson.
- **1814** Publication in English of the first volume of the *Personal Narrative of Travels to the Equinoctial Regions of the New Continent.*
- **1829** Crosses almost 16,100 km (10,000 miles) of Russia in six months.
- **1845** Publication of the first volume of *Cosmos*, a multi-part work that drew on all of Humboldt's observations to propose that the Earth is a living organism. James Lovelock's famous Gaia theory, formulated in the 1960s, bears remarkable similarities.

HUMBOLDT'S DRAWING OF *MELASTOMATACEAE*

HUMBOLDT IN HIS STUDIO, PAINTED BY EDUARD HILDEBRANDT IN 1845

Rediscovering Egypt

In 1798, a failed invasion of Egypt led by Napoleon Bonaparte of France fired up European and American passions for the wonders of a forgotten ancient civilization.

▽ **An ancient world**
An illustration from the *Description de l'Égypte*, a collection of observations made by scholars during Napoleon's expedition to Egypt, shows the portico of the Temple of Isis on Philae.

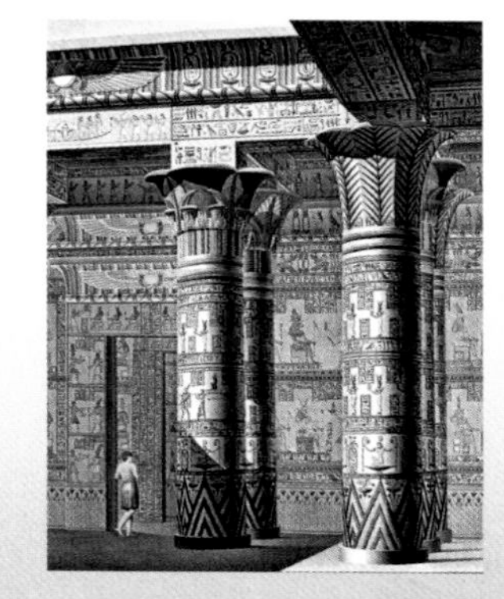

The French First Republic had been at war with Britain and several other European monarchies since 1792. The territory of Egypt was shortly to become a pawn in this game of empires. Napoleon and his advisors saw that by occupying this distant province of the declining Ottoman Empire, they could divide Britain from its colonial interests in India.

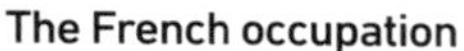

The French occupation

On 1 July 1798, around 40,000 soldiers of the Armée d'Orient put ashore at Alexandria, on Egypt's Mediterranean coast. The occupation was not a success. Early victories against Egyptian forces on land, including the capture of Cairo, were undermined by the British sinking the French fleet at its moorings. Now stranded in Egypt, the French army had to combat local insurgents who were aided and abetted by the British and the Ottoman Turks. Napoleon secretly fled back to France in October 1799. His abandoned and subsequently defeated army was repatriated by the British two years later.

What might be regarded as an inglorious episode was, in time, more than redeemed by the efforts of over 160 scholars, scientists, engineers, botanists, cartographers, and artists who had travelled with Napoleon to Egypt. Their mission was twofold: to bring European Enlightenment ideas, such as liberty and progress, to Egypt, and to study a country which, until then, had only had minimal contact with the West. They were sent out to survey and document all that they saw, both ancient and modern, and to gather specimens and artefacts. It was a mission that represents the greatest scientific undertaking of its kind, the painstaking documentation and categorization of an entire land and its people. It also stands as perhaps the most extreme expression of the colonial impulse to catalogue acquired possessions.

△ **Egyptian fountain**
Napoleon's Egyptian campaign inspired many artists to explore ancient Egyptian styles. This neo-Egyptian statue at 52 rue de Sèvres, Paris, was made by sculptor Pierre-Nicolas Beauvallet in 1806.

Description de l'Égypte

The results were gathered in the encyclopaedic *Description de l'Égypte*, which ran to 23 outsize volumes, published in 1809–29. At the same time, one of the artefacts found by

▷ **Rosetta Stone**
Found in 1799, the Rosetta Stone is a rock stele on which a decree was inscribed in 196 BCE on behalf of King Ptolemy V of Egypt. The decree was written in three scripts, the comparison of which gave modern scholars the key to understanding Egyptian hieroglyphs.

the French in Egypt, a stele containing an inscription in three languages, provided the key for a young French scholar, Jean-François Champollion, to finally unlock the code of the ancient Egyptian script known as hieroglyphics. Together, these laid the foundations for the science of Egyptology, and sparked a fascination in Europe and America for all things Egyptian and Oriental.

For the remainder of the 19th century, what is now the Middle East (but was then better known as the Near East) became an extension of the Grand Tour. Travellers were now given licence to push on from Italy and cross the Mediterranean to Constantinople (now Istanbul), Jaffa in Palestine, and Alexandria in Egypt. The sense of awe engendered by such a journey was perfectly captured by the English poet Percy Bysshe Shelley in his sonnet *Ozymandias*, with its famous opening line: "I met a traveller from an antique land". Further discoveries were soon to follow, including the discovery of the great temple of Rameses II (Shelley's Ozymandias) at Abu Simbel, and, in neighbouring Jordan, the rediscovery of the rock-hewn Nabataean city of Petra.

Egyptian-inspired architecture and decoration became popular across Europe and the US. The Americans even named two new settlements after Egypt's ancient and medieval capitals: Cairo, Illinois, founded in 1817, and Memphis, Tennessee, established in 1819. The fascination proved enduring, peaking in 1922 with the discovery of the tomb of Tutankhamun.

IN CONTEXT
Stocking national museums

The study of ancient artefacts took a great leap forward in the 18th century with the creation of large national museums, notably the British Museum in London in 1759 and then the Louvre in Paris in 1793. Egypt was ruthlessly plundered to help stock such institutions. Many antiquities collected by the French were seized by the British Navy and ended up in the British Museum, including the Rosetta Stone. Treasure hunters continued to cart off whatever they could transport, making a tidy profit by selling them off to the cultural institutions of Europe and then, later, North America.

A HEAD OF RAMESES II IS ROLLED TO THE NILE, EN ROUTE TO BRITAIN, IN 1815

▷ **Napoleon before the sphinx**
The French campaign in Egypt was an attempt by Napoleon Bonaparte, depicted in this painting by Jean-Léon Gérôme c.1858, to obstruct Britain's access to India.

" From the **heights** of these **pyramids, forty centuries** look **down** on us. "

NAPOLEON, IN A SPEECH TO HIS TROOPS BEFORE BATTLE, 21 JULY 1798

Painting the East

Throughout the 19th century, the lands of the Orient provided a rich profusion of exotic subjects for artists of the Western world.

The Orient – which includes Turkey, North Africa, and the present-day Middle East – had been exerting its charm on Western artists long before Napoleon landed in Egypt (see pp.194–95). However, the French expedition certainly increased the fascination with the Orient. Among the painters subsequently inspired were Scotsman David Roberts and Englishman John Frederick Lewis, who captured the region's archaeology and architecture in the 1830s and '40s, respectively. Others, such as the French Romantic artist Eugène Delacroix, who travelled to Spain and North Africa in 1832, shortly after the French had conquered Algeria, were fascinated by the people and their dress. For Delacroix, the Arabs evoked an earlier, purer age: "The Greeks and Romans are here at my door, in the Arabs who wrap themselves in a white blanket and look like Cato or Brutus…".

Other artists went further in their imaginings. The associations between the Bible and the Orient were important for artists such as William Holman Hunt and David Wilkie, who travelled to find authentic settings for their biblical paintings. Jean-Léon Gérôme, who undertook numerous trips to the eastern Mediterranean in the second half of the 19th century, painted large, theatrical canvases redolent with sensuality and violence. However, these were usually made back in his Paris studio with the aid of models and props, resulting in a realistic style of painting that suggested an unwarranted accuracy. One scene he surely never witnessed was a women's bathhouse, a subject that he painted more than once, complete with nude concubines. He was not alone in his fantasy: in 1862, for example, the highly respected French neoclassical painter Jean-Auguste-Dominique Ingres produced his erotic ideal in *The Turkish Bath*, and he had never even visited the East.

The taste for Orientalism – as this style of painting would come to be known – continued until well into the 20th century, with artists such as Paul Klee, Henri Matisse, and Auguste Renoir taking up its themes. The genre came under attack in the 1970s from cultural critic Edward Said, who saw such paintings as a means of exerting Western authority over Arab culture. Ironically, some of the biggest collectors of Orientalist work today come from the Arab world.

◁ ***The Midday Meal, Cairo***
John Frederick Lewis lived in Cairo from 1841–51. On his return to England, he continued to paint scenes of the city, including this peaceful composition of 1875.

Charting the American West

In 1804, Meriwether Lewis and William Clark were sent to find a route across the central United States to the Pacific, creating a pathway for the new nation to spread westward from ocean to ocean.

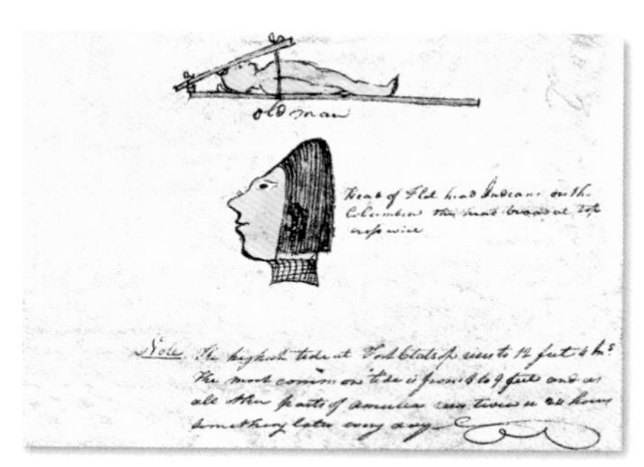

▷ **Flathead Indians**
Sketches in William Clark's diary show the practice among some Native American tribes of altering the shape of their heads by pressing their skulls during infancy.

In the early 19th century, North America was roughly divided into thirds: the eastern third was the fledgling United States, the middle third was French, and the western third was Spanish, apart from an unclaimed region to the northwest, bordering the Pacific and Canada. In 1803, President Thomas Jefferson negotiated the purchase of the whole of what was known as the Louisiana Territory from France's Emperor Napoleon, effectively doubling the size of the United States. He immediately commissioned a survey party to explore the new territory, as well as the unclaimed lands beyond the "great rock mountains" in the west. The hope was that a river route could be found to connect the Mississippi with the Pacific beyond the mountains, making it possible to open the land for settlement.

The Lewis and Clark expedition

Jefferson appointed his personal secretary, Meriwether Lewis, a man of learning and frontier skills, as expedition leader, and selected another frontiersman, William Clark, to help him. Together they assembled a team of 33 men, and in May 1804, this "Corps of Discovery" set off up the Missouri River in a fleet of three boats. They travelled all summer until the first snowfall, then put ashore and built a fort, Fort Mandan, to wait out winter. These were potentially hostile lands, where no non-Native American had ever ventured. A local French-Canadian fur trapper, Toussaint Charbonneau, had purchased or won a Shoshone Native American wife, Sacagawea, when she was just a child. Once they found out she could speak Shoshone, the expedition hired Charbonneau as guide and Sacagawea was required to act as interpreter.

In spring 1805, when the ice on the Missouri broke, the expedition resumed. In the last week of May, the party had its first sighting of the Rockies, the great mountains that it would have to cross. On 13 June, it reached what Lewis described as "the grandest sight I ever beheld" – the Great Falls of the Missouri River, an obstacle that would take a month to pass. At the foot of the Rockies they encountered the Shoshone tribe, with whom they bartered for the horses necessary to cross the mountains. It was an arduous journey, during which the expedition resorted to eating three of the animals. Eventually, they descended to the Clearwater River. The Rockies were behind them, the Pacific Ocean in front.

To the Pacific and back

On 7 November 1805, Clark wrote in his journal, "Ocian in view! O! the Joy", although in fact, they were at the estuary of the Columbia, still 32 km (20 miles) from the coast. By mid-November, however, they had made it to the Pacific. They spent the winter there, building Fort Clatsop, named

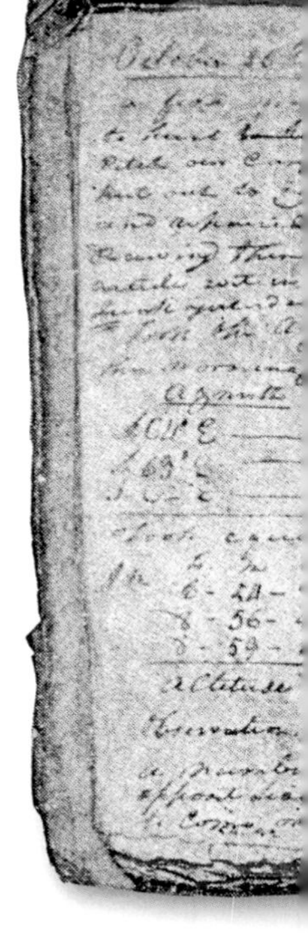

▷ **Expedition journal**
A reproduction of William Clark's diary shows the detail with which he documented the expedition. Thanks to his and Lewis's observations, their adventure became a founding tale of the young United States.

" **Ocian in view!** O! the **Joy**... This **great Pacific Octean** which we been **So long anxious** to **See.** "

WILLIAM CLARK, ON GLIMPSING THE PACIFIC OCEAN, 7 NOVEMBER 1805

▷ **American odyssey**
Lewis and Clark travelled 12,000 km (8,000 miles) by boat, on foot, and on horseback. For part of the return trip they split up to explore more territory.

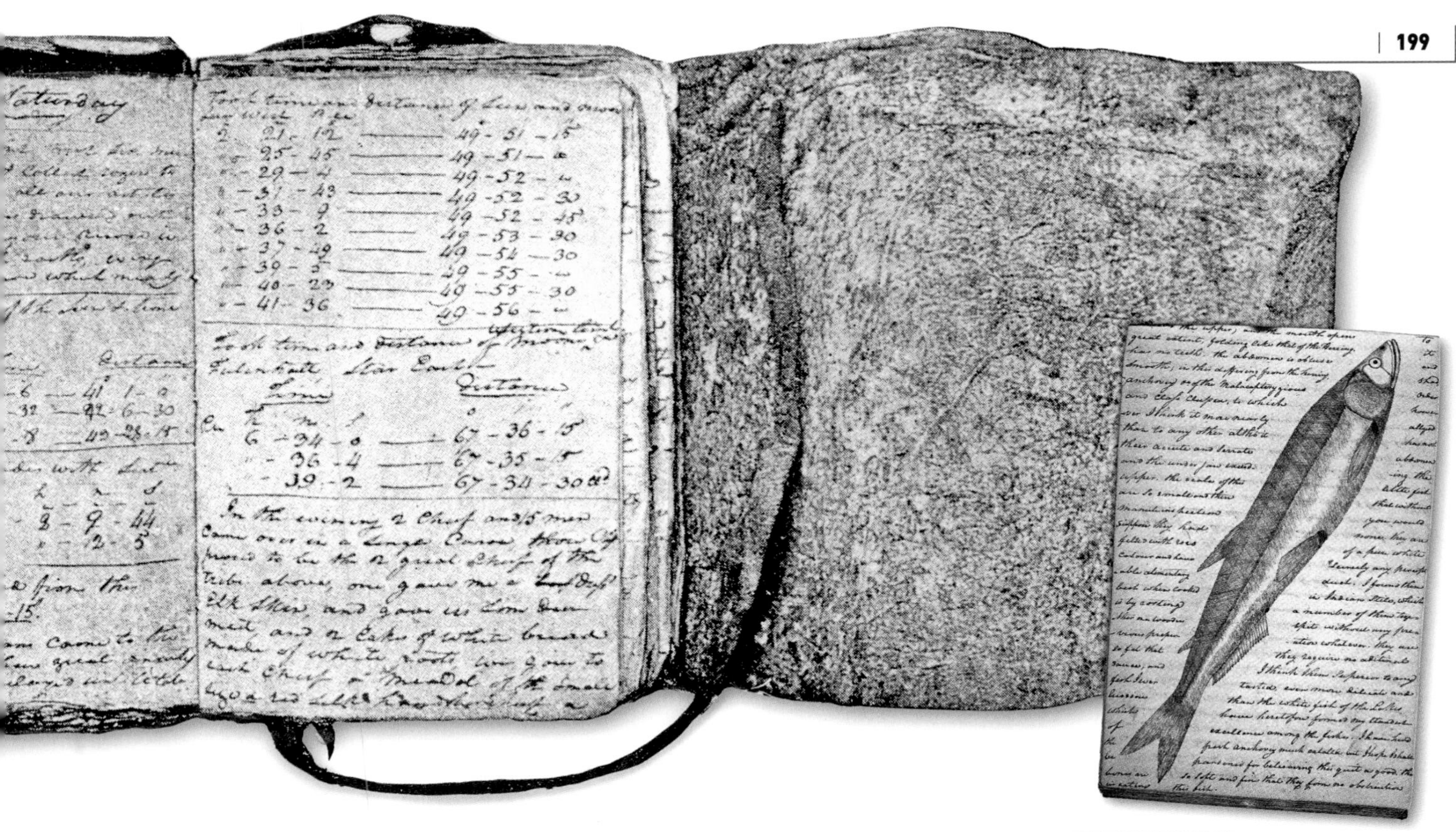

△ **Salmon trout**
This sketch of a white salmon trout by Clark shows the detail with which the expedition catalogued its findings. The picture is surrounded by a transcript of Lewis's diary entries.

after the local Clatsop tribe, and by the third week in March the expedition was ready to retrace its steps back across the continent.

On 23 September 1806, Lewis and Clark arrived back at St Louis, two years, four months, and 10 days after they had left. They had been feared dead, and over a thousand people greeted their return. They carried records of their contact with the Native Americans, maps drawn by Clark, and information on over 300 species of new flora and fauna. Over the forthcoming decades, thousands of Americans followed their lead and crossed the United States, transforming the landscape, displacing wildlife, corralling the tribes into reservations, and furthering Euro-American colonial expansion.

▽ **Heading west**
The Corps of Discovery sets up camp beside the Columbia River. Charbonneau and Sacagawea (carrying their son Jean-Baptiste) stand behind Lewis and Clark, who study the land ahead.

IN CONTEXT
The Lewis and Clark Trail

Today, a series of marked highways, that mostly run parallel to the Missouri and Columbia rivers, serves as an official Lewis and Clark Trail. More in keeping with the original spirit of adventure is the Lewis and Clark National Historic Trail, which is part of the National Trails System and extends some 6,000 km (3,700 miles) from Wood River, Illinois, to the mouth of the Columbia River in Oregon.

Administered by the National Park Service, it provides opportunities for hiking, canoeing, and horse riding. The least changed part is the White Cliffs section of the Missouri River in north-central Montana, a protected area that is only accessible by boat. For Lewis, the sandstone formations here afforded "scenes of visionary enchantment".

FORT MANDAN, THE WINTER HOME OF THE CORPS OF DISCOVERY IN 1804–05

Go west, young man!

In the wake of Lewis and Clark, ever greater numbers of Americans began to migrate west, following an arduous and often deadly route that would enter history as the Oregon Trail.

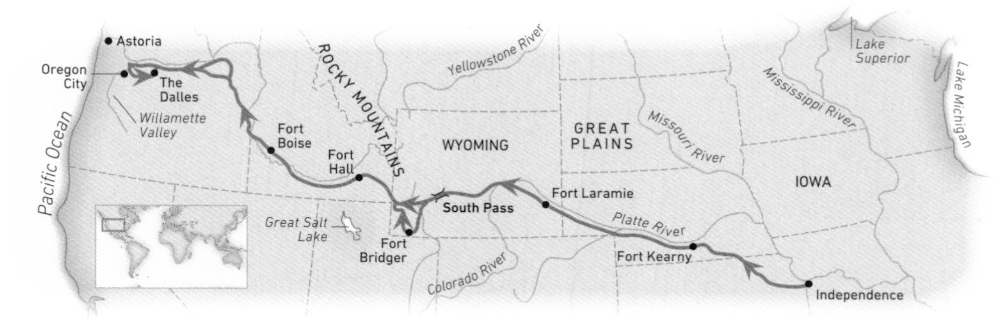

△ **Emigration route**
The route of the Oregon Trail stretches some 3,490 km (2,170 miles) from the town of Independence, Missouri, to the Willamette Valley, Oregon, in the west.

The oft-quoted phrase at the head of this page (usually attributed to newspaper man Horace Greeley) may not have been uttered until the middle of the 19th century, but a route from established settlements in the Midwest to valleys in Oregon was laid by fur trappers and traders between 1810 and 1820. John Jacob Astor, who would become the US's first multimillionaire and the owner of swathes of real estate in New York City, established a fur trading post that he called Astoria on the West Coast in 1811. His agent, Robert Stuart, located the South Pass, which provided a navigable route through the Rockies.

Early emigrants

By the 1830s, the trappers and traders were being joined by missionary groups, who sent back word of the northwest's agricultural potential. The US did not have sovereignty over the territory of Oregon at this time, but settlers were finding it difficult back east, hit hard by economic depression and diseases such as malaria. In 1839, a group set out from Peoria, Illinois, to claim Oregon country on behalf of the US government, and the following year, two families became the first pioneers to make the journey west in wagons. Further wagon trains set out in 1841 and 1842, to follow what the pioneers were now calling the Oregon Trail. In May 1843, in what is known as the Great Migration, a massive wagon train with up to a thousand emigrants, plus cattle, departed from Independence, Missouri, beginning a journey that would end that August in the Willamette Valley, Oregon. Hundreds of thousands more would follow, especially after gold was discovered in California in 1848.

The Oregon Trail

The first section of the 3,490-km (2,170-mile) trail ran through the rolling country of the Great Plains. Obstacles were few, but just a day or two of heavy rain could turn the land into a quagmire and render rivers impassable. In summer, water sources dried up, while snow could close passages through the mountains in winter. Some experienced starvation when they brought insufficient food

◁ **A child's shoes**
Guilford and Catherine Barnard travelled the Oregon Trail in 1852. These hand-made shoes belonged to their two-year-old son, Landy, who died along the trail and was buried in Kansas.

supplies and found it impossible to live off the land. The most dreaded danger was cholera, which claimed many lives. In total, it is thought that at least 20,000 people died along the Oregon Trail. According to the historian Hiram Chittenden: "The Trail was strewn with abandoned property, the skeletons of horses and oxen, and with freshly made mounds and headboards that told a pitiful tale."

On crossing the mountains, the trail became increasingly difficult, with steep ascents and descents over rocky terrain. The pioneers also risked injury from overturned and runaway wagons. There might have been some celebration at South Pass, which was the natural crossing point of the Rockies, from where the land dipped towards the Pacific, but hundreds of miles still lay between there and the wagoners' final goal.

Although traffic declined after 1869, when the Transcontinental Railroad was completed (see pp.236–37), around 300,000–500,000 people made the four-to-six-month journey. It is small wonder that even today, ruts from the wagon wheels remain etched into the soil of the Midwestern landscape.

▽ **Taking a break**
Travellers set up camp beside their wagons, somewhere along the Oregon Trail, c.1890.

" **Keep traveling**! If it is only a few miles a day. **Keep moving**. "

MARCUS WHITMAN, AMERICAN MISSIONARY AND DOCTOR

◁ **Journey's end**
For many, the sight of Mount Hood, in northwest Oregon, signalled that their journey was nearly over. This view of the mountain from Barlow Cutoff was painted by W.H. Jackson in 1865.

IN CONTEXT
Carved in stone

Of the many natural landmarks that served as navigation aids to those on the Oregon Trail, including Courthouse Rock, Chimney Rock, and Scotts Bluff, there is one in particular that has great resonance. A popular campsite for the emigrants travelling along the Sweetwater River in Wyoming was next to a vast, whale-shaped, granite outcrop known as Independence Rock. It was so-called because the schedules of many wagon trains brought them to this place around the Fourth of July. Many of those passing by left their names or initials chiselled into the rock, many of which remain clear and legible to this day.

INDEPENDENCE ROCK, BY THE SWEETWATER RIVER, ON THE OREGON TRAIL, WYOMING

Full steam ahead

After the wheel, the greatest revolution in the way humans travelled came with the exploitation of the power of steam. It delivered the potential to travel further, faster, and in greater numbers than ever before.

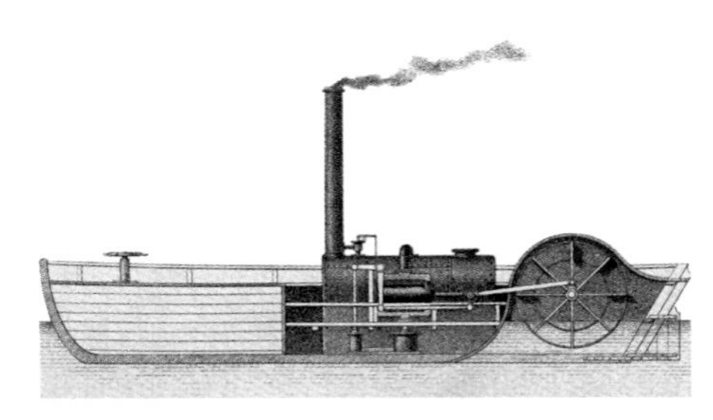

△ **Charlotte Dundas**
A lithograph by C.F. Cheffins shows the mechanics of the *Charlotte Dundas* – brainchild of the steam pioneer William Symington.

On the afternoon of 17 August 1807, a strange-looking vessel, long and sitting low in the water, pulled away from its moorings on the East River off Greenwich Village in New York. As it did so, clouds of smoke billowed out of a chimney set amidships, and two large paddlewheels mounted on the boat's sides began to turn and churn the water. Aboard was Robert Fulton, a former portrait painter who had designed the boat, his patron Robert Livingstone, and a group of their friends. The boat, with the odd hiccup, sailed on up the Hudson River at a speed of about 8 kph (5 mph), reaching Albany, which lay 240 km (150 miles) to the north, the following day. There, it took on two paying passengers and began its return voyage to New York.

◁ **Racing on the Mississippi River**
Steamboats *Baltic* and *Diana* run neck and neck during a race in 1858, as depicted by George F. Fuller.

◁ **Historic voyage**
Robert Fulton's *Clermont* makes its maiden voyage – a 240-km (150-mile) trip from New York to Albany. Built with the backing of Robert Livingstone, it was the first ever steamboat to see commercial service.

Beginnings of the steamboat

The vessel was called the *Clermont*, but although historic, it was not the first ever steamboat. As early as 1787, Fulton's fellow American John Fitch had sailed a steamboat on the Delaware River, and in 1803, Scottish engineer William Symington had built and sailed the *Charlotte Dundas* on the Forth and Clyde Canal, towing two barges. What Fulton proved, however, was that the steamboat was in fact a commercially viable and technically superior alternative to sail.

At the age of 21, Fulton had taken himself off to London and Paris, which were then the centres of the scientific world, to learn more about canals and the recently invented steam engine. While in France, he designed the first working submarine, the *Nautilus*, and in Britain, he pioneered designs of naval mines for the Royal Navy. He returned to the US with an English steam engine, which he adapted for the *Clermont*, and after his first successful trial, Fulton's boat made trips between New York and Albany every four days, sometimes carrying as many as a hundred passengers.

The Mississippi riverboats

Together with Livingstone and Nicholas Roosevelt, Fulton designed a new steamboat, which they called the *New Orleans*. In 1811–12, they sailed it from Pittsburgh, Pennsylvania, along the Ohio and Mississippi rivers, all the way to the boat's namesake city on the Gulf of Mexico. The impact of this was enormous. Steam enabled the settlers who had flooded the Western Plains following the signing of the Louisiana Purchase (1803) to travel upstream on the river as well as downstream. The *New Orleans* was the first of what would be hundreds of steamboats to open the Mississippi and Ohio valleys to trade, and to further open up the continent to exploration, settlement, and exploitation.

In May 1819, steam made an even bigger impact on American history, when the *Savannah*, a sailing ship with an auxiliary engine, crossed the Atlantic Ocean. The ship sailed under steam for only 80 of the 633 hours it took to reach Liverpool, but the achievement encouraged support for the steamship. The following year, a steamship sailed from Liverpool to South America, the first from Europe to cross the Atlantic, and in 1825 the *Enterprise* sailed all the way from England to Calcutta, India. The age of steam on river and at sea was well underway.

IN PROFILE
Mark Twain

Born Samuel Langhorne Clemens in 1835, the man who became Mark Twain, one of America's best-loved writers, left his home in Hannibal, Missouri, at the age of 17. For several years, he travelled the country working as a journeyman printer before taking passage on a boat from Cincinnati to New Orleans in 1857, intending to embark for the Amazon River to seek his fortune in the cocoa trade. His plans changed when he met pilot Horace Bixby. Before reaching New Orleans, the young Clemens's boyhood dream of becoming a steamboat pilot had been revived. Clemens convinced Bixby to take him on as a trainee pilot, and the job gave him his writer's name – "mark twain" being the pilot's cry for a river depth of two fathoms, which was a safe depth for a steamboat.

MARK TWAIN IN 1890

The Romantics

The Romantic movement is usually associated with art, poetry, and philosophy, but it was essentially a different way of looking at the world, and encouraged people to go travelling in search of new experiences.

It seems strange now, but to the majority of 18th-century travellers, particularly those who had set out on the Grand Tour (see pp.180–83), the Alps were a nuisance. At the time, people valued order and symmetry, and humankind's ability to tame nature and shape civilization. The mountains in southern France were simply a massive barrier of rock that stood in the way of getting to Italy. Towards the end of the century, however, a growing number of travellers began to look for beauty and spiritual inspiration in such places.

In search of the picturesque

The change in attitudes was influenced by a group of intellectuals, especially the Swiss philosopher and writer Jean-Jacques Rousseau, whose novel *La Nouvelle Heloïse* (1761) rhapsodized the Swiss landscape and helped to ignite a 19th-century passion for Alpine

△ ***Landscape*, William Gilpin**
Unusually for the time, Gilpin's work was purely aesthetic and focused entirely on the beauty of its subject – nature – rather than trying to express a moral principle. Gilpin described this style as "picturesque".

▽ **Byron in Italy**
Lord Byron did much to popularize Europe as a travel destination, particularly Italy, where he lived for seven years.

▷ **Alone with nature**
Caspar David Friedrich's *Wanderer Above the Sea of Fog* (c.1818) encapsulates Romantic ideas about the isolation of the individual when faced with the sublime forces of nature.

scenery. In England, the works of a minister named William Gilpin were also instrumental in fostering a growing appreciation of untamed nature. His book, *Observations on the River Wye* (1782) sent hordes of tourists to south Wales, and his subsequent books on the Lake District and Scotland did much the same for those regions. Gilpin popularized the term "picturesque", which he defined literally as the kind of beauty that "would look well in a picture".

The sublime

Thomas West's *Guide to the Lakes* (1778) was a bestselling work on where to find the finest views in that famously picturesque part of northwest England. It also capitalized on the growing popularity of the "sublime", a feeling of wonder and awe that people sometimes experience when they see vast, majestic landscapes. This was part of a movement, later known as Romanticism, which renounced the rationalism and order characteristic of the Enlightenment, and stressed the importance of pure emotion and elevating imagination above reason.

An integral part of the sublime was the idealization of raw, rugged landscapes, such as mountains, heaving seas, and wild moors. This was exemplified in the work of the poet William Wordsworth, who had spent his childhood in the Lake District and moved back there in later life. Wordsworth's poetry linked the beauty of nature with the divine, imbuing landscapes with power and mystery.

Byron and Shelley abroad

One reason why English writers and artists had become so interested in the landscapes of their homeland was that two decades of war with France had prevented them from crossing the Channel. When peace was restored to Europe in 1815, Romanticism influenced how travellers perceived the Continent. Among those who took advantage of peacetime to travel were the poets Byron and Shelley, two of the most prominent Romantics. Byron wrote one of the most revolutionary travelogues of the early 19th century, which, unusually, took the form of a poem. *Childe Harold's Pilgrimage* (1818) is an account of a voyage of self-discovery based on an extensive journey that Byron made through Portugal, Spain, Greece, and Albania. It made its author one of the most famous poets in Europe – and travellers took to using *Childe Harold* like a guidebook, its passages of high drama heightening the romance of the lands through which they passed.

> " Man is **born free**, and everywhere he is in **chains**. "
>
> JEAN-JACQUES ROUSSEAU, *THE SOCIAL CONTRACT*, 1762

The glories of Italy

Both Byron and Shelley made Italy their home. They delighted in its ruins, natural beauty, and glorious weather (Byron delighted even more in its women), and exalted all in their work. Although their antics brought the poets notoriety (it was famously said of Byron that he was "mad, bad and dangerous to know"), it was ultimately their strong belief in the importance of nature that had the more lasting effect.

△ ***Childe Harold's Pilgrimage***
This frontispiece, illustrating Canto I, is from the 1825–26 edition of *Childe Harold's Pilgrimage* by Byron. The engraving is by I.H. Jones.

◁ **Shelley in Italy**
Percy Bysshe Shelley wrote many of his greatest poems, including *Prometheus Unbound*, in Italy. He moved from one Italian city to another, and was finally buried in Rome.

The voyages of the *Beagle*

In 1831, a small ship left Plymouth, England, with a young naturalist on board, taken along in part to keep the captain company. His keen observations made during the voyage would later reshape our understanding of humankind.

▷ **Charles Darwin**
This watercolour by George Richmond shows Charles Darwin in 1840, aged only 31. His voyage on the *Beagle* turned him into an eminent natural scientist.

Launched in May 1820, the HMS *Beagle* was just one of more than a hundred ships of its type, a modestly sized twin-masted warship. Originally outfitted for battle with ten guns, she was never called to action and instead was adapted as a survey vessel, sent off in 1826 to explore the coastline of South America. Along the way, the ship's captain sank into depression and committed suicide, possibly from the isolation of the lengthy voyage. His successor, Robert FitzRoy, assumed command and returned the ship to England. There he was asked to take the ship on a second voyage to continue the work of the first. FitzRoy decided he would like to take along a scientist to help with the surveying and, perhaps more importantly, provide some company on the long journey.

A surprise invitation

Charles Darwin (1809–82) was a student with a passion for natural history who had just finished his final exams and was intending to return to Cambridge for theological training. A letter from one of his professors, inviting him to join the *Beagle*, changed everything. Lacking experience and scientific credentials, Darwin was proposed for his enthusiasm and enquiring mind, but his presence on board ultimately made the *Beagle* one of the most famous ships in history. The expedition left England on 27 December 1831, and reached South

▽ **The *Beagle* in the Murray Narrow**
HMS *Beagle* anchors in one of the deep fjords of the Tierra del Fuego, at the southern tip of Patagonia, in this 1836 painting by the ship's artist, Conrad Martens.

> " The **voyage of the *Beagle*** has been by far the **most important event** in my life, and has **determined my whole career**. "
>
> CHARLES DARWIN, *THE AUTOBIOGRAPHY OF CHARLES DARWIN*, 1887

▷ **Discovering new species**
Darwin preserved specimens of numerous species that were new to science – including this parrot fish from the Pacific.

America the following February. Darwin was able to spend considerable time on land, collecting specimens and making observations in notebooks. At the beginning, Darwin, a keen hunter, enjoyed shooting the birds and other wildlife he encountered, before coming to the realization that "the pleasure of observing and reasoning was a much higher one than that of skill and sport".

Intimations of evolution

In September 1835, the *Beagle* reached the Galápagos Islands, where Darwin was intrigued to learn that the shells of giant tortoises differed from place to place, and that these differences allowed locals to tell which specific island a creature came from. He collected samples of mockingbirds, and noted that these birds too were slightly different on each island.

The expedition sailed on, dining on Galápagos tortoises, and arrived at Tahiti in November 1835, then New Zealand in late December. In January 1836, they reached Australia, where they explored coral reefs, and by the end of May 1836, they passed the Cape of Good Hope at the southern tip of Africa. Rather than making straight for Europe, Captain FitzRoy wanted to make further hydrographic surveys, which involved a return trip across the Atlantic to Bahia in Brazil. It was not until 2 October 1836 that the *Beagle* finally put into the port of Falmouth in Cornwall.

"As far as I can judge of myself, I worked to the utmost during the voyage from the mere pleasure of investigation, and from my strong desire to add a few facts to the great mass of facts in natural science," noted Darwin. In the following years, he wrote extensively on his travels, contributing the third volume of the *Narrative of the Voyages of H.M. Ships Adventure and Beagle*. Due to its popularity, this single volume was republished many times, first as Darwin's *Journal of Researches*, published in 1839, and later as *The Voyage of the Beagle*, printed in 1905. The observations recorded in the book would ultimately lead him to formulate his theory of evolution, outlined in his masterwork *On the Origin of Species*, which was published in 1859.

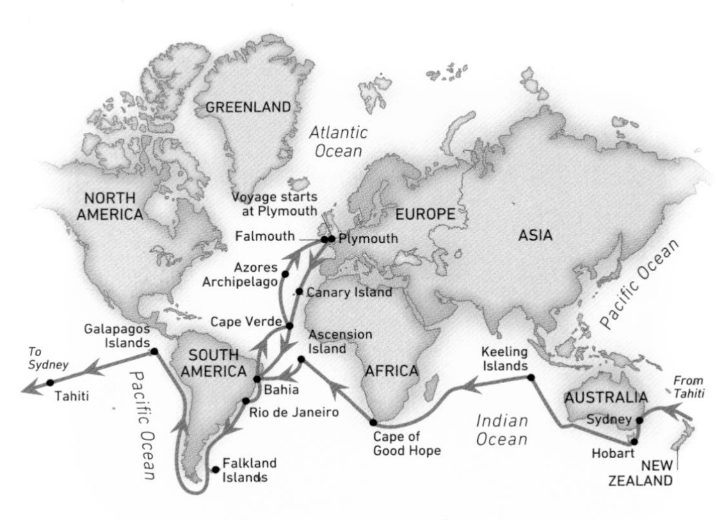

△ **Exploring the natural world**
This map shows the route that the *Beagle* expedition took around the world. The journey took nearly five years.

IN CONTEXT
Darwin's notebooks

Charles Darwin kept a detailed account of his travels on board the *Beagle*. He filled out 15 small notebooks, writing around 116,000 words, and drawing roughly 300 sketches. However, Darwin's account of his journey is anything but dry science. His writing is lively and charming, with vivid descriptions of the places and people he encountered. In places, it is even funny.

Darwin also carefully recorded the various types of flora and fauna he found. The records of the species he came across on his journey were crucial to his later work on evolution. Visiting South America, Darwin encountered and sketched a bird called a "rhea", which he learned had "a very close general resemblance" to another ostrich species. Such observations led him to conclude that "one species does change into another."

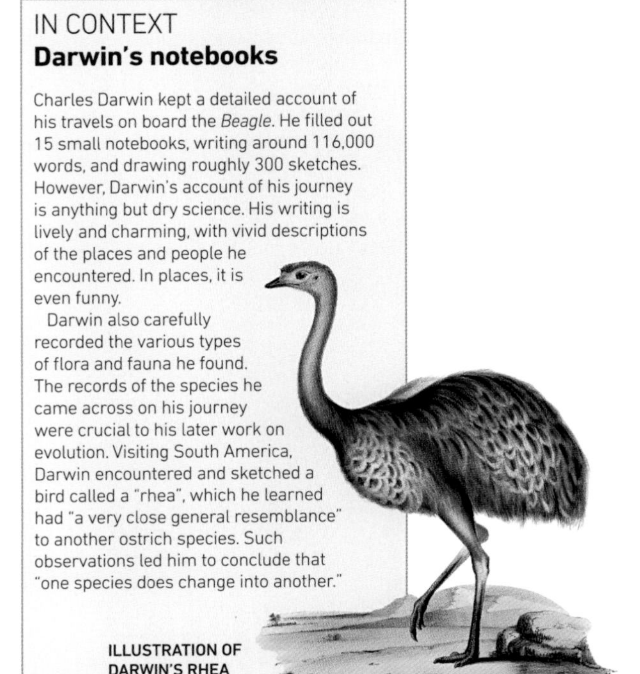

ILLUSTRATION OF DARWIN'S RHEA

Travellers' tales

In the 19th century, explorers' records of their journeys took on a new slant. As well as documenting scientific or geographical discoveries, many travellers began to keep more informal accounts.

Even if he had never written his groundbreaking work about evolution, *On the Origin of Species,* Charles Darwin would probably still be remembered as a notable travel writer. Following his voyage on the *Beagle,* he contributed to the official scientific reports, but he also wrote his own informal account, full of perceptive sketches of the places he visited and observations about the people he had met. This account was typical of travel writing of the age, which was based largely on the journals kept by scientists. Sometimes a personal narrative would emerge and prove quite thrilling.

One such account is John Franklin's *Narrative of a Journey to the Shores of the Polar Sea* (1823), which details his 1819–22 mission to chart the north coast of Canada, in which more than half of his expedition of 20 died. His account almost revels in the hardships: "We enjoyed the comfort of a large fire for the first time since our departure from the coast," he writes. "There was no *tripe de roche,* and we drank tea and ate some of our shoes for supper."

Francis Parkman's majestic account of the American migrant route through the Rockies, *The Oregon Trail* (1849), provides a similarly enthralling personal account. "A month ago," he comments, "I should have thought it rather a startling affair to have an acquaintance ride out in the morning and lose his scalp before night, but here it seems the most natural thing in the world."

Outdoing them all, though, was explorer Henry Morton Stanley. His lively accounts, including *In Darkest Africa* (1890), are full of violent clashes with native tribes, episodes of flogging disobedient porters to death, leaving the lame to die on the trail, and hanging deserters in trees.

The lure of the Orient

While Stanley was on his mission to subdue a continent, a new breed of more cerebral travel writer was emerging, one whose response to visiting foreign lands and cultures was to question civilization, beliefs, and cultural identity. Many of the Victorian travel classics were produced by such writers. Egypt and the Islamic world were of great interest at the time, and featured in many books, notably *Eothen* (1844), Alexander Kinglake's graceful account of a trip from Belgrade to Cairo. Gérard de Nerval's *Journey to the Orient* (1851) offered a more whimsical account of the East, heady with the fumes of hashish – but then its author was an eccentric who enjoyed walking his pet lobster in the gardens of Paris.

In contrast to the sensualist Nerval, English explorer Isabella Bird was driven to travel and write by her evangelical Christian beliefs. She rode – often alone – on horseback across Persia, Japan, Korea, and many other countries, taking her own photographs. Although dismissive of the "false creeds" that she encountered, she showed what was, for the time, a rare empathy with the people she met.

The travelling novelist

Travel writing had become so popular by the mid-19th century that many authors of note felt compelled to write

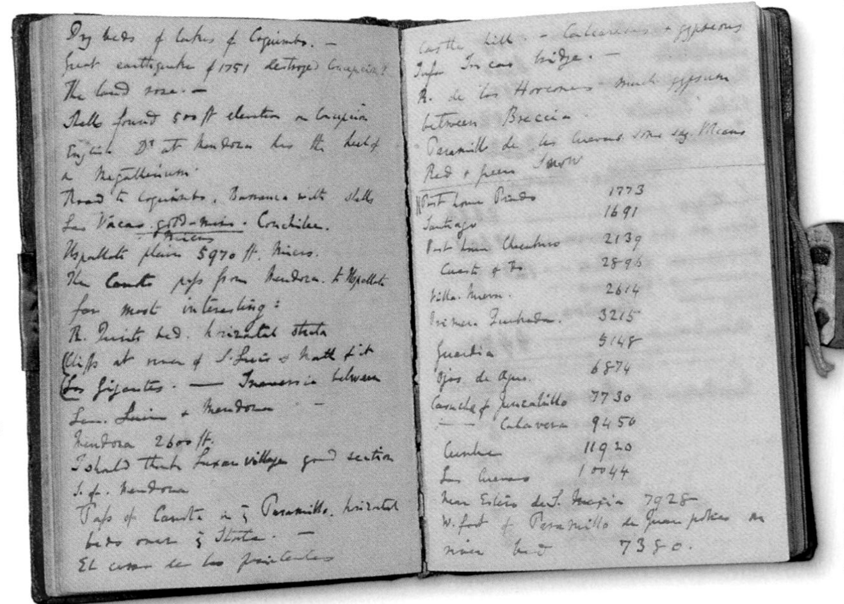

▷ **Book-keeping**
During the voyage of the *Beagle*, Charles Darwin recorded his observations in small notebooks such as this one. He copied up his notes every evening, forming a 750-page journal that would become the basis of his published work.

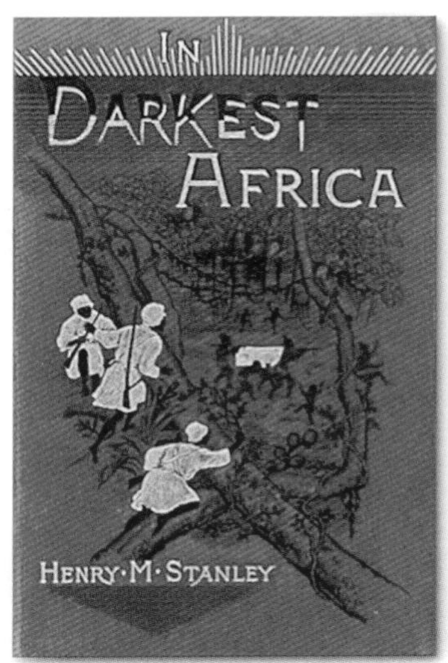

◁ **Imperial adventurer**
In his day, Henry Morton Stanley's feats in Africa enthralled the public. However, he later fell from favour and was seen as a ruthless imperialist who hacked and shot his way across Africa.

◁ **Harsh terrain**
This engraving from John Franklin's *Narrative of a Journey to the Shores of the Polar Sea* brings to vivid life the forbidding landscapes through which the author travelled.

their own travel accounts. In 1844, Charles Dickens set off for southern Europe, a pleasurable jaunt related in *Pictures from Italy* (1846). Henry James turned tourist for *A Little Tour in France* (1884), and the *Treasure Island* author, Robert Louis Stevenson, became a pioneer of outdoor literature with his wry account of a walk taken to recover from heartbreak, *Travels with a Donkey in the Cévennes* (1879).

The most entertaining book of all, perhaps, is Mark Twain's *The Innocents Abroad* (1869), in which he humorously recounts the weeks he spent on board the chartered ship *Quaker City*, visiting the sights of Europe and the Holy Land along with a group of fellow American travellers. "The gentle reader will never, never know," he writes, "what a consummate ass he can become until he goes abroad." With timeless observations like this, it is little wonder that more than a century later, *The Innocents Abroad* is still one of the most successful travel books of all time.

> "Truly a **good horse**, good ground to gallop on, and **sunshine**, make up the sum of **enjoyable travelling**."
>
> ISABELLA BIRD

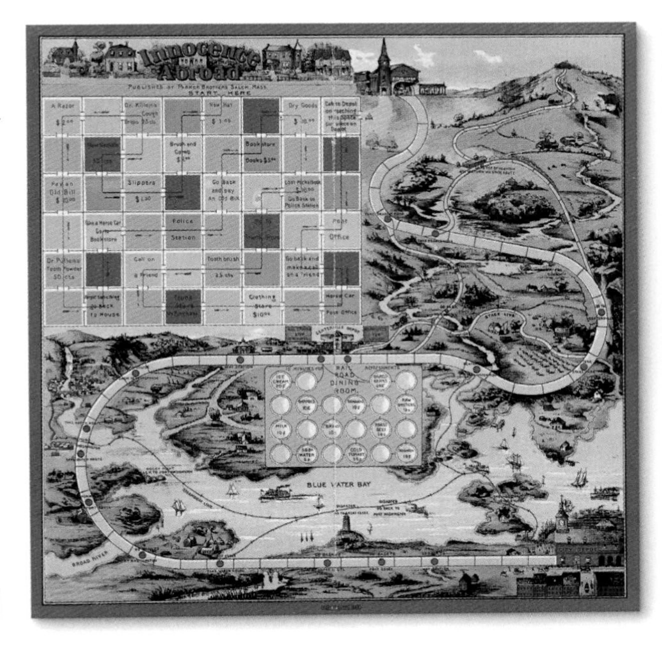

▷ **The Game of Innocence Abroad**
Capitalizing on the popularity of Mark Twain's *The Innocents Abroad*, in 1888 Parker Bros produced this slyly titled game. The board illustrates activities that tourists might engage in as they travelled around Europe.

△ **Mark Twain**
Perhaps to Mark Twain's surprise, his account of a trip to Europe and the Holy Land was his bestselling book – more popular, even, than *The Adventures of Tom Sawyer* and *Adventures of Huckleberry Finn*.

Shooting the world

Soon after photography was invented, cameras went on sale to the public and people began to go on journeys with a single goal in mind – using this incredible new machine.

△ **Ancient Egypt**
The temple of Abu-Simbel was one of hundreds of sites photographed by Maxime Du Camp and Gustave Flaubert on their two-year tour of the Middle East.

The two leading pioneers of photography, Frenchman Louis Daguerre and Englishman Henry Fox Talbot, made their first photographic images in the late 1830s, using a process that became commercially available in 1839. The following year, India hosted its first photography exhibition, and in 1841 Captain Lucas of Sydney took the first ever picture of Australia. Such was the speed with which photography was embraced.

Landscapes of the Middle East

One of the most celebrated early photographers of travel was a man now better known for his novels – Gustave Flaubert, the future author of *Madame Bovary*. In 1849, he set off with his wealthy Parisian friend Maxime Du Camp to tour and photograph the ancient monuments of the Middle East. The two of them travelled through Egypt and Palestine, and published the results in *Memories and Landscapes of the Orient* (1852) – the first travel book to be illustrated with photographs. Egypt also provided rich subject matter for photographer Francis Frith, who visited three times, and Francis Bedford, the official photographer of the Prince of Wales on his 1862 royal tour of the Middle East.

Capturing the unknown

In the mid-19th century, the standard photographer's kit was a very large, heavy camera, a tripod, glass plates, plate holders, a tent-like portable darkroom, chemicals, and tanks for developing the pictures. It was cumbersome, to say the least. However, advances in technology soon

▷ **Glimpse of Japan**
The Samurai, portrayed here by Felice Beato in the 1860s, were Japan's elite warriors. Beato had privileged access to Japanese society, which was little known to Westerners at the time.

◁ **Mammoth plate camera**
Initially, prints had to be the same size as negatives, so in 1861, Carleton E. Watkins commissioned this huge camera so he could do justice to scenes of the American West.

enabled studios, particularly in Europe, to develop large numbers of images for photographers, who were then able to leave their darkrooms behind.

In 1857, the French government commissioned Claude-Joseph Désiré Charnay to go to Mexico. He spent four years there and was the first to photograph the Mayan ruins. Another Frenchman, Émile Gsell, worked in Indo-China, and took the first pictures of the temple of Angkor Wat, in what is now Cambodia. The Italian-British photographer Felice Beato became one of the first to work in China. He took pictures of the Second Opium War and of Japan during its isolationist Edo Period – images that were not only of immense scientific interest at the time, but which also now provide an invaluable insight into a bygone age.

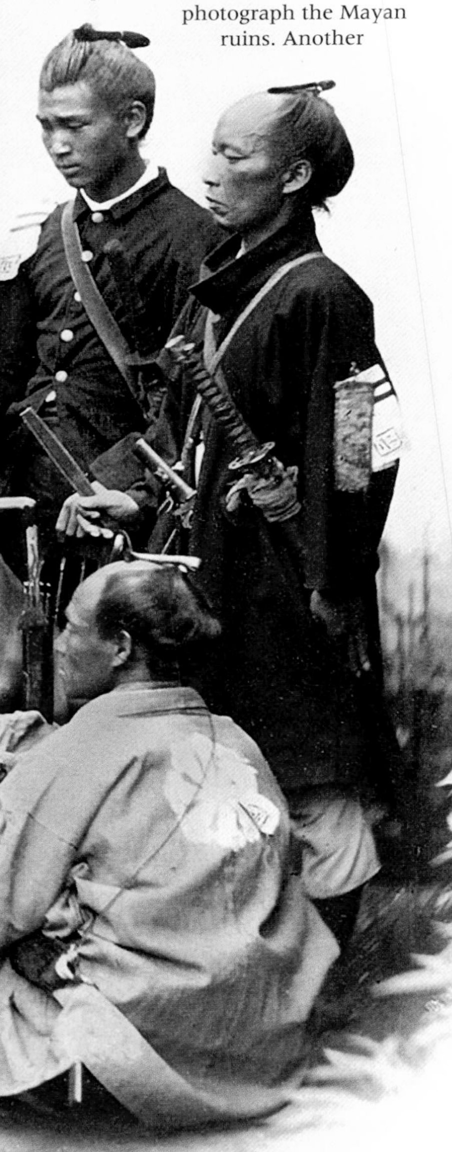

The American West

In the United States, the invention of photography gave many citizens the opportunity to see what was in the rest of their own country for the first time. Frontiersman John Charles Frémont made several failed attempts to use a camera on his early expeditions, so for his fifth crossing of the continent, in 1853, he took along Solomon Nunes Carvalho, who was possibly the first person to photograph the American West and its native peoples. Others swiftly followed, including William Bell, John K. Hillers, Carleton E. Watkins, and Timothy H. O'Sullivan. Each of them returned with plates which, when developed, produced such beautiful photographs that they helped to inspire the establishment of the United States' first national parks.

△ **Cathedral Rocks**
This 1865 picture of a landmark in Yosemite Valley was taken by Carleton E. Watkins when he was working for the California Geological Survey.

" There is no **effectual substitute** for **actual travel**; but it is my **ambition** to **provide** for those to whom **circumstances forbid** that **luxury**... "

FRANCIS FRITH, PHOTOGRAPHER

IN CONTEXT
On the road

At first, travel photography was very much a group enterprise. Francis Frith, for example, travelled with a large entourage of guides and assistants, and set out across Egypt in a caravan of carts and wagons. He was one of the first to experiment with glass negatives, introduced in 1851, but he had to process them on-site, either in a stuffy, dark tent or in an ancient tomb – despite the danger of using explosive materials such as liquid ether and gun cotton. The English war photographer Roger Fenton used a similar technique, but developed his pictures in a specially rigged-up, horse-drawn darkroom. He also used volatile chemicals, even on the battlefields of the Crimean War.

ROGER FENTON'S ASSISTANT, MARCUS SPERLING, WITH THEIR PORTABLE DARKROOM

Into Africa

At the beginning of the 19th century, the European grasp of African geography was confined mainly to the coast. This deficit in knowledge was only slowly eroded by a succession of ambitious European explorers.

As far back as the 15th century, the Portuguese had mapped the outline of Africa as they extended sea routes into the East. It was the Portuguese, too, who initiated the transatlantic trade in African slaves, soon joined by other European powers. By the early 19th century, however, the slave trade was abolished, in part due to the abolitionist movement, Enlightenment ideals of the natural rights of man, and uprisings by enslaved people. As interest turned from slavery to exploration, Africa became the focus of intellectual curiosity, again partly due to Enlightenment concepts of reason and scientific investigation, but also with a predatory eye to potential exploitation opportunities for trade or colonization. In 1884, the Berlin Conference was an attempt to regulate the colonization efforts of various European empires, which became known as the Scramble for Africa. By then, much energy had been focused on exploring the interior of the continent, particularly in establishing the sources of two of Africa's greatest trading arteries, the Niger and Nile rivers.

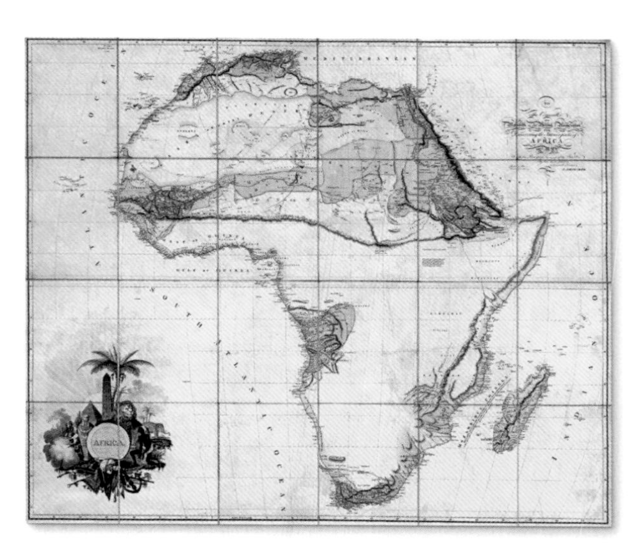

▷ **Uncharted territory**
This map by Aaron Arrowsmith, from 1802, shows the extent to which the centre of Africa was a great unknown for European explorers.

▷ **Mungo Park**
Scottish-born Park had been expected to enter the Church, but a fascination for the natural world led him to voyage to Africa in search of the River Niger.

The search for the Niger

The African Association was founded in London in 1788, with the aim of locating the origin of the Niger. A young Scotsman named Mungo Park, who had trained in medicine, but now had the patronage of the influential botanist Joseph Banks (see pp.176–77), was dispatched to West Africa in May 1795. Putting ashore at Pisania in Gambia, Park made for the interior accompanied only by a local guide, a porter, a horse, and two mules. At one point, he was thrown into prison for four months by a local Muslim ruler, but he managed

◁ **Into the interior**
In December 1795, Park left Gambia on his journey to find the Niger, which he located in July of the following year. It took a further 11 months to make the return journey.

to escape with a horse and compass. He eventually located the Niger, becoming the first European to set eyes on the middle stretch of the river.

Park then returned to the Gambian coast and to England, where his diaries, under the title *Travels in the Interior of Africa*, were published to acclaim. In 1805, he accepted a second mission to the Niger, this time with 40 men. Tragically setting off at the height of the hot season, most lost their lives to illness. The five survivors who actually reached the Niger, including Park, took to the river in a boat, but were attacked by locals defending their territory, and Park drowned while trying to swim to safety.

In 1818, British naval officer George Francis Lyon was sent to chart the Niger and locate the fabled city of Timbuktu, known only from historic accounts. Inadvisably, his party started out from Tripoli, in what is now Libya, meaning they had to first cross the Sahara. The mission was a failure.

◁ **Locating Timbuktu**
Réné Caillé was the first to reach Timbuktu, in 1828, followed by the German Heinrich Barth in 1853. This illustration, by Johann Martin Bernatz, is based on a sketch made by Barth.

Success at last

It was not until 1830 that the Niger was successfully mapped. Richard Lander was a born adventurer. In 1825, he was an aide to the Scots explorer Captain Hugh Clapperton, who went on a trade mission to Sokoto, in what is now Nigeria. Once again, illness wiped out almost the whole group, and Lander had to make a seven-

»

▽ **Mungo Park's second expedition**
A second expedition to the Niger, which sought to assess its potential for European settlement, ended in the death of the whole party, including Park himself.

» month walk on his own to reach the coast and a ship home. Undaunted, he returned to Africa in 1830 along with his brother John, and together they finally succeeded in visiting the Niger Delta and mapping the route of the river.

The prize for being the first European to visit Timbuktu – a literal prize of 10,000 francs offered by the Paris-based Société de Géographie – had been claimed two years earlier by a Frenchman, René Caillié. A Scotsman, Alexander Gordon Laing, had found and visited the city in 1826, but was killed by local tribesmen as he journeyed home. Mungo Park may also have reached the city, but he died before telling anyone. Caillié travelled alone, disguised as a Muslim, and was able to return home safely to claim both glory and the prize money.

Tracing the source of the Nile

In East Africa, finding the elusive source of the Nile proved an irresistible lure to many explorers. The Royal Geographical Society in London commissioned Richard Burton and John Hanning Speke to locate it. Burton was a swashbuckling adventurer and a brilliant linguist who had gone on the *hajj* to Mecca disguised as an Arab, and although more conventional, Speke was also an experienced explorer, having already travelled with Burton on an earlier African expedition.

The pair set off from the coast to cross present-day Tanzania in June 1857, following a map supplied by a recently returned missionary. The going was arduous, and along the way they both contracted tropical infections: Speke lost his hearing in one ear, Burton became unable to walk, and both went temporarily blind. Even so, by June 1858, they had succeeded in reaching Lake Tanganyika. They backtracked to the town of Tabora to recuperate and stock up on provisions, but Burton was too ill to travel, so Speke headed north with a party to investigate claims of another great lake. On discovering the waters of what locals called the N'yanza, he renamed it Lake Victoria, in honour of his queen. Speke was convinced that this was the source of the Nile. Burton disagreed, and the two fell out publicly. Back in England, in 1864, Speke was killed in a hunting accident the day before a debate appointed to bring the two men and their theories together.

"... if I **ever** travel **again**, I shall **trust** to **none but natives**, as the **climate** of **Africa** is **too trying** to **foreigners**."

JOHN HANNING SPEKE, ON AFRICA'S CLIMATE

△ **Speke's sketchbooks**
John Hanning Speke filled sketchbooks with drawings of the flora and fauna he encountered, creating a remarkable record of his travels.

▷ **Captain John Hanning Speke**
This portrait depicts Speke standing in front of Lake Victoria, which he claimed was the source of the River Nile. His theory was confirmed in 1875 by Henry Morton Stanley.

David Livingstone

Perhaps the most famous explorer of all was David Livingstone, the first European to cross the continent from west to east. He arrived at the southern tip of Africa in 1841, and became

fascinated with the African interior. With a young family in tow, he first went to Lake Ngami in present-day Botswana. Continuing north, he reached the Zambezi River, which he followed west before striking off to reach the coast at Luanda, the modern capital of Angola. Returning to the Zambezi, he then headed east to find out whether the river was navigable for trade, and discovered its greatest natural obstacle, the Victoria Falls, making him the first European to set eyes on it.

After 1866, he became obsessed with finding the source of the Nile, a task that filled the final seven years of his life. He was out of contact with the outside world for so long at one point that an American newspaper dispatched correspondent Henry Morton Stanley to find him, resulting in their famous encounter in the jungle.

Livingstone was ill for the last four years of his life. When he died of malaria in 1873, two African assistants carried his body a thousand miles to the coast so it could be returned to Britain for burial. It was sent with a note: "You can have his body, but his heart belongs to Africa." The heart had been cut from the body and buried where he died.

◁ **David Livingstone's cap**
As a British consul, Livingstone wore the Consul's Cap. He was wearing it on the day he met Stanley.

△ **Livingstone at Lake Ngami**
David Livingstone became Victorian England's most famous explorer, even if, ultimately, he failed in his quest to discover the source of the Nile.

IN PROFILE
Mary Kingsley

Captivated by her father's tales of his trips overseas as a travelling doctor, ethnographer and explorer Mary Kingsley set off for her own adventure in 1892, at the age of 30. Her ambition was to discover new species of animals. She headed for Angola and then into the Congo, travelling on the River Gabon. She returned to England in 1893 with a collection of tropical fish for the British Museum.

On her second sojourn to Africa, she spent time with a tribe that was known to practise cannibalism and fell into a game pit, but was saved by her voluminous underskirts. Her subsequent book, *Travels in West Africa* (1897), became a Victorian bestseller. She died of typhoid in South Africa in 1900.

MARY KINGSLEY ON THE FRONTISPIECE OF HER SECOND BOOK, *WEST AFRICAN STUDIES* (1899)

The Railway Age

Until the 19th century, most people lived their lives in the area in which they had been born, and no-one had ever travelled faster than a horse could gallop. The arrival of the railways changed everything.

Before the 19th century, steam engines were cumbersome objects, used only in industry and, latterly, in boats (see pp.202–03). However, in the early 1800s, they became small enough to be mounted on wheels. Experiments in putting steam engines on rails culminated in the first railway to run solely on steam power, the Liverpool and Manchester line, which opened in 1830. It was masterminded by engineer George Stephenson and employed an engine, *Rocket*, designed by his son, Robert. Robert would go on to construct a far longer line connecting London and Birmingham, and assist in constructing lines in Belgium, Spain, and Egypt. When the line celebrated its 150th anniversary in 1980, the then-chairman of British Rail, Peter Parker, observed: "The world is a branch line of the Liverpool and Manchester".

△ **Early rail travel**
This woodcut of the 1830s shows passengers travelling along the Baltimore and Ohio Railroad in a horse-drawn carriage.

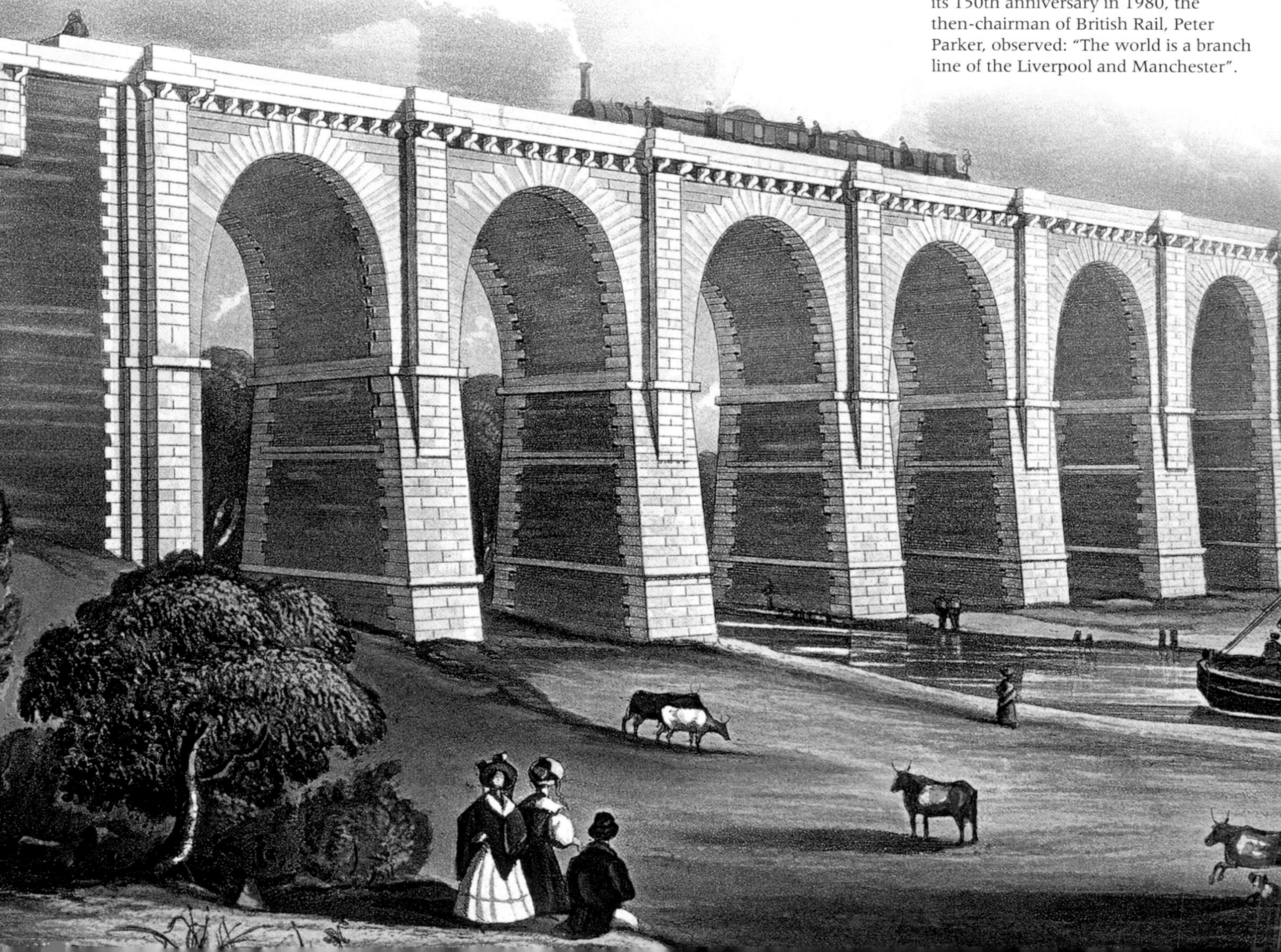

◁ **Steam vs horse**
The *Tom Thumb* races a horse-drawn carriage in 1830. It lost the contest due to a mechanical failure, but proved that steam locomotion was both a viable and particularly fast form of transport.

The first American railroads

The United States was quick to embrace the railway, or railroad, as they would name it. The country was vast and travel was slow. Steamboats were the best form of transport, but they only provided access to certain parts of the country. Otherwise, it was still horse and wagon. The original American railroad pioneer, Colonel John Stevens, was a steamboat operator. The first US railway carriages were pulled along the tracks by horses, as had also been the case in England.

It took a legendary race, between a locomotive called the *Tom Thumb* and a horse in 1830, to convince sceptics that the future lay with steam. Trains started running on completed stretches of the Baltimore and Ohio and Charleston and Hamburg lines by the end of that same year. Within a decade, 4,425 km (2,750 miles) of railroad were in operation all across the US. The spur to build was freight, but passengers soon began to benefit from this new form of transport. Every town wanted to be connected to the railroad as it was considered vital for their future prosperity.

> “ A bell indicates the start and the **locomotive begins to groan** and the **wheels revolve first slowly** and then **faster and faster**, and then **the train flies**. What **fun** it is to travel now. ”
>
> PUBLISHER KARL BAEDEKER, ON HIS FIRST TRAIN TRIP, 1838

Tracks across Europe

Europe was not far behind England and the US. By 1832, France was running locomotives on a line that stretched between Saint-Étienne and Lyon. Originally intended to convey coal, it quickly expanded to carrying passengers. In 1834, building started on a line in Belgium, and Germany's first railway opened the following year. The German line was only 6.5 km (4 miles) long, but unlike most of the early lines, it was built specifically for passengers. This was also the case with the inaugural line in Holland, laid to connect Amsterdam and Haarlem, which opened in 1839.

The rapidly growing railway networks across Europe and the US did not just simplify travel, but also cut travel costs, as increased speed and better routes meant less time and money spent on travelling. Early entrepreneurs, who had expected the rails to be used for goods, were surprised to find that passenger traffic accounted for more than half of their revenue. The bulk of this traffic came from the lower and middle classes, the sort of people who had never before had the opportunity to travel. The arrival of the railway amounted to nothing less than a revolution.

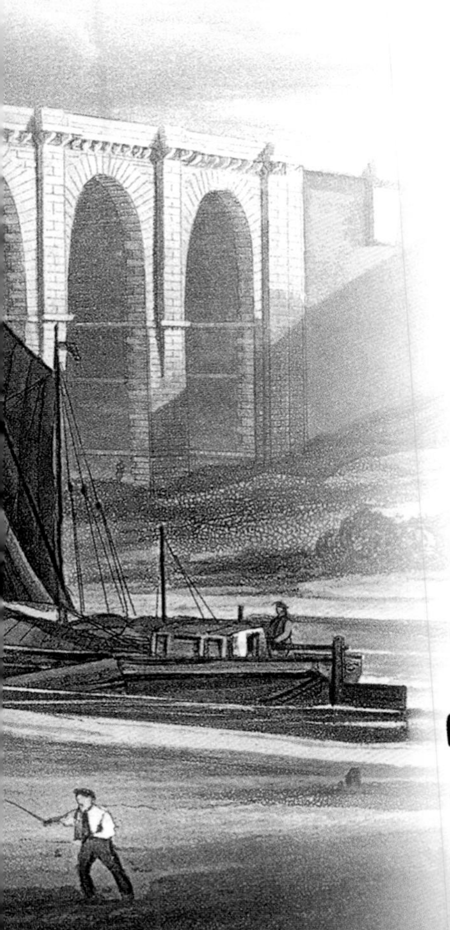

◁ **Changing landscapes**
This English viaduct was erected for the Liverpool and Manchester Railway, under the direction of George Stephenson. Made of brick with stone facings, its 15 m- (50 ft-) wide arches rise from slabs of local sandstone.

▽ **The *Adler***
The first successful steam locomotive to operate in Germany, the *Adler* (meaning "eagle" in German) was built by pioneers George and Robert Stephenson in 1835.

Trains

It was a simple idea – to use steam to turn wheels that ran on tracks – but it was one that transformed transport.

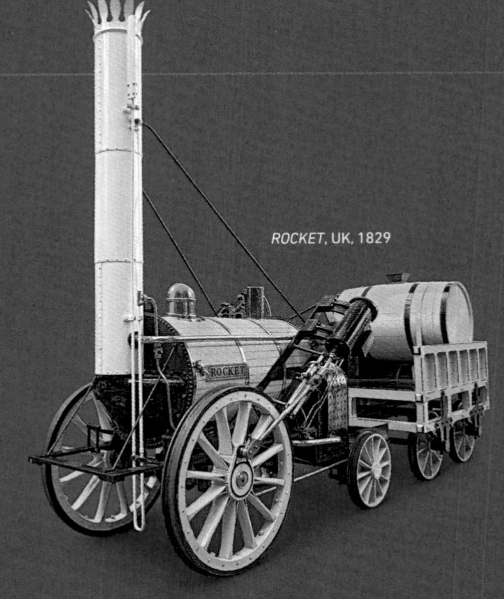

ROCKET, UK, 1829

The steam engine was the great invention of the Industrial Revolution, and its greatest application was the creation of the railways. Robert Stephenson's *Rocket* (1829) was not the first steam locomotive, but its design, which featured a blastpipe, was the most advanced of its day. When it won the Rainhill Trials, held by the world's first railway (the Liverpool and Manchester), the *Rocket* became the template for locomotive design. Its top speed was 56 kph (35 mph), which was amazing for its time.

Trains were powered by steam for more than 130 years, during which time they became increasingly large and powerful. There were many modifications to the basic design: American locomotives, for example, were given huge smokestacks that prevented sparks from escaping into the air, as well as "cowcatchers" at the front to clear obstacles from the track. The steam engine perhaps reached its peak with the streamlined *Mallard*, which achieved a speed of 203 kph (126 mph) in 1938, a record that stood for decades.

Although electricity had been used to power small trains as early as the 1880s, it did not begin to supplant steam until the 1950s. Since then, trains have become more streamlined – notably the Japanese *Bullet*, which borrowed elements from aircraft design – and have been given tilting mechanisms to improve performance. These and other innovations have created trains that are fast, economical to run, and can carry a vast number of passengers.

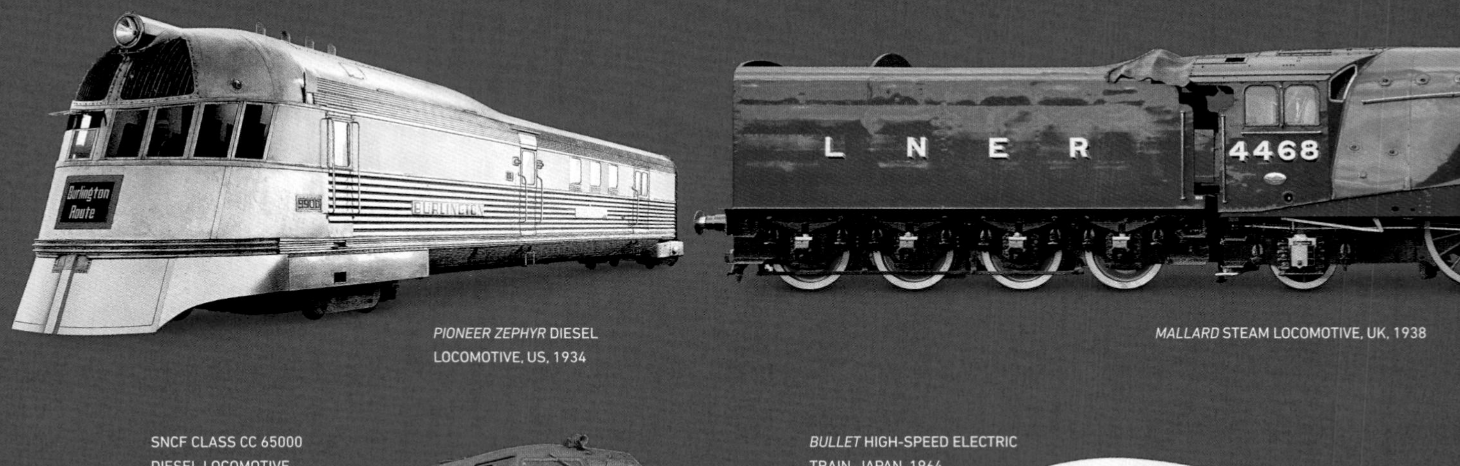

PIONEER ZEPHYR DIESEL LOCOMOTIVE, US, 1934

MALLARD STEAM LOCOMOTIVE, UK, 1938

SNCF CLASS CC 65000 DIESEL LOCOMOTIVE, FRANCE, 1957

BULLET HIGH-SPEED ELECTRIC TRAIN, JAPAN, 1964

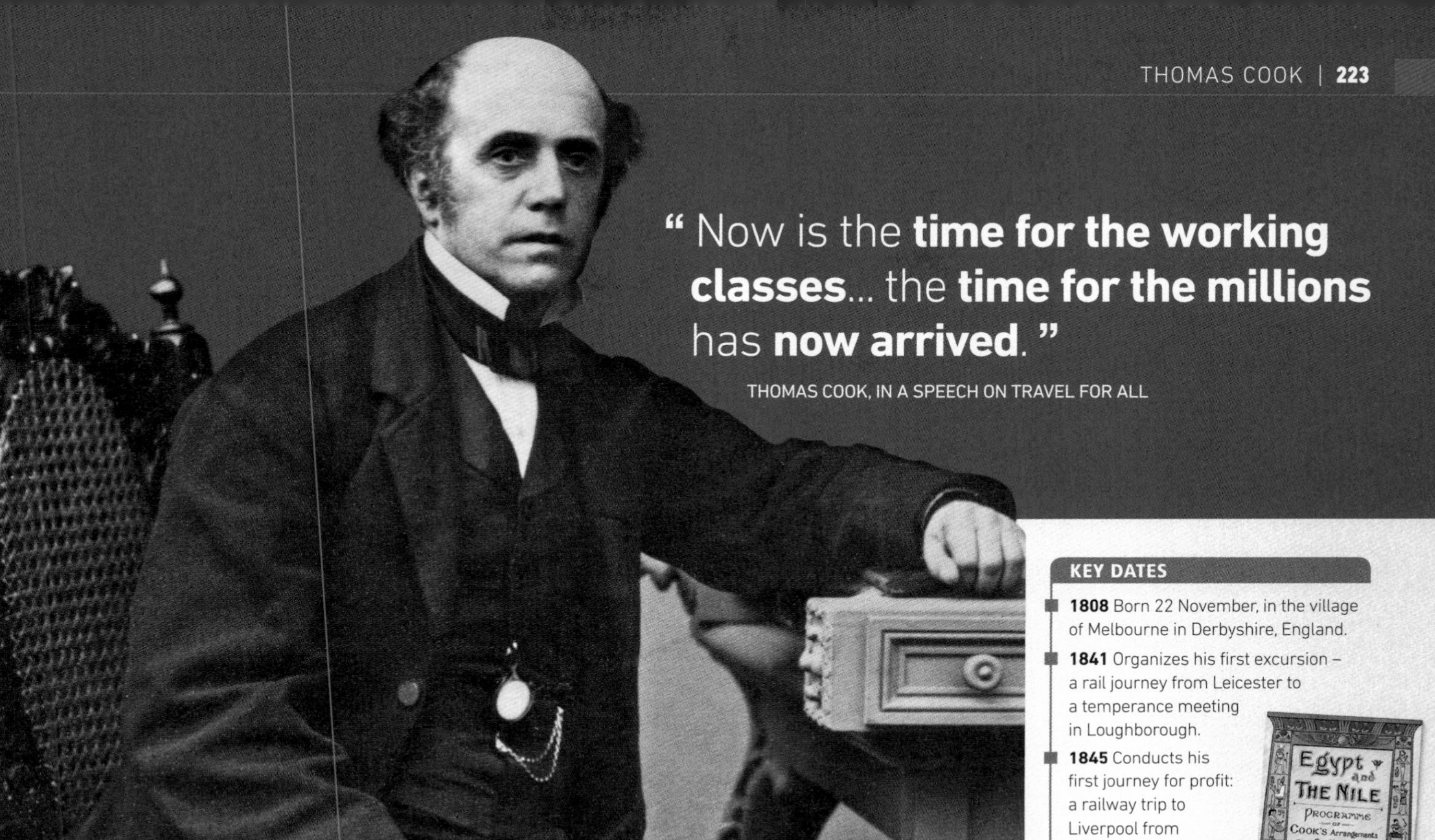

> “ Now is the **time for the working classes**… the **time for the millions** has **now arrived**. ”
>
> THOMAS COOK, IN A SPEECH ON TRAVEL FOR ALL

KEY DATES

- **1808** Born 22 November, in the village of Melbourne in Derbyshire, England.
- **1841** Organizes his first excursion – a rail journey from Leicester to a temperance meeting in Loughborough.
- **1845** Conducts his first journey for profit: a railway trip to Liverpool from Leicester, Nottingham, and Derby.
- **1851** Promotes trips from the English Midlands to the Great Exhibition in Hyde Park, London.

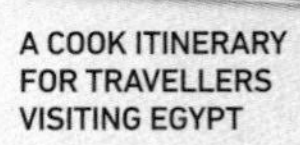

A COOK ITINERARY FOR TRAVELLERS VISITING EGYPT

- **1855** Launches the first continental tour, which takes in Belgium, Germany, and Paris for the International Exhibition.
- **1863** Conducts his first party of 62 people to Switzerland, via Paris.
- **1868** Introduces a system of hotel coupons in an attempt to get fixed, discounted prices for accommodation at selected hotels.
- **1872** Leads the first round-the-world tour, which lasts 222 days and covers more than 40,000 km (25,000 miles).
- **1892** Dies, aged 83.

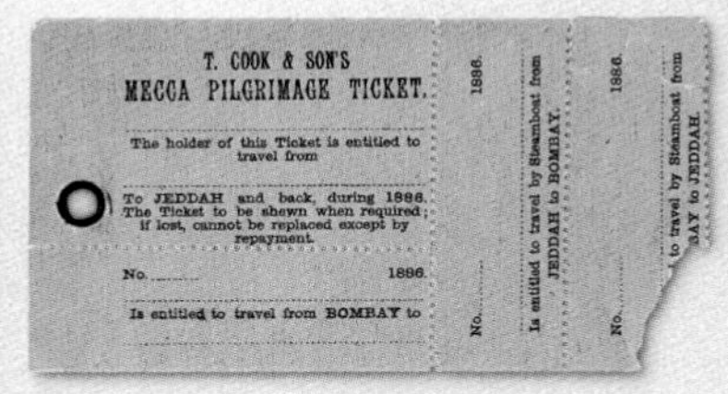

T. COOK & SON'S
MECCA PILGRIMAGE TICKET.
The holder of this Ticket is entitled to travel from
To JEDDAH and back, during 1886.
The Ticket to be shown when required; if lost, cannot be replaced except by repayment.
No. 1886.
Is entitled to travel from BOMBAY to
1886.
Is entitled to travel by Steamboat from JEDDAH to BOMBAY.
No.
1886.
Is entitled to travel by Steamboat from BOMBAY to JEDDAH.
No.

A TICKET FOR A THOMAS COOK PACKAGE TOUR TO MECCA, 1886

Spas

From the ancient Greeks to 19th-century invalids, people have always travelled in search of healing waters – from hot-water springs to spas.

Although 19th-century historian Jules Michelet condemned the Medieval period as *un mille années sans un bain*, or "a thousand years without a bath", he was not entirely correct. Christian morals ended the Roman tradition of communal bathing, but thermal springs attracted health pilgrims throughout the Middle Ages. Well-known hot-water springs existed at Bath in England, Aix-les-Bains in France, and Spa in what is now Belgium. At the latter, according to one 17th-century account, the waters could "extenuate phlegm; remove obstructions in the liver, spleen, and the alimentary canal; dispel all inflammations; and comfort and strengthen the stomach".

The popularity of these institutions grew massively during the period of the Grand Tour (see pp.180–83), when recuperative pit-stops developed as an essential part of the itinerary. This gave rise to the grand spa towns of Central Europe – places like Baden-Baden, Bad Ems, Bad Gastein, Karlsbad, and Marienbad. Here, doctors administered regimes of imbibing water from mineral sources, bathing in thermal pools, and exercising.

Given the social nature of the Grand Tour, the spas became more akin to exclusive country clubs. In time, they offered not only luxurious accommodation but also gardens, theatres, dance halls, casinos, and racecourses, as well as busy programmes of concerts, operas, grand balls, and fêtes. In their heyday, which was virtually the whole of the 19th century, the grandest spas were bywords for glamour, and they competed to attract royalty, heads of state, and figures of political and cultural eminence. For 20 years, the German writer Goethe spent around four months annually at a spa, famously encountering Beethoven during one of his stays at Karlsbad. Baden-Baden, which promoted itself as "Europe's summer resort", was where Brahms wrote his *Lichtenthal Symphony*, and was favoured by Russian novelists Fyodor Dostoyevsky and Ivan Turgenev. A stay at a spa was considered not only good for health, but also beneficial for mental wellbeing and excellent for one's social standing.

◁ **Spa fountain, Karlsbad**
Named after Charles IV, Holy Roman Emperor, who visited this site of hot springs in the 1350s, Karlsbad became the most visited spa town in what is now the Czech Republic. Here, attendants pose for a photograph in 1910.

Going by the book

The Grand Tourists of old had unlimited time and money, and travelled with servants who attended to their needs. In the 19th century, most travellers were less well-off and had to journey more efficiently, so a new kind of guide emerged to help them.

▷ **Murray's handbooks**
London publisher John Murray launched its first handbook in 1836, setting the template for modern-day guidebooks.

▽ ***The Honeymoon***
This 19th-century engraving by Roberto Forell shows a couple holding a Baedeker guidebook. These books not only acted as a guide to local attractions, but also advised on how to behave when abroad.

In E.M. Forster's novel *A Room with a View*, Lucy Honeychurch, the heroine, faces difficulty when she has to venture out in Florence without her Baedeker guidebook. It is a name that is now forgotten, but for almost a century, the words "travel" and "Baedeker" went hand in hand, so much so that for many people, it was unthinkable to visit anywhere without the famous travel guide.

Guidebooks, like general travel literature, have existed in some form since the time of the ancient Greeks. They flourished in the era of the Grand Tour (see pp.180–83), when numerous travellers had written, advising those who followed on places to visit. However, towards the middle of the 19th century, when tourism became a significant industry, driven by rapid improvements in transport and the increasing rise of a wealthy middle class, publishers saw the need for a new kind of travel book.

Handbooks for travellers

John Murray was renowned as the publisher of Lord Byron, among others. In 1820, the company issued *Travels on the Continent* by Mariana Starke, an Englishwoman who had grown up in India and who now lived in Italy. Whereas previous travel guides aspired to the literary, Starke's book was essentially practical. It included advice on what to pack and how to deal with bureaucracy, as well as on what things cost. Starke also introduced a ratings system of exclamation marks to highlight key sights. Her book was very popular with British travellers, and ran to eight editions, each updated by the author (in one, she notes the introduction of streetlamps in Italy, which, she says, has put a stop to "the dreadful practice of assassination").

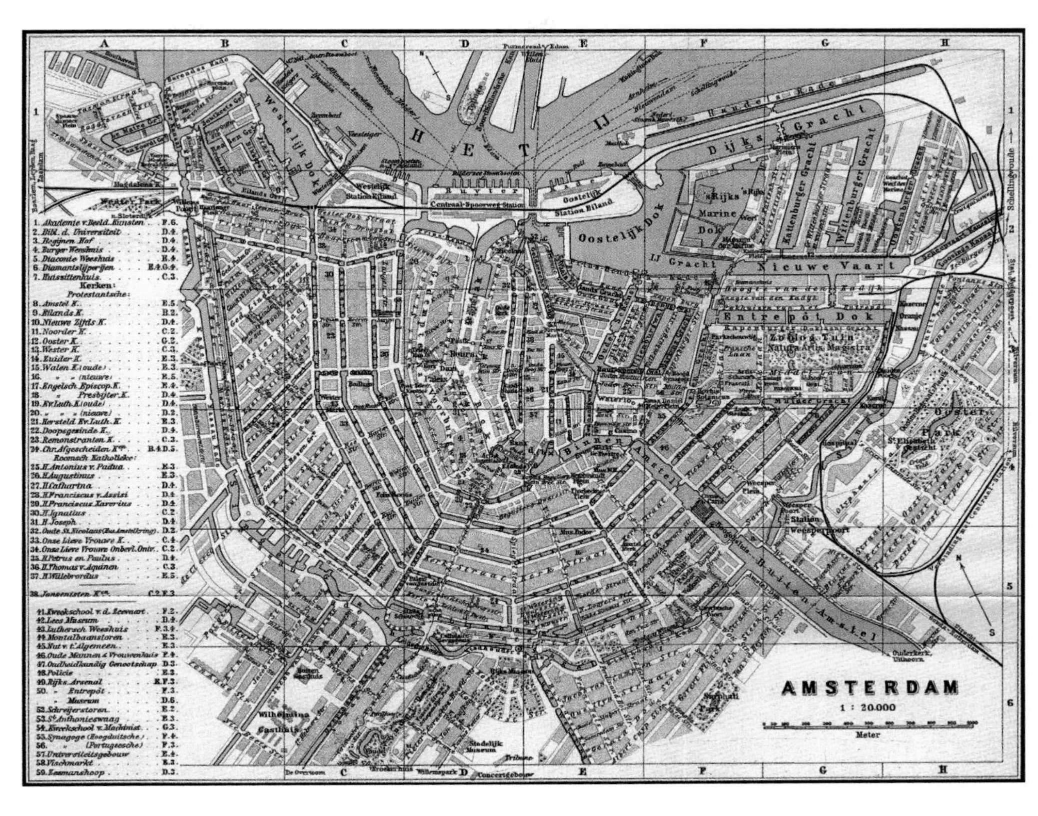

▷ **Baedeker maps**
From the beginning, Baedeker guidebook maps were famed for their detail and reliability. According to English travel writer Eric Newby, they were "made as if by spies for spies".

Arguably, Starke's book set the template for the guidebook as it exists today.

Murray capitalized on the book's success, following it with the *Handbook for Travellers on the Continent* (1836), which covered Holland, Belgium, and north Germany, and was written by the publisher himself. This was quickly succeeded by handbooks to South Germany (1837), Switzerland (1838), and France (1843). The books were standardized, made small to fit in the hand, printed on thin paper to keep down both cost and weight, and were updated on a regular basis. When the *Handbook to New Zealand* was published in 1893, one magazine commented: "Mr Murray has annexed what remained for him to conquer of the tourist world". By this time, however, Murray's handbooks were second in popularity to those published by Karl Baedeker.

Baedekers

Released by the German publishing house in 1839, the first Baedeker guides were modelled on Murray's, and were soon published in French and English as well as in German, all in a distinctive red cloth cover. Over time, they grew in popularity for the quality and concision of their information, their excellent maps, and the star system that told travellers which sites were most important. The books became so popular that Kaiser Wilhelm of Germany was quoted as saying that he stood at a particular window in his palace each noon, because: "It's written in Baedeker that I watch the changing of the guard from that window, and the people have come to expect it."

Less obligingly, in 1942, during World War II, several historic English cities were carpet-bombed in what became popularly known as the "Baedeker Raids", after a spokesperson for the German Air Force declared, "We shall go all out to bomb every building in Britain that is marked with three stars in the Baedeker guides". The following year, Allied bombers destroyed the publishing company's headquarters and archives in the city of Leipzig.

" My book…comprehends **every kind of information** most needful to continental **Travellers**. "

MARIANA STARKE

◁ **Karl Baedeker**
Descended from a line of publishers and printers, Baedeker saw the potential of Murray's handbooks, and soon outsold them with his own guides.

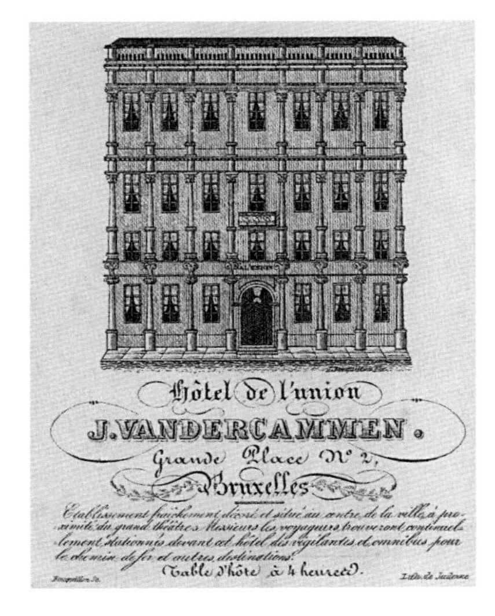

◁ **Advertising**
With the rise of tourism, guidebooks began to feature adverts for specific locations, such as hotels and shops. This engraving is an advert for the Hôtel de L'Union in Brussels.

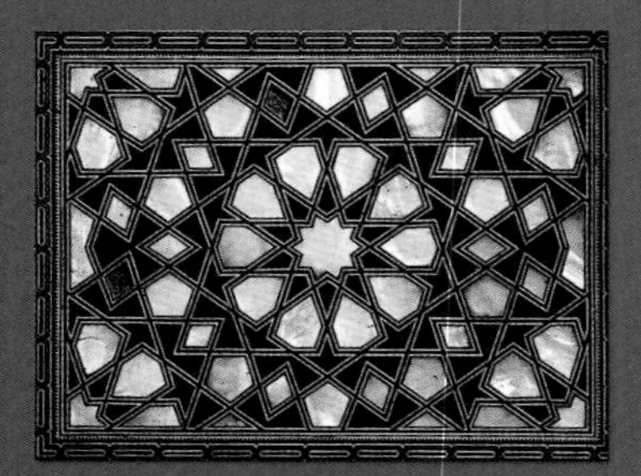

WOOD AND MOTHER-OF-PEARL QUR'AN BOX THAT BELONGED TO SULTAN SELIM II, 16TH CENTURY

PLYMOUTH ROCK FRAGMENT WITH PAINTED INSCRIPTION, US, 1830

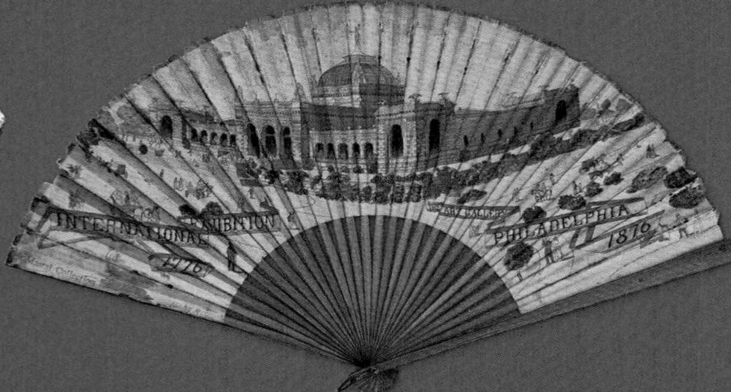

LADY'S FAN COMMEMORATING CENTENNIAL EXHIBITION, PHILADELPHIA, 1876

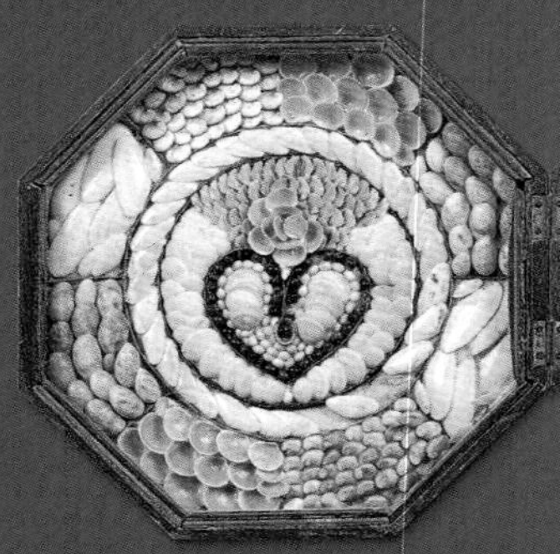

DOUBLE OCTAGON SHELLWORK SAILOR'S VALENTINE, MID-19TH CENTURY

PILGRIM'S BROOCH, EUROPE, 1890–1935

JAPANESE NETSUKE (KIMONO ORNAMENT) OF OLD MAN

BOOKMARK ADVERTISING BUTLIN'S HOLIDAY CAMPS, UK, 1950s

ENAMELLED GOLD SNUFFBOX IN SHAPE OF BUTTERFLY, WITH CARILLON AND WATCH

RUSSIAN NESTING DOLL, MASS-PRODUCED AFTER WINNING AWARD AT WORLD'S FAIR, PARIS, 1900

BABOUCHES (EMBROIDERED MOROCCAN SLIPPERS)

DISH OF PAINTED LACQUERED SHELLS, WITH PHOTO OF A MALTESE HARBOUR, MALTA c.1965–75

NEW YORK CITY NOVELTY PLASTIC SUNGLASSES

WORLD'S FAIR SILVER SPOON WITH IMAGE OF BERTHA PALMER (AMERICAN BUSINESSWOMAN AND SOCIALITE), CHICAGO, 1893

GLASS VIAL HOUSING COLOURED SANDS FROM ALUM BAY, ISLE OF WIGHT, UK, 19TH CENTURY

CHINA JUG DEPICTING KINEO HOUSE AT MOOSEHEAD, LAKE MAINE, US, 1890s

TRAVEL GUIDES TO ITALIAN CITIES OF FLORENCE, CAPRI, AND MILAN, c.1926

SWEDISH DALECARLIAN HORSES, TRADITIONAL WOODEN TOY MASS-PRODUCED AFTER WORLD'S FAIR, PARIS, 1937

Souvenirs

Meaning "memories" in French, souvenirs remind the owner of a place or event, and often have great sentimental value.

The first souvenirs were relics taken from a sacred or historic site such as Plymouth Rock, where the Mayflower Pilgrims were said to have first touched American soil. To preserve sites, as what was left of Plymouth Rock had to be fenced off in the 1880s, officials began to sell trinkets instead.

Souvenirs came into their own during the Grand Tour. In the 18th century, wealthy Europeans shipped home ancient sculpture, artefacts from archaeological digs, and views of Venice by the artist Canaletto that were forerunners of the postcard. Soon there was a market for manufactured keepsakes – Mt Vesuvius erupting on porcelain and fans, bronze replicas of the Colosseum, even a "selfie" by Pompeo Batoni, who specialized in painting English lords lounging in classical settings. Starting with the 1851 British Great Exhibition, world's fairs attracted thousands of visitors in the late 19th century, and many of them took home a silver spoon, a china jug, or a penny trinket. Souvenirs could also be gifts, such as a sailor's shellwork valentine shaped with a heart and tender message – "Think of me" or "Home again".

While love tokens, scenic tableware, and commemorative spoons are still popular, today's mass-produced souvenirs are a far cry from the handmade Qur'an and snuff boxes of earlier days. Russian dolls, snowglobes, slippers, novelty sunglasses, and bookmarks all make cheap and cheerful holiday mementos to bring home.

SNOW GLOBE (INVENTED IN VIENNA IN 1900) DEPICTING NEW YORK CITY

The works of all nations

Called variously Great Exhibitions, Expositions Universelles, and World's Fairs, a series of itinerant celebrations of industry, technology, and arts inspired those who visited to marvel at all the world had to offer.

△ **Javanese dancers**
Exhibitions always featured ethnic attractions, ranging from exotic foods to costumes and dance, as here in Paris, 1889.

The full title of the very first international exhibition, which was held in London in 1851, described it as "The Great Exhibition of the Works of Industry of All Nations". This spectacular event, held in a glittering glass-and-steel "Crystal Palace" designed by Joseph Paxton, was a celebration of the achievements of the Industrial Revolution, as well as the success of the British Empire. It was the first of many exhibitions that, as well as aiming to glamorize the host city, became a platform for promoting wares from countries all around the world.

The world comes to town

In the 1851 Great Exhibition, there were exotic goods from Europe, the Near and Far East, the British colonies, and the US. Exhibits from the latter included Colt revolvers, false teeth, and a large model of the Niagara Falls. It is not recorded how many visitors were inspired to cross the Atlantic by such a display, but people from all across Great Britain flocked to the capital for the exhibition. It provided a huge boost for the new business of Thomas Cook & Son (see pp.222–23), who organized the tickets, transport, and accommodation for 165,000 visitors from the Midlands.

Cook had a similar success when he launched his first Continental tours to the Paris Exposition Universelle of 1855. A second Paris exposition in 1867 was the first of its kind to introduce the idea of national pavilions, as well as a parade of ethnic restaurants where customers could experience different cuisines served by staff in national costume. One of its most popular exhibits was a large-scale model of the Suez Canal, complete with ships, and the Italians brought a model of the entrance to the Mont Cenis Tunnel, which was due to open the following year as part of the world's first mountain railway.

No expense spared

By this time, a pattern for Great Exhibitions had emerged: a large park was filled with a small city's worth of buildings that boasted every conceivable type of innovation, commodity, and activity. Each country was invited to take part, at great expense, and then after six months, the exhibition area was razed to the ground, with maybe just the odd landmark surviving.

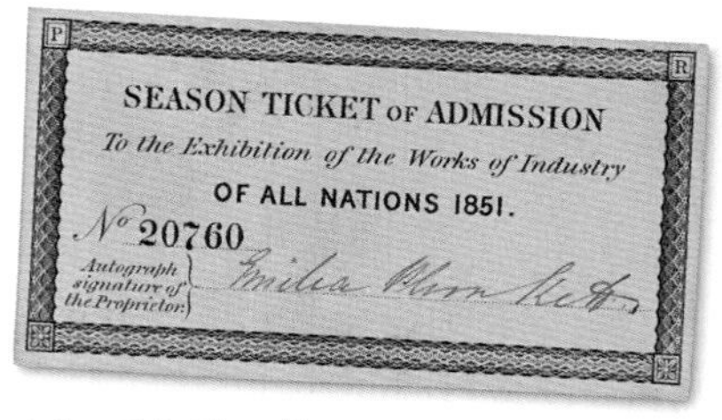

△ **One ticket for all**
Right from 1851, the Great Exhibitions drew enormous crowds of visitors on a scale that was only rivalled by the Olympic Games.

The 1878 Paris Exposition introduced a *Rue des Nations* (street of nations), a long promenade lined with facades of buildings exemplifying the architecture of numerous countries. In the 1889 Paris Exposition, this was developed into a large colonial section, complete with native villages, reproductions of parts of the Angkor Wat temples, and an entire Cairo street.

This street was so successful that the massive World's Columbian Exposition, which marked the 400th anniversary of Christopher Colombus's discovery of America and was held in Chicago in 1893, also featured an Egyptian Street, where Thomas Cook displayed models

▷ ***Rue des Nations***
In Paris, in 1900, participating countries were invited to build national pavilions on the banks of the River Seine.

Ocean to ocean

The completion of America's first transcontinental railroad was a landmark not just in transport but in nation building.

The United States had been quick to grasp the benefits of the railroad. Even so, in 1860, 30 years after the first lines had been laid, the only way to get from one coast to the other was still either by wagon or to sail round South America. To remedy this, in 1862, President Abraham Lincoln signed the Pacific Railroad Act. The legislation granted financial subsidies and land rights to railway companies in order to construct a continuous transcontinental railroad between the eastern side of the Missouri River at Council Bluffs, Iowa, to the Sacramento River in California.

In the west, four businessmen came together to back Theodore Judah's Central Pacific Railroad Company, which won the contract to build the line out of California. Progress was slow, but by 1867, the toughest engineering challenge was overcome with the completion of a passage through the Sierra Nevada mountains at the Donner Pass, 2,160 m (7,086 ft) above sea level. In the east, work was delayed by the ongoing Civil War, but once this ended in 1865, the Union Pacific Railroad Company, which held the contract for that end of the line, was boosted by the addition of ex-soldiers and freed slaves to its army of railway builders.

To build the railroad, an advance party first surveyed the route, then graders smoothed out the terrain, levelling earth, clearing rocks, and creating embankments and bridges. Lastly came the tracklayers, who put down the sleepers and rails. The government contracts paid for each mile of track completed, so both companies competed to lay as much as possible. Eventually, Promontory Summit in Utah was agreed on as the place to join the tracks – so it was there that company representatives took turns hammering a golden spike into the final tie on 10 May 1869. It was not just a railroad that had been united, but a country, and a journey that had previously taken six months was now possible in just a matter of days.

▷ **The great event**
A poster announces the opening of the railroad connecting the Pacific and Atlantic oceans. The service opened the West to settlers of all kinds, and to people travelling for pleasure.

The grand hotel

Around the middle of the 19th century, hotels blossomed from basic resting places to grand, and often luxurious, destinations in their own right.

△ **Grand arrival**
People arrived at the grand hotels in style. In this 1931 photo, hotel porters at the Bayerischer Hof in Munich load up new arrivals' suitcases and bags.

On 5 May 1862, Empress Eugénie of France was invited to the inauguration of the Grand Hôtel, Paris. Occupying a vast triangular plot next to the site on which Garnier's spectacular opera house would soon rise, the Grand was possibly the largest hotel in the world at the time, boasting 800 rooms and 65 salons. It had a spectacular glass-roofed courtyard, a cellar holding a million bottles of wine, décor by a host of well-known artists, and a vast ballroom. It was, said the Empress, "like home".

Until the middle of the 19th century, hotels in the modern sense were rare. Travellers generally stayed in boarding houses or rented lodgings in cities, and put up at coaching inns while on the road. However, as the numbers of travellers increased and stays became shorter, it became impractical to rent lodgings – what people needed were modern hotels.

Two of the first great hotels of note were built in the rapidly expanding cities of North America. The Tremont House in Boston, built in 1929, was soon followed by the first luxury hotel in New York City, Astor House, which opened in 1836. Both establishments offered unimaginable luxuries, such as indoor plumbing, heating, gas lighting, locked rooms for guests, free soap, and dining rooms with extensive menus.

◁ **Swiss bellboy**
Many fine early hotels were established in Switzerland, where winter sports kept them busy year round. The staff were immaculately dressed, in keeping with the status of the hotels.

Beds for the crowds

Some of the earliest large hotels in Europe were built next to stations by railway companies. The first railway hotel was completed at Euston in London in 1839, and even larger hotels were constructed soon after at Paddington in 1854, Victoria in 1861, and Charing Cross in 1864. Several hotels were also built specifically to accommodate the crowds anticipated for the Great Exhibitions (see pp.230–31),

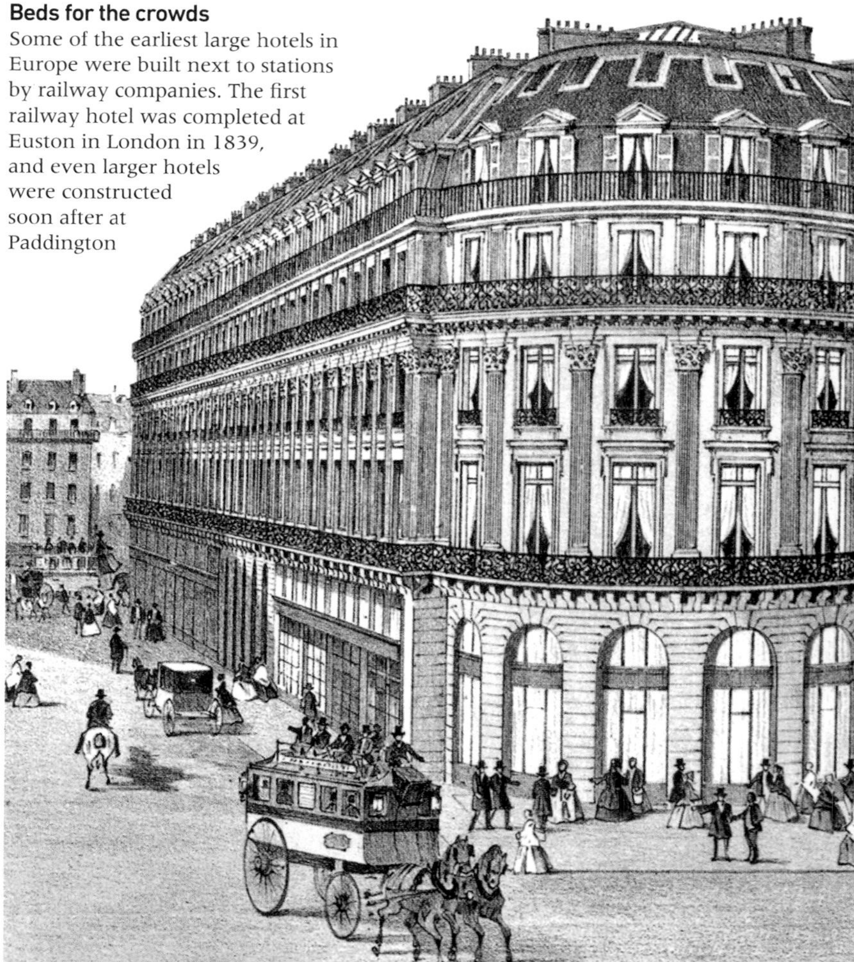

▷ **The Grand Hôtel, Paris**
The luxurious Grand accommodated royalty and a host of celebrities, including actress Sarah Bernhardt, composer Jacques Offenbach, and opera singer Enrico Caruso.

◁ **The Waldorf Astoria**
American hotels led the way in terms of comfort and amenities. New York's Waldorf Astoria was the first hotel to be completely electrified and offer each guest a private bathroom.

notably the Grand Hôtel du Louvre, which was built for the 1855 Paris Exposition, the Langham Hotel (London, 1862), and the Continental (Paris, 1878).

Initiated by the splendours of the Paris Grand, the late 19th century issued in a golden age of hotels. Benefitting from new inventions such as the lift, first publicly demonstrated at the New York World's Fair of 1854, and then electricity, hotels became even larger and more grand. The Americans led the way – European travellers to the United States often remarked on the size and comfort of the hotels they found there. Hotels in Europe, in turn, attempted to attract American visitors by improving amenities, notably the number of bathrooms. When London's Savoy was under construction in the 1880s, its financier, Rupert D'Oyly Carte, requested one bathroom for every two bedrooms, leading his contractor to ask if D'Oyly Carte was expecting his guests to be amphibious.

One of D'Oyly Carte's smartest moves was to enlist the services of Swiss-born César Ritz as manager of his new hotel. In the late 1890s, Ritz left the Savoy to launch his own eponymous hotel, situated on the fashionable Place Vendôme in Paris. Behind an 18th-century façade, it boasted a private bath for each room, setting a standard for luxury that was to become the benchmark for every grand hotel that followed.

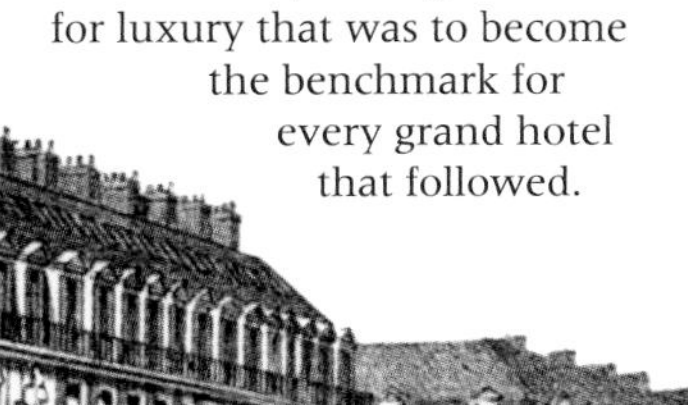

IN PROFILE
César Ritz

Born in Switzerland's upper Rhône Valley in 1850, César Ritz went to Paris to look for work during the 1867 Exposition. He worked in a succession of restaurants before returning to Switzerland to work at the Hotel Rigi-Kulm, near Lucerne. In 1877, he moved to the Grand Hôtel National in Lucerne, which, although the most luxurious hotel in Switzerland, was suffering financial difficulties. In a single season, he turned it around. He struck up a partnership with chef Auguste Escoffier, and together they established the dining room as the hotel's social centre, a destination not just for guests but for all of fashionable society.

CÉSAR RITZ IN 1900, TWO YEARS AFTER OPENING THE HOTEL RITZ IN PARIS

SAVOY HOTEL

Menu.

PETITE BOUCHÉE WLADIMIR
CONSOMMÉ AUX NIDS D'HIRONDELLES
SUPRÊME DE SOLE NANTUA
CAILLE ROYALE AU RAISIN DE MUSCAT
NOISETTE D'AGNEAU À LA MASCOTTE
PÂTÉ DE FOIE GRAS DE STRASBOURG
SALADE ISABELLE AUX ASPERGES
PLUM PUDDING FLAMBÉAU RHUM
FANTAISIE PARISIENNE
FRIANDISES.

New Years Eve 1908. Savoy Hotel, London.

◁ **Haute cuisine**
This 1908 New Year's menu from the Savoy Hotel Grand shows that the grand hotels were not just places to stay, but venues where high society could dine, dance, and be seen.

> “ When I **dream of** afterlife in **heaven**, the **action** always **takes place** in the **Paris Ritz**. ”

ERNEST HEMINGWAY, IN A LETTER TO A.E. HOTCHNER

GRAND HOTEL DE LYON, FRANCE

KYOTO HOTEL, JAPAN

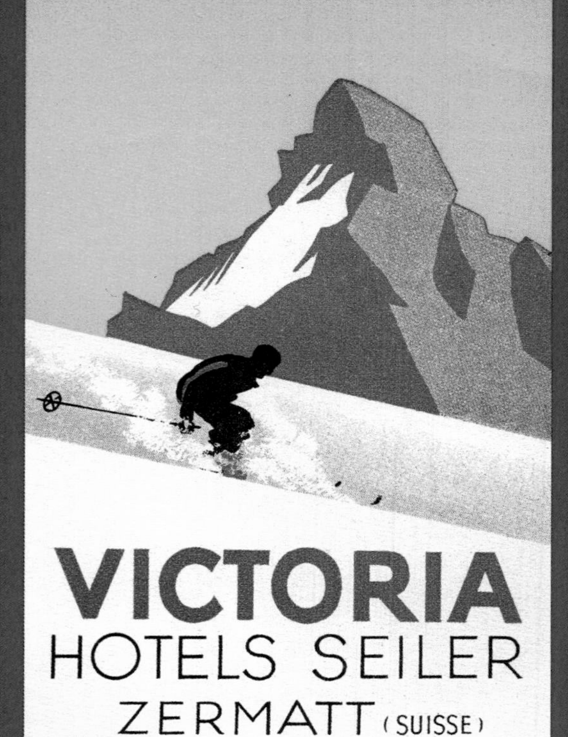

VICTORIA HOTEL, SWITZERLAND

HOTEL LUNA, ITALY

VILLARS PALACE, SWITZERLAND

HOTEL DE LA MAMOUNIA, MOROCCO

CAISTER HOTEL, SOUTH AFRICA

HOTEL TIMES SQUARE, US

SEA VIEW HOTEL, SINGAPORE

HOLDEN'S AIR SERVICES, NEW GUINEA

STATION HOTEL, MALAYSIA

OVERLAND DESERT MAIL, MIDDLE EAST

GRAND HOTEL RICHTER, FRANCE

COSULICH LINE, ITALY

HOTEL REGIS, MEXICO

Luggage labels

Beautiful and practical, luggage labels indicated where a traveller's luggage was to go, and where it had been.

Labels were first issued by shipping companies as an aid to dockside handling, indicating which items were wanted in cabins and which could be stored in the baggage hold. They were later adopted by the hotel industry. Porters at the railway station or harbour stuck them onto travellers' luggage to make sure all items were delivered to the correct hotels. In later years, hotels handed them out to guests as souvenirs of their stay, and as a cheap and effective early form of advertising.

The artists who designed hotel labels typically sought to encapsulate the exoticism and romance of the local area, so they usually included landmarks set against dusky sunsets or brilliant blue skies. Some of these labels are mini masterpieces, resembling classic travel posters from the golden era of travel – unsurprising, given that the same artists produced designs for both posters and labels.

A well-stickered piece of luggage also advertised its owner's worldliness, wealth, and status. "Travelling in a compartment, with my hat-box beside me, I enjoyed the silent interest which my labels aroused in my fellow-passengers," wrote essayist Max Beerbohm in 1909.

The advent of flight forced travellers to travel with far less luggage, and the tradition of the luggage label came to an end.

HOTEL RUHL, FRANCE

HOTELS ESSENERHOF AND SCHLICKER, GERMANY

Measuring India

In India, the British undertook the most arduous map-making expedition ever launched. It took nearly 70 years to complete, and cost more lives than many wars, yet it is all but forgotten today.

By the end of the 18th century, the British East India Company was in control of large swathes of India, but knowledge of the country was limited. Frontiers, districts, and infrastructure had yet to be established, but this was difficult without good maps. So, shortly after a victory over Tipu Sultan, the ruler of the Kingdom of Mysore in southern India, a scientific survey of the country was initiated.

The Great Trigonometrical Survey

What became known as the Great Trigonometrical Survey of India began at Madras, on the east coast, on 10 April 1802, under the command of surveyor William Lambton. It involved taking the most precise measurement possible along the ground between two fixed points, and then sighting a third fixed point from either end of this line. The distance to the third point could then be calculated using trigonometry. This new line then served as the baseline for another sighting, and so on until a given area was measured. All calculations were checked against the positions of the stars, using a huge custom-made theodolite (a kind of telescope), which weighed half a ton. By this rigorous and arduous process, the survey made its way, triangle by triangle, westwards from Madras across the newly acquired territory of Mysore, via Bangalore in central India, to Mangalore on the Arabian Sea.

Logistical problems

This first portion of the survey took four years, and established that the width of India at that latitude was 580 km (360 miles), which was 64 km (40 miles) shorter than had previously been thought. The survey then worked south from Bangalore to the southern tip of India, and north up the spine of the country to Delhi and beyond, in a

◁ **Lambton's theodolite**
Weighing over 508 kg (1,100 lb), this huge piece of surveying equipment was taken to India by William Lambton in 1802. Its sturdiness helped surveyors make accurate readings.

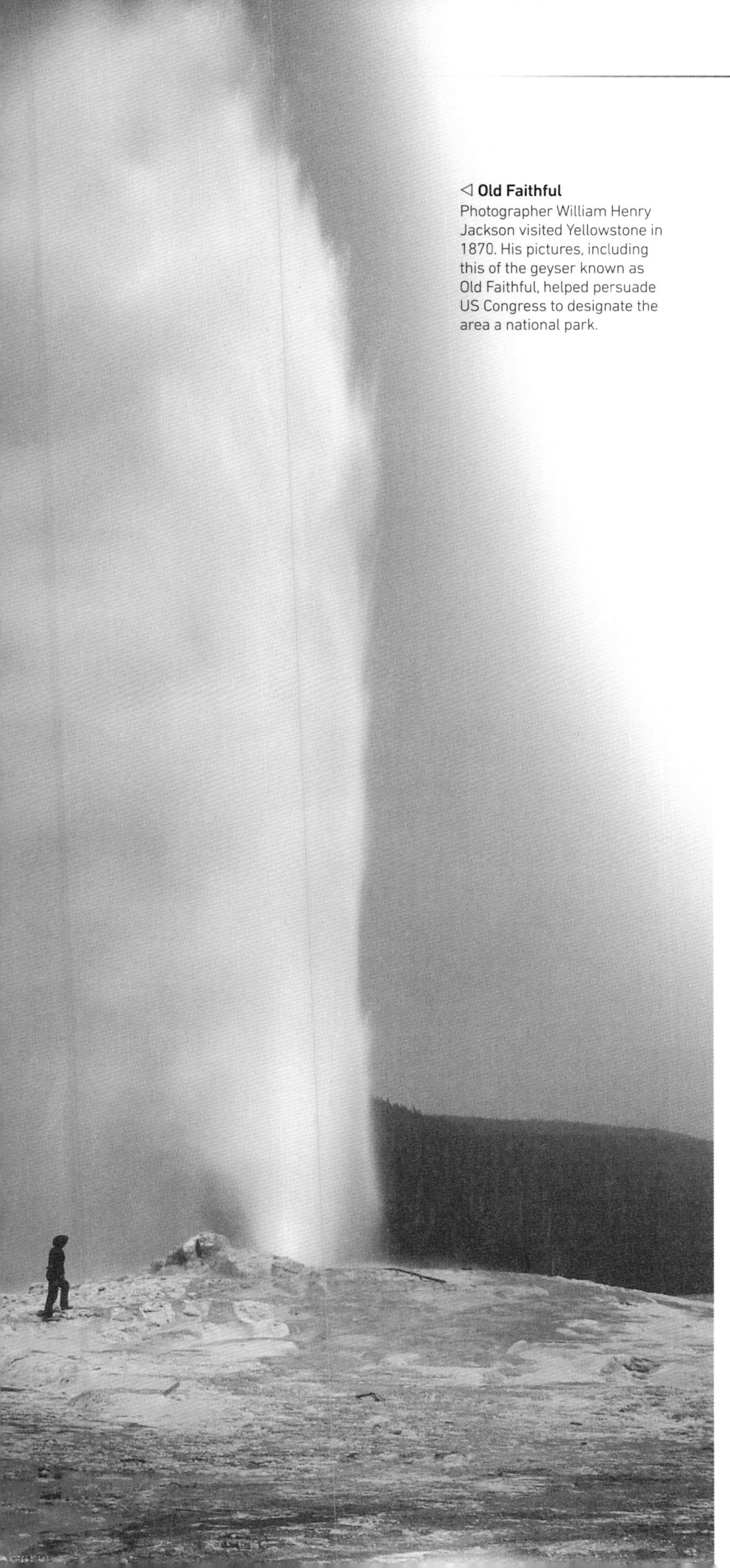

◁ **Old Faithful**
Photographer William Henry Jackson visited Yellowstone in 1870. His pictures, including this of the geyser known as Old Faithful, helped persuade US Congress to designate the area a national park.

△ **Yosemite**
Glacier Point, photographed here in 1877, was one of the rugged wonders that stunned Frederick Law Olmsted during his visit to Yosemite.

set out to investigate tales of a place in northwest Wyoming, at the headwaters of the Yellowstone River, where water and steam spouted out of the ground. The party eventually found what it was looking for. After struggling through dense forest and snow, they came across a large clearing where they witnessed the eruption of an immense geyser, causing them to toss up their hats and shout for joy.

The following year, the area was visited by a party of scientists who prepared a report that was submitted to Congress. On 1 March 1872, President Ulysses S. Grant signed the bill, creating Yellowstone National Park. Unlike Yosemite, which was administered by the state of California, this would be a national park, the first in history.

Following campaigns by naturalist John Muir, who was dismayed by the Californian state's inability to look after Yosemite, this area of outstanding national beauty was also elevated to the status of national park in 1890.

▷ **Cook's Tours**
As early as 1873, Thomas Cook was personally leading groups of tourists on trips around the world. This poster dates from 1890.

Around the world

By the end of the 19th century, thanks to improved transport and the opening of the Suez Canal, intrepid tourists were able to travel right around the world.

△ **Ida Pfeiffer**
Ida Pfeiffer kept meticulous diaries of her travels that were full of thoughtful cultural and behavioural observations.

Ida Pfeiffer was an extraordinary woman. Born in Vienna in 1797, a time when women were meant to devote themselves to "Kinder, Küche, Kirche" (children, kitchen, church), she travelled the world alone, not once but twice. What is more, she paid for her travels with her writing.

Her early life was fairly conventional. Although she grew up with several brothers and was treated as one of the boys by her father (she was even dressed in boys' clothes), she was married at the age of 22 and raised two sons. It was only after she and her husband had separated and the sons had grown up and left home that Ida, then aged 45, decided to travel.

She went first to the Holy Land and Egypt, ostensibly on a pilgrimage, knowing that this would meet with less disapproval from family and friends. She followed her foray into the Middle East with a trip to Iceland and then, in 1846, with a round-the-world tour that took in South America, China, India, Iraq, Iran, Russia, Turkey, and Greece. She travelled with almost no luggage – just a leather pouch for water, a small pan for cooking, and some salt, rice, and bread. She returned home in 1848, staying just long enough to write a bestselling account of her voyages. Then she sallied forth again, heading for South Africa and then on to Asia.

Ida was not just unique in being a woman travelling on her own. In the mid-19th century, few people travelling for pleasure ventured further than Europe, the Middle East, and the eastern part of America. Beyond these familiar places, there was little in the way of scheduled transport or tourist accommodation, and travelling further afield required fortitude and the ability to cope with the unexpected, the uncomfortable, and even the outright dangerous. Ida lodged with local people, slept outside on the decks

▷ **Media darling**
Nellie Bly's youth, sex, and daring led to fame. Here, she is depicted on the cover of an 1890 "Round the World" board game.

of crowded ships, dined with cannibals, and was even imprisoned in Madagascar, where she was accused of conspiring against the queen.

World travel

By the end of the 1860s, this situation was beginning to change. The Suez Canal opened in 1869, enabling direct steamship services between Europe and Asia. The opening of Japan to foreign trade in the 1850s had already led to similar services across the Pacific.

Unsurprisingly, the man who had taken advantage of the new steam railways to launch his guided-tour business, Thomas Cook (see pp.222–23), was quick to spot the possibilities of world travel. By 1872, his company had already escorted parties all around Europe, and to America, Egypt, and the Holy Land. In spring that year, he announced an ambitious around-the-world excursion that he would lead himself. His route went across the Atlantic and the United States to Japan, China, Singapore, Ceylon, India, and Egypt. Travellers then took a steamer across the Mediterranean, and crossed Europe by rail to return to London. In all, the tour covered 40,234 km (25,000 miles) and took 222 days. It was such a success that it became an annual event from then on.

> " In exactly the **same manner** as the artist feels an **invincible desire** to paint ... so was I hurried away with an **unconquerable wish to see the world**. "
>
> IDA PFEIFFER, IN *A WOMAN'S JOURNEY AROUND THE WORLD*, 1850

◁ **Tourist in Shanghai, 1900**
For the well-travelled who had been to Europe, the Middle East, and America, the cities of the Far East provided novelty and new experiences. London publisher John Murray published his first "handbook" to Japan in 1884.

Over the coming decades, the route established by Cook became the standard itinerary for world travellers. It remained expensive, but thanks to ever-improving and better-connected transport systems, travelling further afield became more feasible. By the end of the 19th century, tours of Japan and India were becoming popular as journeys in their own right. Japan's fear of foreign influence on its culture meant that it continued to restrict access for tourists, but it was still possible to visit Yokohama, Tokyo, Kobe, Osaka, Kyoto, and Nagasaki. By the 1890s, Cook's itineraries included Australia, New Zealand, and South Africa.

Record journey

In January 1873, *Around the World in Eighty Days*, an adventure novel written by Jules Verne, was published in France to great acclaim. It was an entertaining work of fantasy about an impossible journey. Just 16 years later, in 1889, Elizabeth Cochran, a 22-year-old American journalist, who wrote under the name of Nellie Bly, decided to follow in the footsteps of Phileas Fogg, Jules Verne's protagonist. She set herself the goal of travelling around the world in just 75 days and did so by ship, train, donkey, and any other means possible. She made it back home in 72 days, setting a world record for the time. Travel seemed to have shrunk the world.

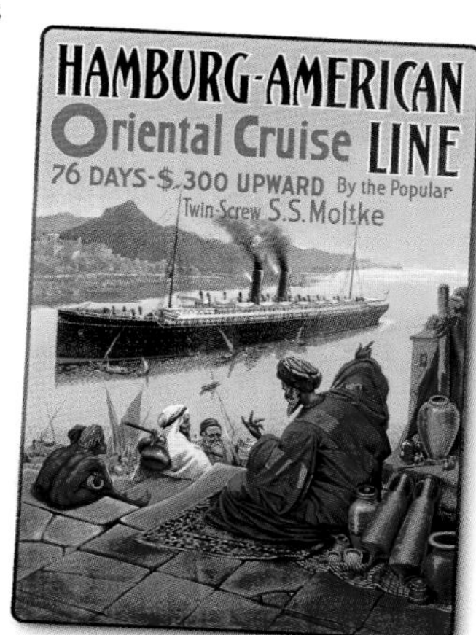

△ **The Orient lines**
By the early 20th century, round-the-world travel no longer meant hardship, and several shipping lines offered comfortable passage to far-flung destinations.

Mapping the oceans

Modern oceanography began with the *Challenger* voyage of 1872–76. It was the first expedition organized specifically to gather data on the oceans and marine life, and to investigate the geology of the ocean floor.

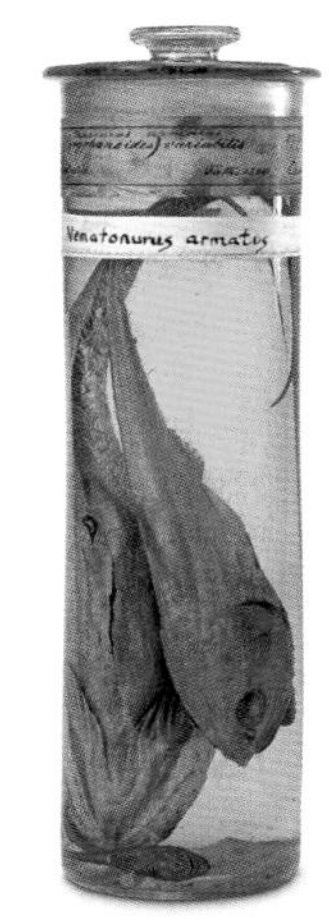

△ **Rattail fish**
This rattail fish, kept in a specimen jar, was discovered by the *Challenger* team in waters south of Australia in 1874.

As recently as the mid-19th century, the oceans and seas were regarded as vast, empty spaces that had to be crossed to get from one place to another. Naturalist Charles Darwin had even referred to them as "a tedious waste, a desert of water". Little was known of the ocean depths, but that began to change when the first transatlantic telegraph cable was laid across the seabed between western Ireland and Newfoundland in 1858. Engineers planned to lay more cables beneath the world's oceans, giving rise to an interest in what lay beneath the waves.

Into the unknown

In 1870, prompted by Scottish natural historian and marine zoologist Charles Wyville Thomson, the Royal Society of London, in collaboration with the Royal Navy, raised an expedition that would put Britain at the forefront of oceanic exploration. The venture was lent a Royal Navy frigate, HMS *Challenger*, a three-masted, square-rigged ship with auxiliary steam engines that had had all but two of its 17 guns removed to make room for two laboratories full of the latest scientific instruments. Some of the equipment had been specially invented or modified to suit the requirements of the expedition. The commanding officer was Captain George Nares, and as well as the crew of around 240, there were six scientists on board, including Wyville Thomson, naturalists John Murray and Henry Mosely, and the official artist, John James Wild.

▷ **On-board laboratory**
The *Challenger* was fitted with two fully equipped laboratories, one for chemistry (shown here) and another for biology.

▷ **Pioneers of oceanography**
This photograph shows the *Challenger*'s scientific staff, led by Wyville Thomson (third from the left). John Murray (on the floor) was the main author of their report.

▷ **HMS *Challenger* in Antarctica**
During its three-and-a-half-year voyage, HMS *Challenger* visited every continent on the planet. This contemporary engraving depicts the vessel in Antarctica in February 1874.

The ship set sail from Sheerness, Scotland, in December 1872, starting a journey that would last three and a half years. The *Challenger* sailed some 127,600 km (68,890 nautical miles) during that time, touring the North and South Atlantic and Pacific oceans, and entering the Antarctic Circle. At regular intervals, the ship would make a sample stop (362 in all), to measure the exact depth of the water and to take temperature readings at different depths. The scientists used special devices that trawled the sea floor to gather samples of rocks, sediment, and seabed fauna, and nets to capture marine life. They also collected samples of the water at different depths, and recorded atmospheric and other meteorological conditions.

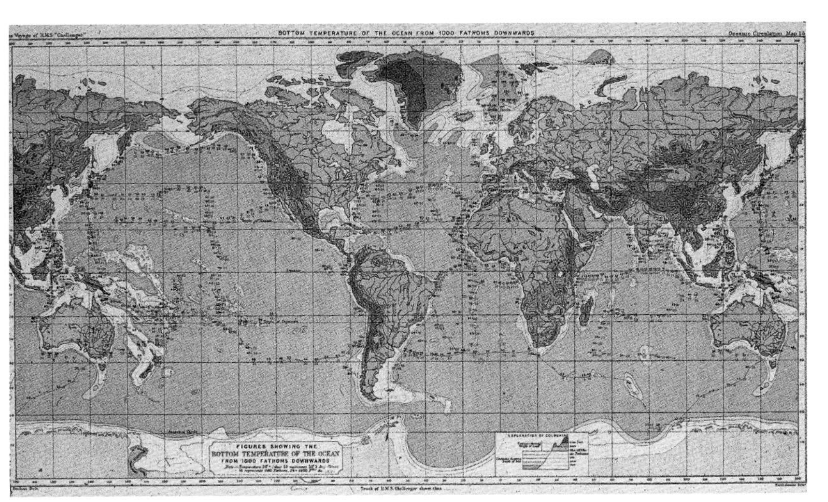

◁ **Thermal map of the oceans**
This map of HMS *Challenger*'s historic circumnavigation shows temperature readings of the oceans at different depths. It was the first time that such comprehensive data had been compiled.

A new science

The findings of the voyage were finally presented in a mammoth 50-volume, 29,500-page report that took 19 years to compile, and which was published between 1877 and 1895. It proved definitively that life existed at the bottom of the oceans – 4,700 new species of plants and animals were discovered. It also investigated the astonishing topography of the ocean floor. It recorded, for the first time, one of the deepest parts of the ocean, the Marianas Trench in the western Pacific, which is more than 6 km (4 miles) deep. The expedition made countless contributions to many different branches of science – hydrography, meteorology, and geology, as well as botany and zoology. It marked the birth of the modern science of oceanography, a term that was invented to describe the work of the *Challenger*'s scientists. It is fitting that NASA named both the 1972 Apollo 17 lunar module and its second space shuttle after the humble Royal Navy frigate that took its crew so far into the unknown.

“ The **greatest advance** in the **knowledge** of our **planet** since the **celebrated discoveries** of the **fifteenth and sixteenth** centuries. ”

NATURALIST JOHN MURRAY ON THE FINDINGS OF THE *CHALLENGER* EXPEDITION

◁ **New life forms**
Over 4,700 new species of marine animals were discovered by the *Challenger* expedition. Many were painstakingly painted and then published in the expedition's report.

Fantastic voyages

As new technologies pushed the possibilities of travel ever further, fiction writers faced the danger of being overtaken by current events.

Writing in 1869, a Sacramento newspaper noted: "Now that the Pacific Railroad is complete, few of our readers are aware that a journey around the world can be made in eighty days". Was this statement the inspiration for French author Jules Verne, who just three years later published *Around the World in Eighty Days*? Perhaps, except that Verne did not usually write about the possible – he preferred the fantastical. He favoured tales of wildly imaginative exploration, such as *Journey to the Centre of the Earth* (1864), *From the Earth to the Moon* (1865), *A Floating City* (1870), and *Twenty Thousand Leagues Under the Sea* (1870). In all of these, Verne celebrates the 19th-century mastery of science and technology, and looks ahead to future possibilities. Today, we call this genre "science fiction". We have to suppose that Verne's readers would have treated *Around the World in Eighty Days* as one of his usual, amusing conceits, not something that could really happen. Except the challenge was soundly met and bettered just 17 years later, by a young woman called Nellie Bly (see pp.248–49).

Given the swift evolution of technology, new vehicles were needed to make the fantastic voyage, and they needed to reach ever more fantastical places. In the last few years of the 19th century, English author H.G. Wells took up the challenge with *The Time Machine* (1895), which opened up tourism to the limitless vistas of the future. Wells also sent men to the moon in *The First Men on the Moon* (1901), using a device that reversed gravity. Two decades earlier, in *Across the Zodiac* (1880), English writer Percy Greg took a spacecraft to Mars, propelled by a force called "apergy". Wells also thought of these voyages inversely: in *The War of the Worlds* (1898), the Martians visited Earth – and liked our planet so much they decided to take it. Luckily, some of these early science fiction tales were wider off the mark than others.

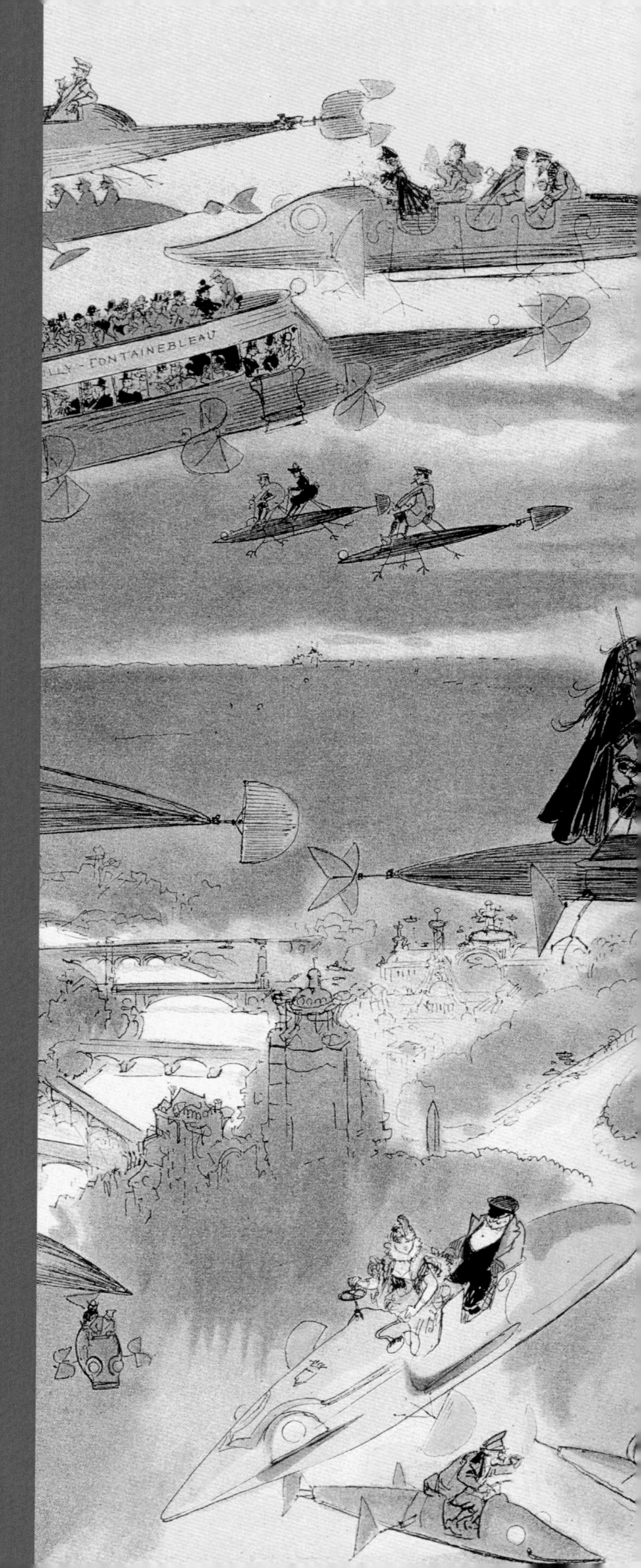

> " How much **further** can we **go?** What are the **final frontiers** in this **quest for travel?** "
>
> JULES VERNE, FRENCH AUTHOR

Imaginary landscapes
In this lithograph made in 1882, French artist Albert Robida imagines travel a hundred years on, and shows upper-class Parisians, dressed in fashionable French attire, leaving an opera in the year 2000. Robida creates a city skyline filled with futuristic flying vehicles, from private taxis and limousines to buses and police cars.

THE GOLDEN AGE OF TRAVEL
1880–1939

THE GOLDEN AGE OF TRAVEL, 1880–1939

Introduction

By the end of the 19th century, the only parts of the world still left to explore were those that were extremely remote and inhospitable, such as the frozen expanses of the polar caps and the burning heart of the Arabian Desert. But people did go to them, with journeys such as the race to the South Pole yielding stories as thrilling as any in the history of exploration. It was in 1912 that British explorer Robert F. Scott trudged to the South Pole, only to find that Norwegian Roald Amundsen had beaten him to it. On his journey back, Scott perished, just 18km (11 miles) short of a cache of supplies. Fourteen years later, Amundsen also reached the North Pole, albeit in an airship (the *Norge*), making him the first person to reach both poles.

Even on more familiar terrain, travel remained a considerable undertaking. Railways connected major towns and cities, but most journeys were made on foot, or at best in a horse-drawn coach. Dedicated travellers were venturing as far afield as Japan and New Zealand, but reaching the other side of the world often involved hardship and a willingness to rough it on occasion. Beyond Europe, some guidebooks still advised travellers to pack a gun for self-defence.

Bicycles, cars, and aeroplanes

For all its simplicity, the bicycle revolutionized people's lives when it first appeared. People living in cities could get out into the country easily, and those in rural areas could travel from one village to another. The motorcar liberated people even more, especially when Henry Ford started mass-producing the Ford Model T cheaply. The greatest freedom of all, however, came when two bicycle manufacturers from Ohio, Orville and Wilbur

THE BICYCLE WAS LIBERATING, GIVING PEOPLE THE FREEDOM TO TRAVEL UNDER THEIR OWN STEAM

WITHIN 10 YEARS OF ITS INVENTION, THE AEROPLANE HAD REVOLUTIONIZED TRAVEL

THE FORD MODEL T MADE IT POSSIBLE FOR MANY PEOPLE TO BUY A CAR FOR THE FIRST TIME

> " We find ourselves still **dreaming** of impossible **future conquest**. "
>
> CHARLES LINDBERGH

Wright, built and flew the world's first real aeroplane. They made their historic, inaugural flight in 1903, and within 20 years, airlines launched the first passenger flights.

The early aviators were fêted as heroes as they smashed one record after another. One of the first pilots was Alan Cobham, who flew from London to Australia and back in 1926. Amy Johnson made the flight from England to Australia in 1930. In 1927, Charles Lindbergh became the first aviator to fly solo across the Atlantic, followed by Amelia Earhart in 1930. These young aviators, male and female, gave flying a glamour that cast its spell on passenger flights too. For those who could afford it, flying became a heady world of refreshments served at 1,200m (4,000ft) in the sky, games of bridge in the lounge, and sleeping overnight in comfortable bunks.

Inequality

Luxury travel characterized the age, whether in the opulent carriages of the Orient Express or the ballrooms and tea salons of Cunard, White Star, and other transatlantic liners. Few people, however, travelled in such comfort and style. For every passenger in a cabin above decks on a liner bound for the US, six or more were hunkered down in the hold, enduring the most primitive conditions. Many of them, driven by poverty, had fled their homes forever, and were hoping for better lives in America.

Such radical inequalities in the nature of travel continued after World War I, which proved to be a mere hiatus in the traditional order of things. This was not true of World War II, however, which triggered massive changes in society, transport, and the way in which people travelled.

ROBERT F. SCOTT'S *TERRA NOVA* TEAM WAS ONE OF THE FIRST TO EXPLORE ANTARCTICA

IN 1926, ROALD AMUNDSEN FLEW OVER THE NORTH POLE IN THE AIRSHIP *NORGE*

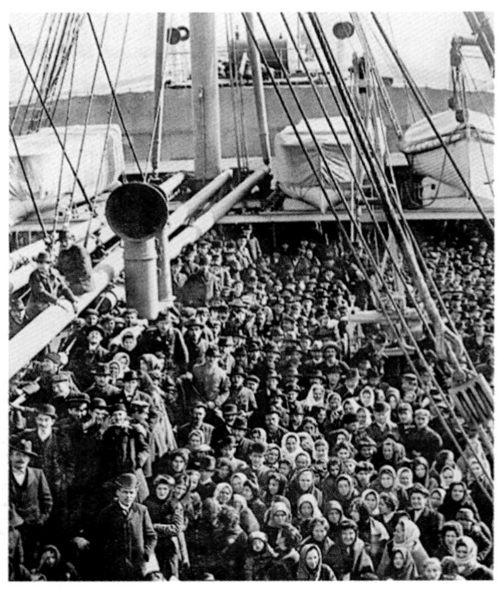

THOUSANDS OF EUROPEANS TRAVELLED TO AMERICA IN SEARCH OF A BETTER LIFE

Central Asia

Even in the late 19th century, there were still areas of the Eurasian landmass that were nothing but white spaces on European maps, challenging explorers to try and fill in the blanks.

△ **Ármin Vámbéry**
The Hungarian (1832–1913) was the first European to undertake the long and arduous journey to the interior of Asia.

The outskirts of an area – the points furthest east, west, north, and south – are usually the last to be explored. A territory identified as "central" would normally be charted first. This was never the case with Central Asia, which, for a long time, existed as a blank in the middle of European maps. During the era of the Mongols, this ill-defined no-man's-land had provided a conduit for trade. But with the decline of the Mongol Empire after 1259, and particularly once Europeans had discovered a sea route to the Far East, Central Asia became depopulated and splintered into small Islamic city-states, with the steppes, deserts, and mountains roamed by nomadic tribes.

From around the 17th century, both Russia and China began to expand into Central Asia (or Turkestan, as it was also known). By the 19th century, both countries had asserted some element of control, but most of the land remained untamed frontier territory.

△ **Bronze-age vessel**
This 3,500-year-old tripod-footed vessel was unearthed in central Iran by Aurel Stein, who donated it to the British Museum.

Heading into unknown territory

This was still the case in March 1863, when Hungarian Ármin Vámbéry made his way overland from Tehran into the heart of Central Asia, where few Westerners had set foot since the 1600s. To prepare for this expedition, he had spent many years in Constantinople learning over 20 Turkic dialects and studying the Qur'an. He disguised himself as a Muslim pilgrim, fearing that if he were to be recognized as a European, he would be imprisoned or even executed. He aroused suspicion on several occasions and was brought before the authorities, accused of being a spy, but each time he managed to talk his way free. Vámbéry spent a year in Central Asia, visiting Khiva, Bukhara, and Samarkand, before returning to Budapest, where he recounted his experiences in *Travels in Central Asia* (1865).

Sven Hedin

As Vámbéry published his work, the life of another Central Asian explorer was just beginning. The Swede Sven Hedin witnessed as a child the triumphant return of Adolf Erik Nordenskiöld after his navigation of the Northeast Passage, and this inspired him to become an explorer. He travelled in Russia, the Caucasus, and Iran while he was a student, before setting off on the first of three major expeditions to Central Asia in 1893. From 1894 to 1908, he explored and mapped parts of Chinese Turkestan (officially Xinjiang) and Tibet, much of which was completely unknown to Europeans. He nearly died in the Taklamakan Desert after gravely miscalculating his party's water supplies, but went on to discover the site of the lost city of Taklamakan, where he looted hundreds of artefacts.

Sven Hedin's writings about his explorations went on to inspire another Hungarian, Marc Aurel Stein

◁ **Tibetan tribesmen**
Photographs such as this hand-tinted slide by Hedin, of mounted tribesmen, are some of the earliest images of Tibet, but apply a colonial gaze and were often taken without consent.

△ **Shigatse Dzong**
In 1907, Hedin slipped unseen into the city, outraging the Tibetans by violating and disrespecting their rules. He painted this image based on earlier sketches of the city.

△ **Hedin's route**
The Swede's first visit to Central Asia took him from China, through Mongolia, into Russian Turkmenistan, and finally to the Taklamakan Desert. On his second visit, he travelled along the desert's northern edge to reach the Lop Nor Lake.

> "I was content with nothing less than **to tread paths** where **no European had ever set foot**."
>
> SVEN HEDIN

(1862–1943). In 1900, Stein set out on the first of his own expeditions to Chinese Turkestan, travelling to places Hedin had not reached.

Journeys of rediscovery

During the 30 years Stein spent exploring, however, his most significant achievements were actually rediscoveries. He uncovered evidence of a lost Buddhist civilization and long-forgotten stretches of the Chinese Great Wall, and also established how the routes of the ancient Silk Road (see pp.86–87) were connected. In this respect, Hedin's journeys set a template for much of the exploration that was to come in the following century, with a focus on retracing where people had gone before. These journeys would take the form of archaeological discoveries, such as those of Machu Picchu in 1911 and of Tutankhamun's tomb in 1922.

▷ **Disguised as a pilgrim**
Like Vámbéry, Hedin used deception to evade discovery, aware that Europeans were not welcomed. Hedin's disguise was as a Buddhist pilgrim to attempt entry into Lhasa, while Vámbéry's was that of a Muslim dervish.

On skis across Greenland

In 1888, Fridtjof Nansen became the first man to cross Greenland on skis. In doing so, he paved the way both for polar exploration and for the advancement of winter sports.

According to Roland Huntford, the author of *Two Planks and a Passion: The Dramatic History of Skiing*, it is the Norwegians who should be thanked for the staple of winter sports. It is they, he writes, a nation of native skiers, who invented modern skiing, developing both the equipment and techniques used today. The first known public ski races were held near Tromsø in northern Norway in 1843, and the Norwegians were the first to organize ski marathons and ski touring. Huntford also claims that skiing was first introduced to central Europe by young Norwegians studying in Germany – and it is widely known that Norwegian immigrants were the first to use skis in North America, possibly as early as the 1830s.

◁ **Early skis**
Skis have been used in Scandinavia since the Middle Ages. In the late 19th century, they were refined by the Norwegians, who introduced many refinements to create the modern ski.

Taking to the skis

The greatest milestone in Norwegian ski history, however, came in 1888. Five years earlier, a Finnish-born Swede named Adolf Erik Nordenskiöld had attempted to cross Greenland on foot, and failed. With him had been two Lapps on skis who were able to traverse the snowy terrain far more easily than Nordenskiöld, suggesting that this was a better mode of transport. Fridtjof Nansen, an accomplished Norwegian skier, took up the challenge. He had already twice skied across the mountains from Bergen to Christiana (Oslo), a distance of over 500 km (310 miles). The distance across Greenland was similar, but the only settlement was on the coast at the far side, and the terrain and temperatures were likely to be a good deal worse than in Norway. One critic in the press commented, "the chances are ten to one that he will uselessly throw his own and perhaps others' lives away". Nevertheless, on 10 August 1888, Nansen and his small team of six put ashore on the east coast of Greenland and began a 150-km (93-mile) trudge uphill to the summit of the ice cap. It was only there that they could finally don skis, and the going still proved hard, as they had to haul their sledges behind them. On downhill slopes, they rigged sails made from their tents to the sledges, so that they only had to steer them.

▽ **Ready for adventure**
Fridtjof Nansen poses with his Greenland expedition team: three fellow Norwegians and two Lapps from Finland, Samuel Balto and Ole Nielsen Ravna.

IN CONTEXT
The first ski resort

Isolated high in a Swiss valley, the settlement of Davos found itself booming in the 1860s as word spread of the recuperative qualities of its climate for those suffering from tuberculosis. Before long, visitors began to outnumber the inhabitants, and Davos became the first alpine winter spa. The first clients were German, but in the late 1870s, the English began to arrive and soon colonized the place. The invalids were supplanted by healthy holidaymakers, lured to the spot by accounts of novel new sports such as tobogganing, horse-drawn sledging, and skiing. When Arthur Conan Doyle, journalist and creator of Sherlock Holmes, visited in 1893, Davos was well on its way to becoming Europe's first winter-sports resort.

TRAVEL POSTER FOR DAVOS, 1918

▷ **Dogskin gloves**
In *The First Crossing of Greenland*, Nansen described his expedition clothing, including gloves made of dogskin, with the hair on the outside.

The value of skiing

On 21 September 1888, the skiing came to a stop when the group reached the fjords of the west coast. By early October, they had arrived at their destination, Godthaab (now Nuuk), the Danish settlement that served as the capital of Greenland. Having missed the last boat out, they had to spend the winter there, and didn't return to Norway until the following spring.

"Without skis," wrote Nansen, "this expedition would have been an absolute impossibility. We would never have returned alive." By this, he meant that the speed and economy of effort provided by the skis saved them from running out of supplies.

From then on, skis became vital in polar exploration. News of Nansen's exploits spread around the world and he became the spiritual father of skiing, popularizing the sport from the Alps to the Rocky Mountains of North America.

◁ **Hostile terrain**
Nansen proved that skis could make the difference between life and death in the Arctic.

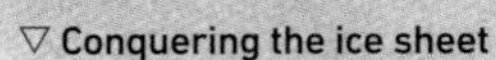

▽ **Conquering the ice sheet**
Nansen photographed his team struggling up the Greenland Ice Sheet, sledges in tow, in August 1888. They could not use their skis until they reached the summit.

" For the first time, the polar explorer had revealed to him a **new means of transport** [skis], which ought to **ease his task** fantastically... "

ADOLF ERIK NORDENSKIÖLD, ARCTIC EXPLORER

△ **Penny-farthing**
This kind of bicycle, developed from 1869 onwards, was also known as a "high-wheeler" or an "ordinary", to distinguish it from the safety models that followed.

The bicycle craze

For the vast majority of the population who did not own horse-drawn vehicles, the invention of the modern bicycle meant that they had their own personal means of transport for the first time.

The distinguished historian Eric Hobsbawm believed that the bicycle was one of the greatest inventions ever. It gave people true mobility in a way that was beneficial to all classes, not just the rich. He also thought it was one of the rare inventions that had no bad side effects and could not be turned to any malign use.

Although the bicycle had existed in one form or another since the early 19th century, it had to evolve through a series of technical improvements before it became popular. By the 1870s, it had reached the stage of the so-called "penny-farthings", bicycles that had an oversized front wheel and a tiny rear wheel, but these were heavy, cumbersome, and dangerous. From the mid-1880s, these awkward contraptions were gradually replaced with what was called the "safety bicycle", a model that had most of the hallmarks of what was to become the modern bicycle – two equal-sized wheels on a diamond-shaped frame, a chain-driven rear wheel, and inflatable tyres (invented by the Scot John Dunlop in 1886).

Increasing popularity

These improvements made cycling less hazardous and more enjoyable. As cheaper models became available, bicycle fever took hold. In Britain, it was estimated that by the 1890s, around one and a half million men and

▽ **Bicycle clubs**
The boom in cycling led to the formation of hundreds of clubs across the US. The first to form was the Boston Bicycle Club in 1878. This picture shows the Brighton Bicycle Club in Cincinnati, lined up at the start of an annual race.

> " In the near future **all women**...will **bestride the wheel**, except possibly the **narrow-minded**... "
>
> ANNIE "LONDONDERRY" COHEN, 1895

▷ **The new woman**
The practicalities of cycling led to new fashions for women. Restrictive layers of petticoats and trailing dresses were replaced by loose knickerbockers and protective headwear.

women were cycling on roads that had yet to be monopolized by the motor car. In the US, in 1897 alone, more than two million bicycles were sold.

Cycling was not confined to Britain and the US – it was a worldwide craze. An edition of *The Egyptian Gazette* in January 1894 reported on an excursion undertaken by about 30 cyclists, who rode from central Cairo out to the Pyramids one morning. "Amongst them was a lady riding astride and garbed in female knickerbockers," the correspondent noted. "She rode very well."

Social change

The impact the bicycle had on society was remarkable. In rural communities, it dramatically increased the distances people could travel and the places they could visit. It also increased the numbers of other people one could meet, including potential marriage partners. Steve Jones, a respected biologist, argued that the bicycle played an important role in combating genetic disorders, because people from different towns were marrying, which led to a greater and healthier genetic mix. Contemporary commentators also variously linked the bicycle to changing fashions, a drop in church attendance, and even a decline in piano playing. However, perhaps the most notable social change brought about by the bicycle was in the lives of women. Many people initially found the idea of a female sitting astride a machine "unbecoming", but that did not stop young women taking to the road in their thousands.

Clothes for cycling

As well as the mobility that the bicycle offered, it also led to more personal freedom of movement. The accepted clothing of the era – long, heavy skirts worn over stiff petticoats – was totally unsuitable for cycling, which required something less restrictive, such as the knickerbockers mentioned by the Egyptian newspaper. These were baggy trousers, sometimes called a divided skirt, that were cinched at the knee. Indeed, the women's movement of the 1890s and the cycling craze became so closely intertwined that in 1896, women's rights activist Susan B. Anthony told Nellie Bly of the *New York World* (see p.249) that cycling had "done more to emancipate women than anything else in the world".

THE ROVER SAFETY BICYCLE (PATENTED).
Safer than any Tricycle, faster and easier than any Bicycle ever made. Fitted with handles to turn for convenience in storing or shipping. Far and away the best hill-climber in the market.
MANUFACTURED BY
STARLEY & SUTTON,
METEOR WORKS, WEST ORCHARD, COVENTRY, ENGLAND.
Price Lists of "Meteor, "Rover," "Despatch," and "Sociable" Bicycles and Tricycles, and the "Coventry Chair," Illustrated, free on application.

▷ **Rover Safety Bicycle**
The safety bicycle was developed in the 1880s as a "safer" alternative to the unwieldy penny-farthing. The "Rover" created by John Kemp Starley was one of the most commercially successful designs.

△ **Open country**
Part of the appeal of the bicycle was that it put the countryside within easy reach of city dwellers, as illustrated in this poster for a "bicycle-camp" on the banks of the Connecticut River.

IN PROFILE
Annie Londonderry

Cycling the world seems a very modern challenge, but it was a feat completed for the first time in 1885, by Englishman Thomas Stevens, riding a penny-farthing. Two Boston gentlemen bet that no woman could do it. Annie Cohen Kopchovsky, mother of three, took them up on it. She had barely learned to cycle when she set out from Boston on 27 June 1894. To finance her attempt, she wore a placard advertising Londonderry Lithia Spring Water Company and even agreed to change her name to promote the business. On 23 March 1895, she arrived back in the US at San Francisco and cycled west, reaching Chicago in September, to collect her $10,000 winnings.

ROUND-THE-WORLD CYCLIST ANNIE "LONDONDERRY" COHEN

Escape to the open air

As industry wreathed cities with increasing amounts of smog, a movement arose that encouraged working men and women to head for the open air. It was the beginning of a craze for camping.

Of all the travel literature published towards the end of the 19th century, there is probably no work more popular than Jerome K. Jerome's *Three Men in a Boat* (1889). In late-Victorian England, there was a fashion for recreational boating, particularly on the Thames, and Jerome's book was originally intended as a river guide. In fact, it ended up as a comic novel. There were many reasons for its success, including that it was extremely funny, but most of all it was topical. It dealt with scenarios that were increasingly familiar to readers, such as boating, holidays in the countryside, and camping.

In Britain, in particular, the rapid growth of cities and the often harsh conditions in which

◁ **Camping pioneer**
Thomas Hiram Holding, author of *Cycle and Camp* (1898) and *The Camper's Handbook* (1908), is regarded as the founder of modern recreational camping.

△ **Liberating technology**
Numerous innovations freed people to go camping. After the bicycle, none was more important than the portable Primus stove, invented in Sweden in 1892.

▷ **Camping club**
Camping was already popular in Britain in the 19th century, as is clear from this picture from the 1890s. In 1901, the Association of Cycle Campers was set up. It soon had Robert Scott, the future Antarctic explorer, as its president.

people worked inspired many to seek the simplicities of rural life. It was always possible to get away by train, but the newly invented bicycle brought real countryside within easy reach of the average worker.

Camping fever

The man who popularized what became known as cycle camping was Thomas Hiram Holding, born in England in 1844 to Mormon parents. The family emigrated to Salt Lake City, Utah, in 1853, and Thomas's first experiences of camping came on the long trek across the US. Tragically, two of the Holding children died en route, and the family returned to England. Thomas became a tailor, but in his spare time he travelled and camped. He took a canoe and tent off to the Scottish Highlands, then a bicycle and tent to Ireland. In 1901, he formed the Association of Cycle Campers, which later became the Camping and Caravanning Club. He cemented his position as the father of camping with the publication of *The Camper's Handbook* in 1908. Camping fever spread across the classes and professions – and the sexes. The *Handbook* included a guest essay by a Mrs F. Horsfield titled "Ladies and Cycle Camping". Holding advised on how to assemble lightweight, easy-to-transport kits for weekends away, and his campers often made their own tents and gear. This did not always mean kit was modest. Writing of a visit to "an ideal family camp", Holding notes it consists of "the eating tent, the sleeping tent, the servants' tent, the cooking tent for wet weather and the overboat tent".

◁ **The Thames Rowing Club**
Camping in England evolved in part from boating for leisure, particularly on the River Thames. Boaters carried tents so they could spend the night on the river bank, as shown in this illustration of 1878.

The American wilderness

During the 19th century, American settlers were mostly too busy living off the land to sit back and enjoy its finer qualities. Nevertheless, Ralph Waldo Emerson's seminal essay "Nature" (1836) had a profound influence on how Americans perceived the wilderness. So did the writings of his friend Henry David Thoreau, even if, prior to publishing *Walden, or Life in the Woods* (1854), Thoreau accidentally destroyed 300 acres of prime timber with a campfire.

A guide to camping in the Adirondacks was published in 1869, but it was far ahead of its time. Those who ventured into the mountains at this early date found no infrastructure, but plenty of blackfly, mosquitoes, and dangerous animals. A turning point came in 1903, when President Theodore Roosevelt spent two weeks camping in Yellowstone National Park. In his wake, thousands of campers were drawn to the national parks, and their numbers rapidly increased as the century progressed. The first camping club, calling itself the "Tin Can Tourists", was formed in 1910, and by 1912, the Forest Service was reporting 231,000 campers in the parks.

> " We **pitch** our tents **far apart** so that **our hearts stay closer together**. "
>
> BEDOUIN PROVERB

▷ **Middle-class pursuits**
Pleasure-boating, cycling, and camping were pastimes reserved for those with money and leisure time. Manufacturers catered to these hobbyists with specialist products, such as this Edwardian picnic basket.

Far-flung railways

The last quarter of the 19th century was an adventurous period for railway building that saw extraordinarily ambitious lines and services created across Asia, Africa, Europe, and the US.

The Trans-Siberian Railway
Lamentable transport connections between European Russia and the tsar's Far Eastern territories provided the impetus to build a railway across Siberia. It was by far the most ambitious railway project ever attempted.

It was all very well for the tsar in St Petersburg to lay claim to all the lands from the Baltic Sea to the Pacific Ocean, but Russia's ability to govern and protect its Far Eastern territories was questionable. The distances and conditions involved made communication almost impossible. For half of the year, the rivers that served as transportation highways were frozen, and most of the rivers in eastern Siberia ran from north to south, rather than from east to west. By road, along the roughshod Sibirsky trakt, it could take up to nine months to reach Irkutsk from Moscow, and that was barely two-thirds of the way to Vladivostok, Russia's main Pacific port. The US had linked its two coasts by railroad in 1869 (see pp.236–37), and it was obvious that Russia needed to do the same.

△ **Hard labour**
Workers lay tracks for the central portion of the Trans-Siberian Railway in the Krasnoyarsk region. The labourers included Russians, Iranians, Turks, and even Italians.

The Trans-Siberian Railway

A railway across Siberia had been discussed as early as the 1850s, but the cost and scale of the undertaking were prohibitive. However, by the 1880s, fear of American influence in Russia's Pacific region prompted the tsarist government to go ahead with the scheme. Even then, thanks to stifling bureaucracy, five years passed before anyone stuck a spade in the ground.

The difficulties facing the railway's builders were immense. These included the sheer length of the route, below-freezing temperatures for a good part of the year, and a dearth of local people to use as labour. Also, in addition to two mountain ranges, the vast Lake Baikal in the middle of Siberia had to be negotiated. Looping around the north of the lake was too much of a detour, but the southern way was blocked by rocky terrain. Initially, the two tracks (from east and west) stopped at the lake and a ferry transferred the trains across the water. Then a series of shelves and tunnels were blasted through the southern route to carry the railway round.

It was not until 1916 that the final stretch was completed, forming a continuous 9,250-km (5,748-mile) track that crossed seven time zones between Moscow and Vladivostok. In the early days, the journey could take up to four weeks, due to problems with the track, but no-one doubted its value. It was a national symbol – an iron ribbon binding sovereign Russian territory.

An Imperial dream

Similarly, imperial ideals of stamping authority on a map and accessing the resources of a continent fuelled an even more ambitious plan in Africa. Cecil Rhodes, the Prime Minister of Cape Colony (in what is today South Africa), dreamed of linking the whole of Africa with a railway that travelled through British colonies alone. Its terminals would be Cape Town at the southern tip of Africa and Cairo in Egypt, from where a line already ran to the port city of Alexandria on the Mediterranean.

Initially, the track was laid at an astonishing rate of 1.6 km (1 mile) a day, heading north. In 1904, it reached the Zambezi River, where the railway crossed the Victoria Falls via a 200-m (656-ft) suspension bridge. The bridge had been built in northern England, taken to pieces, shipped to Africa, and reconstructed in situ. »

▷ **The Lake Baikal train ferry**
Until the 180-km (110-mile) Circum-Baikal stretch of railway was finished, passengers (and train) had to take an ice-breaking ferry across Lake Baikal.

△ **Bridging Africa**
This suspension bridge over the Zambezi River at Victoria Falls was the brainchild of Cecil Rhodes, as part of his Cape-to-Cairo railway scheme. He instructed engineers to build the bridge "where the trains, as they pass, will catch the spray of the Falls".

» However, in 1891, Germany secured large tracts of East Africa for its empire, and thereby blocked the path for the railway. Forced to rethink their route, the builders turned west into the Belgian Congo (where the British had permission to build), but here the project foundered, owing to difficulties negotiating the mountainous terrain. By this time, Rhodes had died, and Britain's influence in Africa was waning, so the idea of a trans-African railway was abandoned. However, 100 years on, the concept is not entirely dead and has been discussed by African leaders in recent times.

The Orient Express

The great railway-building projects in Russia and Africa were spurred on by politics and commerce. In Europe, however, the most significant development was all about service. By the late 19th century, every European nation was crisscrossed by its own national network of railways. What Belgian engineer Georges Nagelmackers wanted was to link them with a single service that ran *sans frontières*. In 1872, he launched a service that ran from Ostend on the North Sea coast of his native Belgium to Brindisi in southern Italy. It was a success, so he began work on a new line to connect Paris with Constantinople (Istanbul), a 2,989-km (1,857-mile) route that would cross Europe – Germany, Austria, Hungary, Serbia, Romania, and Bulgaria.

To make this happen, Nagelmackers had to deal with the railway companies of six nations, negotiating which locomotives and lines would be used, and making sure that all the tracks were a standard gauge.

The inaugural service of the *Orient Express* left the Gare de l'Est in Paris on 4 October 1883, and was scheduled to take three-and-a-half days to reach Constantinople. The route was undoubtedly unique, but what really captivated the public was the sheer opulence of the train. The sleeping cars were panelled with teak inlaid with marquetry, and had seats that could be opened out into beds with silk sheets. There was a library and a smoking room, and a coach at the rear that had showers with hot water. On this inaugural trip, the last leg of the journey was made by boat because the line was unfinished. Six years later, however, the train went all the way to Constantinople.

The Orient Express became popular because it was quicker and far more convenient than travelling by boat. Alternative routes were soon added, including one via Milan and Venice. At Constantinople, travellers could make connections for destinations further east, including Tehran, Baghdad, and Damascus. And if the patchwork of countries the train passed through was unstable, the occasional hold-up by bandits, kidnapping, or a day-long halt in snow only seemed to add to the trip's romance.

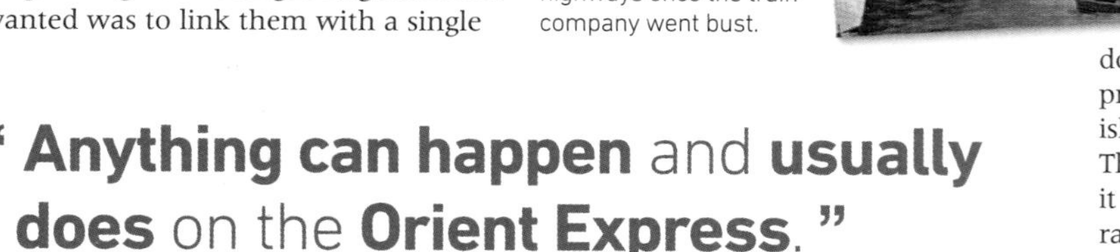

▷ **East Coast Railroad**
A series of viaducts carried the Florida East Coast Railroad across the lagoons to Key West. This souvenir folder shows them converted into highways once the train company went bust.

> " **Anything can happen** and **usually does** on the **Orient Express**. "
>
> MORLEY SAFER, CANADIAN-AMERICAN JOURNALIST

The Florida coast

The United States may have connected the Atlantic and Pacific by rail back in 1869, but at the end of the century there were still vast swathes of the country to be opened up. Oilman Henry Flagler saw potential in Florida and set about unleashing it with a self-financed railway line from New York. He took it all the way down to Biscayne Bay, a beautiful spot protected from the Atlantic by a barrier island and watered by the Miami River. There, he founded a settlement and gave it the river's name. He then extended his railway further south, creating one of the most astonishing lines, which was carried on 27 km (17 miles) of bridges to

reach the far-flung island of Key West. Unfortunately, the railroad went bankrupt in 1932, but Flagler's bridges still support the southernmost section of Highway 1, which remains one of the most breathtaking drives in the world.

◁ **Luxury on wheels**
The Orient Express was the world's most glamorous train service. It operated on several routes, each of which offered lavishly appointed facilities. This wood engraving of 1885 shows passengers in the dining car.

▽ **Legendary train**
The Orient Express was immortalized by many writers, most notably Agatha Christie. It was also the subject of an opera by Oscar Sachs and Henri Neuzillet, which was performed in Paris in 1896, as advertised below.

IN CONTEXT
Great railway literature

Some of the greatest modern travel books are set on trains. *The Great Railway Bazaar* (1975) chronicles Paul Theroux's journey across Europe and Asia, while *The Old Patagonian Express* (1979) sees him on train after train crossing the US and South America.

Eric Newby spent weeks on Russian rails for his *The Big Red Train Ride* (1978), and Jenny Diski months in Amtrak smoking cars exploring the US for *Stranger on a Train* (2002). Meanwhile, Andrew Eames travelled in the tracks of Agatha Christie for his book *The 8.55 to Baghdad* (2004).

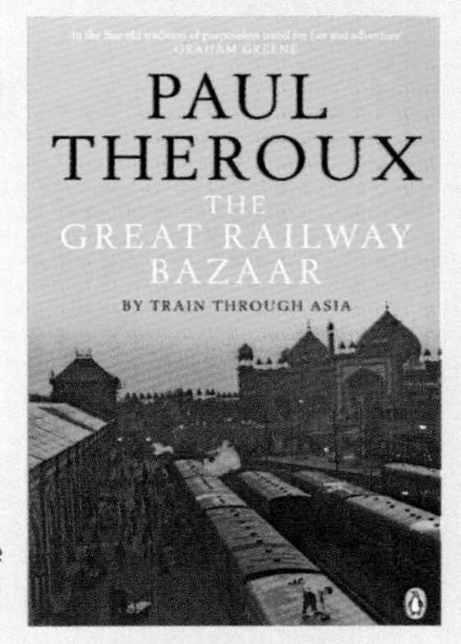

PAPERBACK EDITION OF *THE GREAT RAILWAY BAZAAR*

△ **Lady Liberty**
The Statue of Liberty (officially titled *Liberty Enlightening the World*) was dedicated on 28 October 1886. Placed at the entrance to New York Harbor, she welcomed travellers arriving from abroad.

The American Dream

For over 200 years, the United States has embraced immigrants from all over the world – some fleeing persecution and others seeking riches, but all looking for a better life, regardless of social class.

In 1849, American author Herman Melville sailed out of Liverpool on a ship carrying 500 emigrants bound for New York. The journey he described was horrific. Bad weather made many people ill. Those in cabins could at least suffer in private, but for the poorer passengers – by far the majority – there was no such consolation. Travelling in the hold, or "steerage", they were "stowed away like bales of cotton, and packed like slaves in a slave-ship". Sealed in the darkness, they had nowhere to be sick but their beds. He writes: "We had not been at sea one week, when to hold your head down the fore hatchway was like holding it down a suddenly opened cesspool".

Such vessels, of course, were not designed for passengers. They were cargo ships, bringing cotton, tobacco, and timber to England from the colonies. To maximize profits, their owners filled the holds with emigrants, who served as freight for the journey home. Enduring such voyages for weeks on end, particularly in bad weather, made sickness unavoidable, the worst arising from poor sanitation and meagre food. As a consequence, many died en route. Even so, from 1846 to 1855, emigration to North America boomed. In those 10 years, it is estimated that more than 2,300,000 people crossed to the US and Canada, compared to 1,600,000 over the previous 70 years. The attraction was that anyone could go. All they needed was the steerage fee to board a commercial ship.

Land of opportunity

Melville's mention of slave ships was a reminder that until 1808, when Congress banned slavery, the greatest number of people living in a new place because they were forcibly brought across the Atlantic in chains were Africans. Now, the greatest number were Irish or German, with smaller numbers of English, Scots, and Italians. The main points of arrival were New York and the eastern seaboard ports. There was also San Francisco, on the west coast, which had a huge influx of foreigners, particularly Chinese, when, in 1850, news of the Gold Rush spread around the world.

The Irish came to escape a devastating potato famine at home. From 1845 to 1852, a blight on the country's staple crop claimed the lives of around 1,000,000 people. The same number emigrated, all risking the brutal Atlantic voyage and the uncertainty of what lay ahead. For them, the US and Canada were far more welcoming than Britain. North America needed farmers, carpenters, masons, haulers, and unskilled labourers – the trades of the lower classes – to help construct its countries. It positively welcomed the poor, as long as they were willing to work. »

△ **Poor of all nations**
In the early 19th century, the largest numbers of immigrants came from Ireland and Germany. There were numerous Italians too, such as this family, who were fleeing poverty at home.

◁ **Inspection card**
Immigrants on steamships were subjected to a daily health inspection by the ship's surgeon, who recorded the check-ups by stamping each passenger's health card.

◁ **Huddled masses**
Immigrants to America gather on the decks of the SS *Patricia* in December 1906. The ship carried 408 second-class passengers and 2,143 in steerage.

> "The only **encouragements** we hold out to **strangers**... are **good laws**, **a free government**, and **a hearty welcome**."
>
> BENJAMIN FRANKLIN, ONE OF THE FOUNDING FATHERS OF THE UNITED STATES

△ **Arrival at Ellis Island**
In 1892, Ellis Island opened as New York's federal immigration station, a purpose it served for more than 60 years. Millions of new arrivals passed through the station during that time.

» The Germans came to escape economic hardship, and the political unrest caused by riots, rebellion, and eventually a revolution in 1848. Unlike the Irish, who settled in east coast cities such as New York, Boston, Philadelphia, and Pittsburgh, many Germans had enough money to travel to the Midwest in search of farmland and work. After New York, some of the largest settlements of Germans were in Cincinnati, St Louis, and Milwaukee.

◁ **American liberty**
In this World War I poster, immigrants are reminded that America has given them liberty; now it was their duty to buy bonds to help preserve that liberty.

▷ **Immigrant city**
From 1870 to 1915, New York's population more than tripled. In 1900, when this picture of Mulberry Street in Little Italy was taken, foreign-born immigrants and their children made up 76 per cent of the city's population.

Shaping the New World

Not every immigrant dream came true, but there were enough rags-to-riches stories to keep the influx of labour flowing. One such tale was Paddy O'Dougherty's, as related in the *New York Illustrated News* in 1853. O'Dougherty, the article said, entered the US a pauper but was taken on by the railways. After working hard for nine months, and spending nothing on drink or tobacco, he had enough money to bring over his wife and children and buy a house for the family. "Few more gentlemanly in appearance can this day be seen in Water Street than this same Paddy O'Dougherty," the paper concluded. It was a good advertisement for life in their new home.

As steamships replaced sailing ships, the transatlantic journey became faster and cheaper. Now the hopeful streamed in not just from northern Europe but from all over the world, including eastern Europe, the Mediterranean, and the Middle East. In the 1880s alone, nine per cent of the total population of Norway emigrated to America. The Chinese were not so lucky. They were banned from entering the US by the Exclusion Act of 1882, which remained in force until 1943.

From 1880 to 1930, over 27 million people entered the US. About 12 million of that number passed through the federal immigration station on Ellis Island in New York Harbor after it opened in 1892. Among the masses, there were Jews fleeing pogroms in Russia, Mexicans displaced by revolution, and Armenians fleeing genocide. However, after the outbreak of World War I, American attitudes towards immigration began to change. Nationalism was on the rise, and quotas were introduced to slow the flow of

> " **New York**, indeed, resembles a **magic cauldron**. Those who are **cast into it** are **born again**. "
>
> CHARLES WHIBLEY, IN *AMERICAN SKETCHES*, 1908

Claiming the South Pole

As scientific expeditions from numerous countries probed one of the last remaining unexplored regions on Earth, two headstrong individuals set out on a gruelling, frozen dash for glory.

Antarctica was the last major landmass on Earth to be explored and charted. Following the first expedition there, launched by the Belgian Geographical Society in 1897, some 17 major expeditions were sent to the region by 10 countries in a mere 20 years. It is a period known as the Heroic Age of Antarctic Exploration.

The ultimate prize, of course, was to reach the South Pole. Two expeditions set off in 1910 to achieve this goal: one from Norway, led by Roald Amundsen (see pp.280–81), and the other from Britain, led by Robert Falcon Scott. Scott was a former Royal Navy officer who had left the service to lead the Royal Geographical Society's National Antarctic Expedition of 1901–04 aboard the research ship *Discovery*. This expedition penetrated further south than anyone had ever ventured before. On his return, Scott immediately planned a second venture, this time to the South Pole.

Winners and losers

In June 1910, the *Terra Nova*, Scott's converted whaling ship, left Cardiff, Wales, with a crew of 65, including six veterans of the *Discovery*. For transport on the ice, he took ponies, which he preferred to dogs, and some new and largely untested motorized sledges. The team spent the first winter at the most southerly point they could sail to, where the Ross Sea met the Ross Ice Shelf. They waited out the harshest weather in a hut at Cape Evans (named after one of the group), and set up supply caches on the route to the pole.

By 24 October 1911, the start of the Antarctic summer, everything was ready. It quickly became apparent that the motorized sledges and ponies could not cope with the harsh conditions, so the expedition proceeded without them. On 21 December, they reached the high expanse of the Polar Plateau, where Scott selected four men to accompany him on the final leg of the journey: Henry Bowers, Edgar Evans, Lawrence Oates, and Edward Wilson.

◁ **Scott at Cape Evans**
For three years the *Terra Nova* expedition lived in a large, prefabricated hut at Cape Evans. Here, Scott sits in his private room, pipe in hand, writing his journal.

Slowed down by bad weather and Scott's insistence that the men haul their sledges, they reached the South Pole on 17 January 1912. Awaiting them was a Norwegian flag and marker tent – proof that Amundsen had arrived before them. Their dreams of glory shattered, Scott's exhausted party now faced an 1,300-km (808-mile) journey back to Cape Evans.

Final words

On 17 February, a month into their return trip, Evans died after a fall. On 17 March, Oates, realizing that his frostbitten and gangrenous feet were handicapping the rest of the party, uttered the most gallant words in the history of polar exploration: "I am just going outside and may be some time". He stumbled out of the tent into a blizzard and was never seen again.

On 21 March, with only two days' rations left, Scott, Wilson, and Bowers, faint with hunger and ravaged by scurvy, pitched their tent in a gale. They had walked 1,190 km (740 miles) from the Pole. They were still 225 km (140 miles) short of Cape Evans, but One Ton Depot, where ample food and fuel was cached, was only 18 km (11 miles) away.

"We are weak..."

Eight months later, a search party from Cape Evans found the small green tent. Three frozen corpses lay inside, tucked into their reindeer-hide sleeping bags. Scott's journal was next to his body. "We are weak," he had written, "but for my own sake I do not regret this journey, which has shown that Englishmen can endure hardships...and meet death with as great a fortitude as ever in the past." A sheaf of letters was also found, including one to his wife, which began, "Dear widow".

◁ **Second place**
Battered and frostbitten, Wilson, Scott, Evans, Oates, and Bowers pose for a portrait at the South Pole in January 1912. Behind them stands the marker tent left by Roald Amundsen.

POLAR EXPLORER, 1872–1928

Roald Amundsen

Norwegian explorer Roald Amundsen was the first man to reach the South Pole, the first to visit both poles, and the first to navigate the Northwest Passage between the Atlantic and Pacific oceans.

As a child, growing up a son of shipowners, Amundsen dreamed of being a polar explorer, his imagination fired by the exploits of fellow Norwegian Fridtjof Nansen. His mother did not want him to go to sea and encouraged him to become a doctor. Amundsen dutifully studied medicine, but when his mother died when he was 21, he left university and joined a sealing expedition to the Arctic.

In 1897, he sailed as first mate on an expedition to Antarctica aboard the ship RV *Belgica*. On his return in 1899, he set his sights on the Northwest Passage. Since the 16th century, seafarers had been searching for an Arctic route between the Atlantic and Pacific oceans (see pp.138–39). Among those who had tried and failed to find it were Henry Hudson, James Cook, and John Franklin, whose disappearance in the icy seas of the far north fascinated Amundsen.

▽ The South Pole
Amundsen stands before the Norwegian flag he planted at the South Pole on 14 December 1911. With him are some of the huskies that gave his expedition the edge over his rival, Robert Falcon Scott.

The Northwest Passage

In June 1903, Amundsen and his crew of six set off from Christiania (Oslo) aboard a small fishing vessel called *Gjøa*. After crossing the Atlantic, they set up base on King William Island to the north of the Canadian mainland, where they spent two winters carrying out scientific studies and spending time with the local Inuits. Here they learned Arctic survival skills, including how to build snow huts, drive sledge dogs, and make clothes from animal skins.

In August 1905, the *Gjøa* weighed anchor and sailed on. Amundsen took a westward course through uncharted channels until the boat became ice-bound, forcing them to overwinter just short of Alaska. They finally reached the Alaskan town of Nome in 1906, becoming the first crew to navigate the Northwest Passage. However, the journey had taken over three years, and parts of the route had only just been deep enough to give Amundsen's small fishing boat clearance. The Northwest Passage had no commercial viability.

◁ Pocket watch
This pocket watch, made by US watchmaker Elgin National Watch Company, was used by Amundsen. It is now held in St Petersburg's Arctic and Antarctic Museum.

The South Pole

After his return to Norway, Amundsen began to prepare for an attempt on the North Pole, using Fridtjof Nansen's old polar vessel, *Fram*. Then he received news that Frederick Cook and Robert Peary had beaten him to it, so he switched his attention to the South Pole. Robert Scott had already announced his expedition, so Amundsen telegraphed him with the courtesy message: "BEG TO INFORM YOU FRAM PROCEEDING ANTARCTIC".

The *Fram* arrived in the Antarctic in January 1911, and the party dug in and overwintered until October, when Amundsen judged the time was right to set off. His party consisted of five men and four sledges, with 13 dogs for each sledge. The skills learned from the Inuits proved invaluable, and Amundsen swiftly reached the pole on 14 December 1911 – more than a month before his English competitor.

▷ Called to the wild
This photograph from 1923 shows Roald Amundsen dressed in a fur parka. Unlike others who had attempted expeditions to the poles, Amundsen and his Norwegian crew were used to the freezing climate. They were also more skilled at skiing.

“ With **sufficient planning**, you can almost **eliminate adventure** from an **expedition**. ”

ROALD AMUNDSEN, POLAR EXPLORER

KEY DATES

- **1872** Born into a ship-owning family in Borge, in southern Norway.
- **1897** Joins the crew aboard the RV *Belgica* led by Adrien de Gerlache, which becomes the first expedition to winter in Antarctica.
- **1903–06** Becomes the first to navigate between the Atlantic and Pacific oceans via the Arctic.
- **1910–12** Sails aboard the *Fram* to the Antarctic, where he beats Robert F. Scott in a race to the South Pole.
- **1926** Passes over the North Pole aboard the airship *Norge*, piloted by Italian Umberto Nobile. As previous claims are disputed, this makes them the first verified explorers to reach the Pole. Amundsen becomes the first to visit both poles.
- **1928** Disappears with five crew on 28 June while flying on a rescue mission to find Umberto Nobile. His body has never been found.

A YOUNG AMUNDSEN POSES FOR A STUDIO PHOTOGRAPHER

THE AIRSHIP *NORGE*, WHICH CARRIED AMUNDSEN OVER THE NORTH POLE

> " **Some day** people will be **crossing oceans** on **airliners** like they do on **steamships** today. "

THOMAS BENOIST, PIONEER AEROPLANE BUILDER

△ **First flight**
Wilbur Wright (pictured here on an early test flight of a glider, in October 1902) and his brother Orville made the first ever controlled, sustained flights at Kill Devil Hills, near Kitty Hawk, North Carolina, in 1903.

France was the inventor Louis Blériot, who, the very next year, became the first person to fly across the English Channel, and did so in an aircraft that he had designed himself.

The first airlines

Incredibly, a mere six years after the Wright brothers' international demonstration, the first aircraft capable of carrying several passengers made its debut. The four-engined *Ilya Muromets*, designed by Russian engineer Igor Sikorsky, was the first aircraft to have heating and electric lighting – not to mention a passenger saloon complete with a bedroom, lounge, and toilet.

▷ **Rapid progress**
Fewer than 20 years after the Wright brothers' first flight, there were aircraft that could carry more than 150 passengers. These people are on board a giant 12-engined DO-X flying boat crossing Lake Geneva.

It took off on a demonstration flight on 25 February 1914, with 16 people aboard. In June, it flew from St Petersburg to Kiev in 14 hours and 38 minutes, and returned in less time than that. However, due to the onset of World War I, the plane was never used as a commercial airliner.

The world's first scheduled airline had commenced its operation just one month earlier. The St Petersburg to Tampa Airboat Line began operating across Tampa Bay, Florida, in January 1914. The service only lasted for a few months, but during that time, its two flying boats had flown more than 1,200 passengers across the bay.

Airlines come of age

Technological innovation sped up during World War I, so by 1918, there were many new and improved aircraft that could be adapted for civilian service. On 25 August 1919, the Aircraft Transport and Travel company (AT&T) launched the first regular international service in the world – a daily flight between London's Hounslow Heath Aerodrome and Le Bourget in Paris.

AT&T only flew until December 1920 before it went bankrupt, but within a year, no fewer than six companies were operating a London-to-Paris service. On 17 May 1920, another new airline had its inaugural flight from London to Amsterdam. The Dutch company behind it was Koninklijke Luchtvaart Maatschappij, better known as KLM.

These airlines transformed travel just as radically as the steam engine had 100 years earlier. Journeys that had once taken weeks, or even months, now took a matter of hours. The idea of spending months at sea to travel from one place to another became unthinkable, not just to the rich, but to businessmen and many other people too. By the 1950s, flying had become the principal way to travel long distances.

◁ **KLM**
KLM, the Dutch airline that made its inaugural flight in 1920, has been in service for longer than any other airline. This poster from 1929 states: "The Flying Dutchman – A Legend Becomes Reality".

Adventurers of the skies

Early aviation history is a roll call of leather-hatted and goggled adventurers who risked life and limb – all too often with fatal results – to push back the boundaries of air travel.

In modern times, flight has become merely the fastest way of getting from one place to another. This was not always so. At the beginning of the 20th century, taking to the air was an incredible, almost mystical experience, and airmen were the heroes of the day. The most exciting time was the period between the wars, when engineers were revising aircraft designs almost weekly, and pilots were shattering speed and distance records.

For every success there were many failures, and often deaths, but the pioneers believed that flying was a calling that justified the risks.

Explaining his decision to become an airmail pilot, Frenchman Antoine de Saint-Exupéry (who vanished over the Mediterranean in 1944) could think of no better comparison than entering a monastery. He wrote about his experience in books such as *Wind, Sand and Stars* and *Flight to Arras*.

Death in the air came in unexpected ways. Alan Cobham was a pioneer of long-distance aviation, and in March 1926 he completed a flight from London to Cape Town and back. Three months later he set off again from London, this time bound for Australia. While flying over Iraq, a sandstorm forced Cobham to fly low and the aircraft was shot at by tribesmen. His co-pilot was hit, and although Cobham landed at Basra and got him to a hospital, he died later that night.

Long-distance hero
This photograph shows Alan Cobham landing on the River Thames, in London, on 1 October 1926. He had just made a 42,974-km (26,703-mile) flight from England to Australia and back.

△ **The *Spirit of St Louis***
The aircraft in which Lindbergh flew across the Atlantic was a single-engine, single-seat monoplane, built by Ryan Airlines of San Diego. Today, it is on display at the Smithsonian Institution's National Air and Space Museum.

" Sometimes, **flying feels too godlike** to be attained by man. "

CHARLES LINDBERGH, AVIATOR

Across the Atlantic

In the annals of aviation history, nearly all flights are overshadowed by the one made by Charles Lindbergh on 20–21 May 1927. The 25-year-old military and airmail pilot achieved worldwide fame when he flew solo non-stop from Long Island, New York, to Paris, travelling the 5,800 km (3,600 miles) in 33½ hours. However, he was not the first to cross the Atlantic by air. In May 1919, Lieutenant Commander Read of the US Navy flew from Newfoundland to Portugal by way of the Azores. Two weeks later, Captain John Alcock and Lieutenant Arthur Brown flew non-stop from Newfoundland to Ireland. In fact, by 1927, over 100 other men had flown across the Atlantic. Lindbergh's achievement was to do it solo, and to fly from New York to Paris, thus linking two key cities.

The following year, Australian Charles Kingsford Smith and his three-man crew were the first to fly across the Pacific, from Oakland to Brisbane. The headlines that year, however, were made by Amelia Earhart, who became the first woman to cross the Atlantic by air, albeit as a passenger. "I was just baggage, like a sack of potatoes," she said in an interview after landing – adding, "Maybe someday I'll try it alone." That day came on 20 May 1932, when, five years to the day from Lindbergh's flight, the 34-year-old set off from Newfoundland. Four minutes shy of 15 hours later, she touched down in a field in Northern Ireland, becoming the first woman to fly the Atlantic solo.

Round the world

Ironically, perhaps the greatest aviation achievement of all is the least known. Wiley Post, a native of Oklahoma, worked in the oilfields, spent a year in prison for armed robbery, and lost an eye in an accident before he became the personal pilot of a wealthy oilman. On 23 June, flying the oilman's aircraft (a Lockheed Vega named *Winnie Mae*), Post and his navigator, Harold Gatty, set off from New York, heading west. They arrived back in New York on 1 July, after circling the world in 8 days, 15 hours and 51 minutes.

△ **Amelia Earhart**
In 1932, Earhart became the first woman to fly solo across the Atlantic. She has been an inspiration to aviators ever since.

This smashed the previous round-the-world record of 21 days, which had been set by a Zeppelin.

In July 1933, Post took the *Winnie Mae* around the world again, this time solo, becoming the first person to do so, and in just 7 days, 19 hours. Two years later, his extraordinary career came to an end when the engine of his seaplane failed and he plunged into a lake in Alaska.

MORE PICTURES, MORE NEWS AND MORE SPORT

Daily Mirror 28 PAGES

THE DAILY PICTURE NEWSPAPER WITH THE LARGEST NET SALE

MONDAY, MAY 30, 1927 — One Penny

CROYDON BATTLE TO WELCOME ATLANTIC HERO

▷ **Hero's welcome**
This front page of Britain's *Daily Mirror* records Charles Lindbergh's arrival at Croydon Airport, London, in the *Spirit of St Louis*, following his historic flight across the Atlantic.

Travels in Arabia

Over the centuries, a succession of lone maverick travellers from the West have been drawn to Arabia. They often travelled disguised in Arab dress, and adopted local ways.

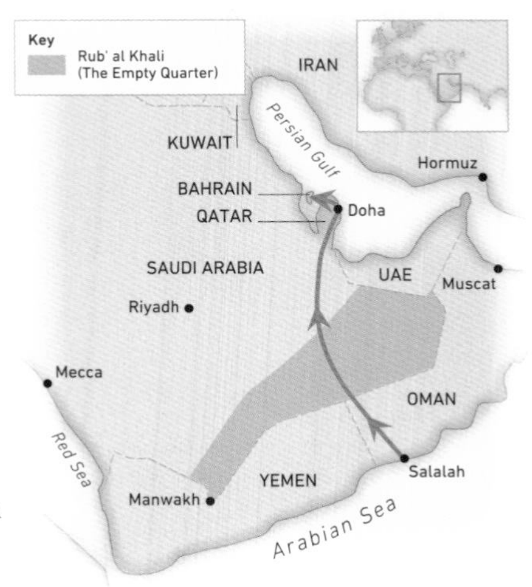

△ **Expeditions of Bertram Thomas**
In 1930, Bertram Thomas set off from Salalah, Dhofar, along with Sheik Saleh bin Kalut. They arrived in Doha – the other side of the Empty Quarter – 60 days later. They then boarded a dhow and finished their journey in Bahrain.

As the heartland of Islam, from the 7th century onwards, Arabia was familiar to Muslim travellers from all over the world as a place of pilgrimage. However, for non-Muslims, Mecca and Medina were part of a list of forbidden cities – including Lhasa and Timbuktu – that for centuries captured the imagination of the West. Although strictly forbidden to non-Muslims, intruders were not uncommon.

By the early years of the 20th century, some 25 Westerners had visited Mecca and written about it.

The first known traveller was an Italian named Ludovico di Varthema. In 1503, he left Damascus, where he had spent two years studying Arabic, in a pilgrim caravan bound for Mecca. He was the first Westerner to describe the rituals of the hajj.

▽ **When in Rome**
Bertram Thomas is pictured here, in 1930, with warriors from the Shahari tribe in Oman. He gained the trust of local people by observing their customs.

across Persia, dodging bandits and police, in her breakthrough book *The Valley of the Assassins* (1934). Some two dozen more works followed, which did much to inspire women to venture into exotic climes. "To awaken quite alone in a strange town is one of the pleasantest sensations in the world," she enthuses in *Baghdad Sketches* (1932).

Although men still outnumbered women when it came to travel writing, Rebecca West's *Black Lamb and Grey Falcon* (1941), a complex portrait of Yugoslavia and Europe on the brink of war, was another of the most significant books of the period.

Most travel writers at this time looked to venture where the Baedeker guides did not, but others combined travel writing with reportage to present new angles on familiar places. The young teacher and essayist Eric Blair wrote about Paris from the vantage point of a kitchen labourer in his book *Down and Out in Paris and London* (1933), which was published under the pseudonym George Orwell. The French poet and pilot Antoine de Saint-Exupéry vividly described the Sahara and South American Andes as seen from his aeroplane as he flew delivering the mail in the classic flying adventure *Wind, Sand and Stars* (1940).

◁ **Antoine de Saint-Exupéry**
Author of the classic book *The Little Prince* (1943), daring aviator Antoine de Saint-Exupéry (left) attempted to break the speed record for flying from Paris to Saigon in 1935.

The end of an era?

Few writers were able to resist the ease with which people could now travel. The mountaineer H.W. Tilman wrote sarcastically in the preface to his book *China to Chitral* (1951), "Comparatively few travellers have visited Chinese Turkestan; which is perhaps just as well because of those fortunate few, not many have refrained from writing a book." However, in 1946, Evelyn Waugh suggested that travel writing had had its day, stating that he did not "expect to see many travel books in the near future". He called his new collection of travel writing *When the Going Was Good*: to him, the joy of journeys was over.

EXPLORER AND NATURALIST, 1884–1960

Roy Chapman Andrews

Considered by many to be the original Indiana Jones, Roy Chapman Andrews was an explorer, educator, and naturalist whose adventures took him to the Gobi Desert, where he became the first person to discover dinosaur eggs.

Roy Chapman Andrews was born in the small town of Beloit, Wisconsin, in 1884. As he wrote in his autobiography: "I was born to be an explorer. There was never any decision to make. I couldn't be anything else and be happy."

After graduating from college in 1906, Andrews went to New York to pursue another thing that he had always wanted: a job at the American Museum of Natural History. When told that there were no jobs available, he volunteered to scrub the floors. He was taken on as a cleaner in the taxidermy department, then as a taxidermy assistant, but he soon moved on to the type of adventurous fieldwork he had hoped for. The museum assigned him to measure and study different whale species, and to collect their skeletons. To do this, he had to travel all around the Pacific, including to Alaska, China, Japan, and Korea. From 1909 to 1910, he sailed around the East Indies on the USS *Albatross*, carefully observing marine mammals and collecting snakes and lizards. In 1913, he went to the Arctic, where he filmed seals.

It was apparently always in Andrews' nature to tackle challenges head-on. "In [my first] fifteen years [of field work] I can remember just 10 times when I had really narrow escapes from death," he wrote. "Two were from drowning in typhoons, one was when our boat was charged by a wounded whale, once my wife and I were nearly eaten by wild dogs, once we were in great danger from fanatical Lama priests, two were close calls when I fell over cliffs, once I was nearly caught by a huge python, and twice I might have been killed by bandits."

◁ Action man
This water bottle is similar to the one Chapman took with him on his adventures. Other items included a ranger's hat and a revolver.

The Gobi Desert

In 1922, Andrews embarked on the first of the expeditions for which he is best known. These were to Mongolia's Gobi Desert, where Henry Fairfield Osborn, the director of the American Museum of Natural History, hoped he might find evidence that supported his theory that humans originated in Central Asia rather than Africa. No such evidence was found, but on the first trip, the team uncovered several complete fossil skeletons of small dinosaurs, as well as parts of larger dinosaurs. A second expedition in 1923 led to even more exciting finds, including the skull of a small mammal that had lived alongside the dinosaurs – few mammal skulls had been found from that period. Even more exceptional, however, was a nest of fossil dinosaur eggs, the first to be scientifically recognized. Until then, no-one had been sure how dinosaurs reproduced. To everyone's great excitement, 25 eggs were found altogether. Andrews even appeared on the cover of *Time* magazine.

Andrews' stories were just as thrilling as his discoveries. He told how, on one extremely cold night in the desert, a huge number of poisonous pit vipers slithered into the Americans' camp in search of warmth. The men killed 47 snakes within just a few hours.

In 1930, the political situation in Mongolia and China forced the museum to suspend expeditions there. This meant that one phase of Andrews' career was over, but another was about to begin. In 1934, he became director of the American Museum of Natural History, a post that he held until 1942, when he retired.

◁ Inspecting fossils
Andrews (right) examines mammal fossils in Mongolia in 1928. The expedition had initially set out to look for traces of "pre-dawn man" in Asia. No such evidence was found, but Andrews did discover a wealth of mammal fossils.

> “ **Always** there has been an adventure **just around the corner** – and the world is still **full of corners**. ”
>
> ROY CHAPMAN ANDREWS

◁ **A shining career**
A determined and resourceful man, Andrews went from sweeping floors at the American Museum of Natural History to becoming its director. During his career, he led many successful expeditions, which he wrote about in bestselling books.

KEY DATES

- **1884** Born in Beloit, Wisconsin, to a father who sells pharmaceutical drugs wholesale.
- **1906** Begins work as a taxidermist's assistant at the American Museum of Natural History in New York.
- **1908** Goes on his first expedition, to Alaska. On this trip, and until 1914, he specializes in the study of whales and other aquatic mammals.
- **1916** Serves as the museum's chief of Asiatic exploration, leading three trips: to Tibet, southwest China, and Burma (1916–17); to Outer Mongolia (1919); and to Central Asia (1921–25).
- **1935** Becomes director of the American Museum of Natural History, until he resigns in 1942 in order to write.
- **1943** Publishes the autobiographical *Under a Lucky Star*, which is followed by *An Explorer Comes Home* (1947) and *Beyond Adventure* (1954).

ANDREWS EXAMINES DINOSAUR EGGS

SKELETON OF A WHALE IN THE AMERICAN MUSEUM OF NATURAL HISTORY

POSTER ADVERTISING CORSICAN RESORT, WILLIAM SPENCER, c.1932

ITALIAN LINE POSTER FOR SOUTH AFRICA, GIOVANNI PATRONE, 1935

HOLLAND-AMERICAN LINE POSTER, WILLEM TEN BROEK, 1936

BOAC AIRLINE POSTER, TOM ECKERSLEY, 1947

Travel posters

The glamour of passenger travel was the perfect subject matter for colourful advertising posters.

At the end of the 19th century, the colour lithographic process transformed graphic art and coincided with the growing popularity of travel for pleasure. Large, bold, brightly coloured posters could be mass-produced, with words and images skilfully combined to entice passers-by to set off on an adventurous trip by boat, train, car, or aeroplane.

Posters advertising travel and exotic holidays catered for every taste and season. Station billboards encouraged passengers to explore the vast expanses of North America by train. Images of chic ski resorts helped make alpine sports popular. For those who preferred sunnier climes, Mediterranean beach hotels and the exotic allure of India and Africa were on offer, as were the cooler charms of autumn in Japan. Railway and airline operators, travel agents, and tour organizers seized on the colourful graphic posters as modern marketing tools.

Artists were quick to spot the potential of commercial art. A.M. Cassandre led the way in Paris in the mid-1920s with his Art Deco posters inspired by the Machine Age. The geometric shapes of his streamlined, speeding trains and looming ocean liners caught the public imagination, and his style influenced the posters of the 1940s and '50s in Europe and the US.

CANADIAN PACIFIC TRAVEL POSTER, 1956

The Long March

In 1934, the Chinese Red Army made an epic trek to escape the clutches of its enemies. The year-long journey that it made created a founding myth for modern China.

In the mid-1930s, China was ruled by Chiang Kai-shek, the leader of the Kuomintang, or Nationalist Party. However, he faced a threat from the Red Army – the armed forces of the Communist Party. It had mounted a rebellion that was centred in the southeastern province of Jiangxi, with a base in the city of Ruijin.

In September 1933, Kuomintang forces encircled the communists in Ruijin. After an 11-month siege, it was clear to the besieged communists that they could not hold out for much longer, so they planned a diversionary attack. Military leader Fang Zhimin charged enemy lines with a small group of troops, taking the enemy by surprise and allowing most of the Red Army to escape. Around 86,000 communists crossed the barrier of the Yudu River on makeshift pontoon bridges built from doors and bed boards.

Over the following year, the Red Army trekked west and north, constantly harried by Kuomintang forces. Numerous battles ended in defeat, such as the Battle of the Xing River, in which it lost 45,000 men – over 50 per cent of its fighting force. However, these defeats were punctuated with the occasional morale-raising victory. In May 1935, for example, the Red Army managed to capture an 18th-century bridge across the Dadu River, in the remote town of Luding. It only managed this, however, because one detachment made a harrowing march of 121 km (75 miles) in 24 hours over mountain roads. Elsewhere, the communists, who were only wearing light clothes and straw sandals, suffered terribly as they marched across the icy mountains in Sichuan.

Mao Zedong, who had become chairman of the party, declared the Long March over when it reached the province of Shaanxi. Mao claimed that his army had covered a distance of around 12,500 km (7,770 miles). Of the 86,000 communists who broke out of Jiangxi, only a few thousand remained, but they had secured some kind of victory simply by surviving this arduous journey. Furthermore, on learning of their heroism, thousands of young Chinese people enlisted in the Red Army. In 1949, Mao came to power and proclaimed the People's Republic of China.

▷ **Mountain-crossing in Sichuan**
It is estimated that the Red Army marched just over 6,400 km (4,000 miles) in the 12-month Long March. Some of their greatest obstacles were the mountains in Sichuan, which exacted a terrible toll on the men.

TWA
TWA

THE AGE OF FLIGHT

1939–PRESENT

THE AGE OF FLIGHT, 1939–PRESENT

Introduction

The early 20th century was a golden age of travel for those who could afford the first passenger flights, luxury cruises, and train journeys across continents. However, everything was changed by World War II. By the time the war broke out, thousands of Jews and ethnic minorities had already fled persecution in central Europe, and millions of refugees were on the move around the world, seeking new homes.

After the war, intrepid adventurers once more ventured to explore the globe. Jacques Cousteau plumbed the ocean depths, Wilfred Thesiger explored the Arabian Desert, and Edmund Hillary and Tenzing Norgay were among the many who risked their lives to try and conquer the highest mountain on Earth. Meanwhile, with the postwar economic recovery gathering pace, travelling was becoming an option for more than just the most wealthy in society. The postwar boom in car manufacturing made cars affordable for many people in the West, releasing them from reliance on public transport. This freedom to drive wherever they liked found unique expression in the US, where the sheer size of the continent led to a passion for the road trip, as evoked by Bobby Troup in his 1946 song, *Route 66*.

The jet age

In 1952, the Comet, the first commercial jet airliner, took to the skies, instigating a revolution in aviation. The jet engine made it possible to build bigger, faster, and lighter aircraft, which could fly higher, and this made flying more economical. It took a while for the cost benefits to filter through to passengers, but keen competition between airlines, the subsequent slashing of air

TRAVEL WRITERS SUCH AS WILFRED THESIGER WROTE ABOUT LIFE IN EXOTIC PLACES

AFFORDABLE CARS, SUCH AS THESE CHEVROLETS, GAVE MANY PEOPLE THE FREEDOM OF THE ROAD

IN 1953, EDMUND HILLARY AND TENZING NORGAY REACHED THE SUMMIT OF EVEREST

> “ I think **humans will reach Mars**, and I would like to see it happen **in my lifetime**. ”
>
> BUZZ ALDRIN, ASTRONAUT

fares, and the advent of budget airlines finally brought the price of tickets down. For the first time, flights and package holidays abroad became affordable for the masses. However, the interconnectedness that air travel made possible was impacted by – and impacted on – the COVID-19 pandemic, and the climate crisis presents further challenges for the travel industry.

Leaving Earth

The Space Race, which began with the launch of the Sputnik satellite in 1957, was a competition between two superpowers – the US and the USSR. It ended in 1969, when astronauts Neil Armstrong and Buzz Aldrin became the first humans to land on the Moon. As Armstrong memorably said, it was a "giant leap for mankind". Since then, space exploration has become more collaborative. By 1994, Russia and the US were collaborating on the Shuttle-*Mir* programme, which led to the construction of the International Space Station (ISS). Completed in 2012, the ISS has since been manned by crews from 10 different countries and hosted visitors from 18 lands. Meanwhile, since 1977, the Voyager programme, often called the greatest feat of human exploration, has sent two probes past the great planets and beyond into the outer reaches of the solar system, from where they are continuing to send back remarkable data. In 2021, space tourism took a step closer to being commercial reality – although still only for the super-wealthy.

Even today, we are constantly exploring new places or discovering exciting information about those places we thought we already knew. Who knows what will be revealed next.

CHEAP FLIGHTS HAVE MADE HOLIDAYS ABROAD AVAILABLE TO MANY PEOPLE

EUGENE CERNAN (PICTURED) BECAME ONE OF TWELVE ASTRONAUTS TO WALK ON THE MOON

AFTER 136 SPACE FLIGHTS, THE INTERNATIONAL SPACE STATION WAS FINALLY COMPLETED IN 2012

The great displacement

Throughout history, war and conquest have resulted in the mass movement of peoples, but conflicts in the 20th century led to forced migrations on an unprecedented scale.

The 20th century was scarcely underway when World War I uprooted millions of European civilians. Meanwhile, revolution and civil war in Russia from 1917 to the early 1920s led to an exodus of more than one million people opposed to the Bolshevik regime. And from 1915–23, the Ottomans set about eradicating 1.5 million Armenians from the territories of the Ottoman Empire. But as far as forced displacement of people goes, all of this was only the start.

The rise of nationalism

When the political map of Europe was redrawn in 1919 after the fall of the Austrian and Russian empires, many countries suddenly found themselves hosting large populations of ethnic minorities who had previously lived in a neighbouring country. This created sectarian tensions and fuelled the rise of nationalism, ultimately setting the stage for World War II, as Adolf Hitler began to claim territory that was inhabited by ethnic Germans. When Hitler's National

▽ **Leaving home**
This photograph from June 1945 shows people packed into every available space of a train leaving Berlin. By May 1945, in the wake of 363 Allied air raids, 1.7 million people (40% of the total population) had fled Berlin.

◁ **Refugee camp**
From 1945 to 1947, more than 700 refugee camps were set up all over Europe. In these, homeless people were given shelter, food, and medical treatment and then repatriated or sent to build new lives elsewhere.

Socialist German Workers' Party gained power in Germany in 1933, it targeted and actively persecuted the Jewish population, resulting in hundreds of thousands of Jews fleeing central Europe to safety elsewhere. Many of them headed for the US, and around 250,000 made their way to Palestine between 1929 and 1939. Some 10,000 children under the age of 17 were also rescued from German and German-annexed territories and taken to Britain on the Kindertransport (children's transport).

The legacy of war

When World War II broke out in 1939, Jews from all over northern Europe fled the threat of German occupation. After the invasion of Poland in 1939, the Germans set about "cleansing" parts of the country of its indigenous population. They deported 2.5 million Polish citizens, resettled 1.3 million ethnic Germans, and killed around 5.4 million Polish Jews. The Soviet Union, similarly, deported tens of thousands of Estonians, Latvians, and Lithuanians to Siberia when it annexed the Baltic States in 1941.

However, it was the end of the war in 1945 that brought about the largest mass migrations in European history. The number of civilians who were displaced from their homes during World War II has been estimated at 11–20 million. They included prisoners of war, workers who had been brought into Germany by the Nazis for wartime labour, survivors of concentration camps, and those who had simply fled the destruction. All of them had to be sent back to their homelands or resettled somewhere else.

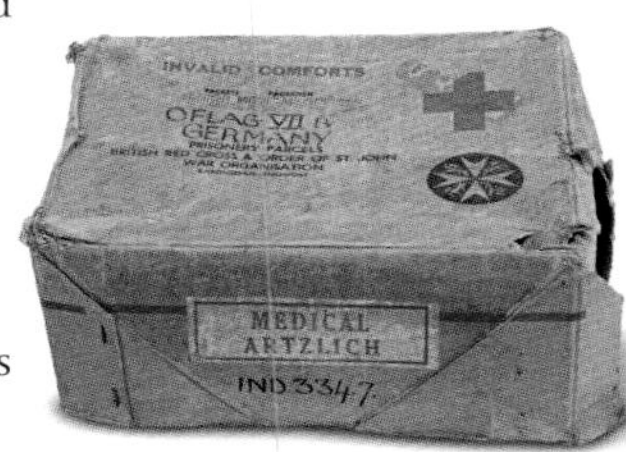

△ **Food parcels**
Red Cross parcels were a lifeline for prisoners of war. Up to 20 million packages were dispatched from Britain, containing medical supplies and basic foodstuffs.

Sent home

Meanwhile, the Allies were instigating their own form of ethnic cleansing. At the Potsdam Conference in July 1945, British, American, and Russian leaders agreed that the Germans who were still living in Czechoslovakia, Poland, and Hungary would have to be sent back to Germany. As a result, more than 11 million people were expelled from their homes and forced to move to Germany over the next five years.

Refugees' rights

The international response to the refugee crisis took legal form with the Universal Declaration of Human Rights of 1948, which guaranteed the right "to seek and to enjoy in other countries asylum from persecution".

Three years later, the 1951 Geneva Convention on Refugees accorded refugees specific rights. Among those, it expressly prohibited the forcible return of refugees to the countries from which they sought to escape. This principle was evoked at the time by many citizens of east European states that had become part of the newly formed Soviet Union. And although the creation of the State of Israel in 1948 finally provided a secure place for the Jews who had been hounded from their homes in central and eastern Europe, around 700,000 Palestinians then became refugees during the war that led to the creation of Israel.

IN CONTEXT
Partition

In August 1947, the British Raj, which had ruled India for 100 years, was dissolved and India was given independence. As a result, the country was divided into two new states along religious lines (Partition): Hindus and Sikhs were to live in India and Muslims in Pakistan. This triggered one of the greatest and bloodiest migrations in history, involving up to 15 million people, as Muslims trekked from India to Pakistan, and millions of Hindus and Sikhs went the other way. There was extensive and appalling sectarian violence, during which it is estimated that between 1–2 million people were slaughtered, and refugee crises were created in both countries.

MAHATMA GANDHI VISITS MUSLIM REFUGEES AT PURANA QILA, DELHI, IN 1947

▽ **Debris of war**
By 1946, bombed German cities were full of piles of rubble. Prisoners, ex-Nazis, and local women volunteers helped to clear it, the latter for extra food coupons.

EMPIRE WINDRUSH

Arriving in Britain
In this photograph taken for the *Daily Herald*, a crowd waves as it approaches the dock. The ship had sailed from Kingston, Jamaica, carrying 490 men and two women.

The *Windrush*

When the *Empire Windrush* arrived in London from Jamaica in 1948, it marked the start of a postwar immigration boom that changed British society.

On the morning of 22 June 1948, the former troop carrier *Empire Windrush* arrived at Tilbury Docks in Essex, near London, having sailed from Kingston, Jamaica. Aboard were 492 immigrants who had arrived at the invitation of His Majesty's Government. When World War II ended, the government realized that reconstructing the British economy would require a large influx of immigrant labour, so the British Nationality Act 1948 was passed to give Commonwealth citizens (citizens of British colonies) free entry to Britain. The *Empire Windrush* arrivals were the first wave of immigrants to come and staff vital jobs at institutions like the National Health Service. Of the arrivals, 236 were temporarily housed in a deep air-raid shelter under Clapham Common. Others, many of whom had served in the British forces during World War II, had organized accommodation for themselves beforehand.

Not everybody, however, was convinced that importing Jamaicans was the correct solution to the country's problems. Prime Minister Clement Attlee was obliged to reassure a group of Members of Parliament that "it would be a great mistake to take the emigration of this Jamaican party to the United Kingdom too seriously". He thought that immigration would be limited by the fact that not many people in the Caribbean would be able to afford the £28 10s fare to England. In fact, over the next 14 years, some 98,000 West Indians made Great Britain their home.

As well as Caribbeans, large numbers of workers and their families from other Commonwealth nations, notably India and Pakistan, emigrated to Britain after World War II. Many hundreds of thousands of people came during the 1950s, not just for short-term work, but to settle for good. Immigration has continued ever since, resulting in a multicultural diversity that would have been unthinkable in 1945.

" **London** is the **place for me**. London this **lovely city**. "

LORD KITCHENER, A FAMOUS TRINIDADIAN CALYPSO SINGER WHO WAS A PASSENGER ON THE *EMPIRE WINDRUSH*, 1948

The *Kon-Tiki* expedition

Norwegian anthropologist Thor Heyerdahl went to extreme lengths to prove his theories of migration, most famously sailing a handmade raft from South America to Polynesia.

△ **Thor Heyerdahl**
In this photograph from 1947, the anthropologist is climbing the mast of the *Kon-Tiki*. The large figure of the Inca sun god on the sail behind him was painted by his crewmate Erik Hesselberg.

As a student with an interest in the civilizations of the South Pacific, Thor Heyerdahl travelled to Fatu Hiva in the Marquesas Islands in French Polynesia in 1937. During his one-and-a-half-year stay, he became convinced that the islands had originally been populated by South Americans rather than by people from Asia who came to the West, as was commonly believed. It was no coincidence, he thought, that the huge stone figures of the mythical Polynesian leader Tiki resembled the monoliths left behind by pre-Incan civilizations, and he concluded that the original Polynesian inhabitants had crossed the Pacific on rafts, 900 years before Columbus traversed the Atlantic.

Heyerdahl presented his theory to a group of leading American academics, who were sceptical. One of them went so far as to challenge Heyerdahl: "Sure, see how far you get yourself sailing from Peru to the South Pacific on a balsa raft!" So that is exactly what Heyerdahl did. He built a raft using only materials and techniques that were available to the pre-Columbians, so no nails, spikes, or wires. Working from illustrations made by the Spanish conquistadors, he used balsawood logs for the base, mangrove wood for the mast, and plaited bamboo for a cabin, then made a roof for it with large banana leaves.

Gone fishing

On 28 April 1947, Heyerdahl and his crew of five, plus a parrot, sailed out from Callao, Peru. They named their raft *Kon-Tiki*, after the Inca sun god. None of the crew, made up of five Norwegians and one Swede, were sailors – Heyerdahl could not even swim. Other than some essential modern equipment, such as radios, watches, shark repellant, and a sextant, they carried nothing that would have been unavailable to ancient sailors. Heyerdahl counted on easterly winds and the Humboldt Current,

◁ **The sole loss of life**
The expedition consisted of Heyerdahl, five other crew, and a Spanish-speaking green parrot named Lorita. Unfortunately, she was washed overboard partway through the voyage.

▷ **Balsawood raft**
Heyerdahl's crew spent around 15 weeks on board the 3.7 sq m (40 sq ft) *Kon-Tiki*. It was built from nine balsawood logs, which were covered with a deck made of bamboo. A bamboo cabin provided some shelter.

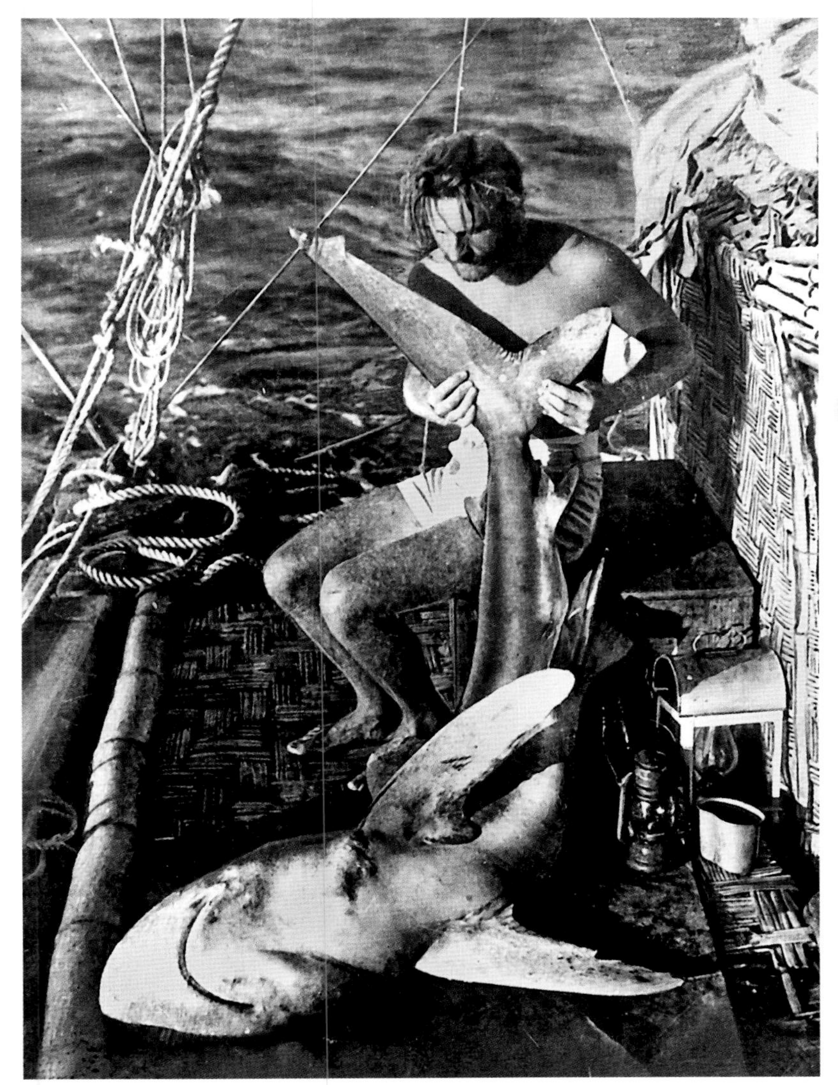

△ **Shark baiting**
Sharks were numerous as the *Kon-Tiki* sailed across the Pacific. The crew even caught some, as shown in this photograph, where Heyerdahl is holding one by its tail.

△ **The *Kon-Tiki* voyage**
From Callao in Peru, the *Kon-Tiki* set off west across the Pacific Ocean. It travelled for 101 days and covered around 6,980 km (4,340 miles), before reaching French Polynesia.

a cold-water current in the southeast Pacific Ocean, to carry the raft in the right direction.

Critics were convinced that the crude *Kon-Tiki* would break up after a week or two, but it proved highly seaworthy. The hand-woven ropes that bound the logs swelled with seawater, embedding themselves into the soft wood and strengthening the vessel rather than weakening it. It was, Heyerdahl wrote in his log, "a fantastic seagoing craft".

According to Heyerdahl's account, when the seas were rough, the crew were sometimes waist-deep in water, and had to hang on to avoid being washed away. To supplement their rations of coconuts, sweet potatoes, and fruit, they caught plenty of fish, particularly flying fish, yellowfin tuna, and bonito. For amusement, they dangled fish overboard for the ever-present sharks to snap at.

Reaching French Polynesia

After 101 days at sea, the *Kon-Tiki* struck a coral reef and was beached on an uninhabited islet off the Raroia atoll. The raft had finally reached Polynesia, and had travelled a distance of around 6,980 km (4,340 miles). After spending a few days marooned, the crew was rescued by inhabitants of a nearby island, and was then collected by a French schooner bound for Tahiti.

Heyerdahl's expedition had successfully demonstrated that South American peoples could, in fact, have journeyed to the islands of the South Pacific by balsa raft. Subsequent DNA tests have shown that the Polynesian people are, in fact, of Asian descent. However, the expedition brought Heyerdahl immense fame, and even triggered a craze for Tiki bars, motels, and cocktails, and a hit track released in 1961 by English band The Shadows.

> " **Borders**? I have **never** seen one. But I have heard they **exist in the minds** of **some people**. "
>
> THOR HEYERDAHL

EXPLORER AND TRAVEL WRITER, 1910–2003

Wilfred Thesiger

Arguably the last of the great explorers, Thesiger was an Englishman who rejected the modern world in favour of the tribespeople of Africa and the Arabian deserts.

Wilfred Thesiger was born in Addis Ababa, in Imperial Abyssinia (modern-day Ethiopia), where his father was the chief British representative. He was sent to England for his education, but was unhappy there. During his first summer holiday from university, he worked his passage on a boat to Istanbul. He spent his second summer on a fishing trawler off the coast of Iceland.

Sudan, Syria, and the SAS

As soon as he was able to, Thesiger returned to Africa. Here, at the age of 23, he decided to explore a remote region of Abyssinia. Two years later, he found work as an assistant district commissioner in the Sudan, where one of his roles was to shoot lions that attacked the farmers' herds. He served in Darfur, where he learned how to travel by camel. It was in Sudan that he had his first real experience of the desert: "I was exhilarated by the sense of space, the silence, and the crisp cleanness of the sand. I felt in harmony with the past, travelling as men had travelled for untold generations across the deserts".

During World War II, Thesiger fought with the British Army against Italian forces in Abyssinia, as part of the Special Air Service (SAS) in North Africa, and against the Vichy French in Syria.

Arabian sands

After the war, Thesiger began to work with the United Nations in Arabia. In 1946, ostensibly on a search for the breeding grounds of locusts, Thesiger made a 2,414-km (1,500-mile) circuit of the Rub' al-Khali – or Empty Quarter – a famously inhospitable region of desert. Although he was not the first to cross it, he was the first to explore it thoroughly, and the first outsider to visit the oasis of Liwa and the quicksands of Umm As-Samim. A second expedition two years later penetrated even further into the desert, during which time he was arrested by the Saudi authorities and became caught up in inter-tribal hostilities.

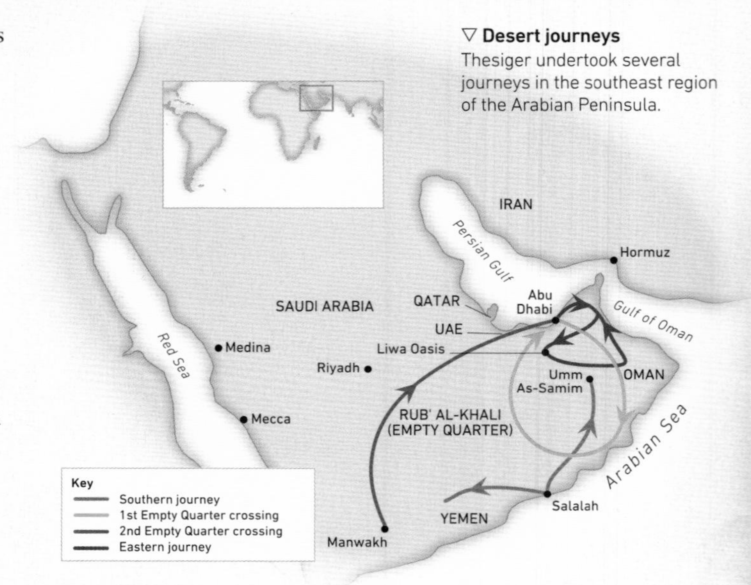

▽ **Desert journeys**
Thesiger undertook several journeys in the southeast region of the Arabian Peninsula.

These journeys represented the last, and possibly greatest, expeditions of Arabian travel. Thesiger later said he found the experience humbling because he rarely matched the standards of endurance and generosity of his Bedouin travelling companions.

In 1950, Thesiger made a new base among the Marsh Arabs in southern Iraq, where he lived for eight years. Every summer, he left to go trekking in the mountains of the Hindu Kush, the Karakorams, or Morocco. In 1959 and 1960, Thesiger undertook two journeys by camel to Lake Turkana, in northern Kenya. He lived and travelled in Kenya for much of the next 35 years.

With his health failing, Thesiger reluctantly returned to England for good in the mid-1990s. However, his heart remained with the tribespeople in whose company he had spent his life. He resented much of modern technology, and regarded cars and aircraft as "abominations". In Thesiger's autobiography, *The Life of My Choice*, published in 1987, he expressed his belief that Western civilization was a corrupting force that had robbed the world of its rich diversity.

◁ **Crossing the Empty Quarter**
One modern invention Thesiger did approve of was the camera. This photograph shows him crossing the Awarik sands in the Empty Quarter.

▷ **Last of his kind**
This portrait of Thesiger was taken in Abu Dhabi in March 1948, during the British explorer's second journey across the Empty Quarter. He wears typical Arab garments, including a *thawb* (shirt) and a headscarf.

" Here in the desert I had found **all that I asked**: I knew that I should **never find it again**. "

WILFRED THESIGER, *ARABIAN SANDS*, 1959

KEY DATES

- **1910** Born in Addis Ababa, Ethiopia, the son of the British minister Wilfred Gilbert Thesiger.
- **1945–46** Makes the first of his legendary journeys, crossing the Empty Quarter of Arabia by camel in the company of four Bedouin.
- **1950** Travels to the southern marsh regions of Iraq where he lives intermittently for seven years.
- **1959** Publishes a book about his travels, called *Arabian Sands*. The book brings him acclaim as a writer and photographer, and becomes an international bestseller.

THESIGER'S 1959 TRAVEL ACCOUNT, *ARABIAN SANDS*

- **1964** Publishes his second great work, *The Marsh Arabs*, another portrait of a world on the verge of vanishing.
- **2003** Dies in England. His collection of 38,000 travel photographs is donated to the Pitt Rivers Museum, Oxford.

THESIGER WITH COLONEL GIGANTES IN THE WESTERN DESERT IN LIBYA, 1942

The jet age

If time is money, then the jetliner made the world a much richer place. It drastically reduced flying times and slashed costs for a new generation of air travellers.

On 2 May 1952, the first commercial jet airliner took to the skies. A de Havilland Comet flying for BOAC (the British Overseas Airways Corporation) took off from London for Johannesburg, South Africa. It was a four-engine aircraft capable of carrying 36 passengers at a cruising speed of 720 kph (450 mph). Truly innovative, it travelled faster and higher than propeller aircraft and gave passengers a quieter and smoother journey. At the time, it was nothing short of revolutionary. In the words of Juan Trippe, founder of Pan American Airways, the jetliner was the most important development in aviation since Lindbergh's transatlantic flight.

However, after 18 months of service, design weaknesses led to metal fatigue, resulting in three catastrophic accidents. The faults were duly corrected, but the Comet's reputation was damaged.

The aircraft that truly began the jet age appeared six years after the Comet's launch. Just as Henry Ford's Model T had popularized the car late in the day, it was the Boeing 707 that brought jet travel to the masses. The 707 was almost half again as fast as the Comet, and carried five times as many people. This made it commercially more viable than any previous aircraft. In October 1958, Pan Am flew the first commercial service of a 707, from New York to Paris. Trippe proclaimed that a trip to Europe was every American's birthright.

▷ **Graphic appeal**
Airlines used contemporary graphics and branding to woo customers. This 1960s poster suggests that the glorious sun and cool modern architecture of Miami are just a quick flight away.

“ In **one fell swoop**, we have **shrunk the earth**. ”

JUAN TRIPPE, FOUNDER OF PAN AM, ON THE INTRODUCTION OF JET AIRCRAFT

▽ **The future is now**
Eero Saarinen's dynamic, wing-like TWA terminal at New York's Idlewild Airport opened in 1962. It was a symbol of the advances in technology and design that jet travel helped to promote.

The Boeing 707

With its sleek body and swept-back wings, the 707 rolled out before a public that was already in love with air travel. Just that year, Frank Sinatra had released his album *Come Fly with Me*, its cover showing him gesturing with his thumb to a TWA airliner on a runway. President Eisenhower himself had already ordered three Boeing 707s to serve as the very first Air Force Ones.

The popularity of the 707, and the new generation of aircraft it spawned, led to developments in almost every aspect of aviation infrastructure and design, from the terminal buildings to cabin crew uniforms. The heady

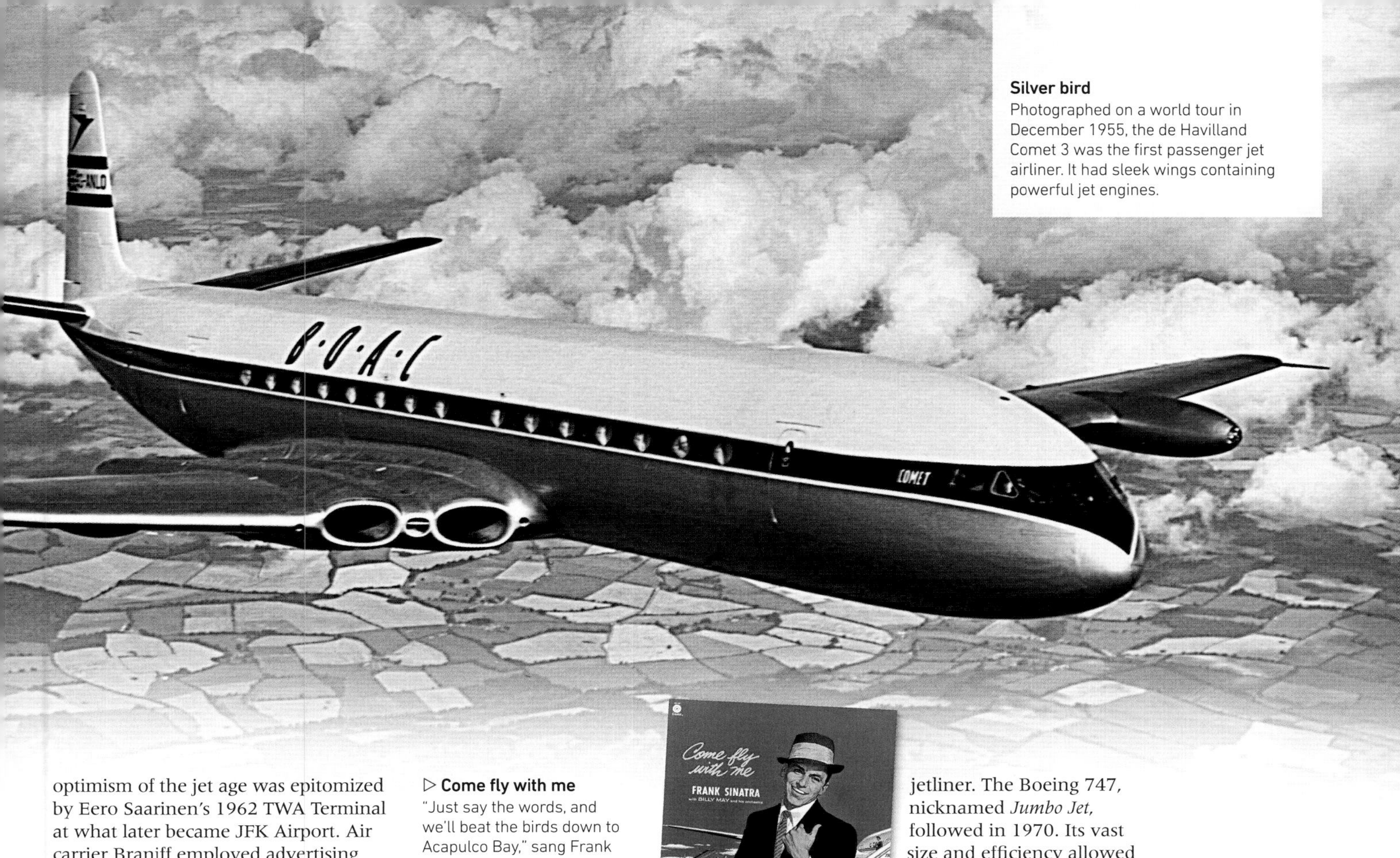

Silver bird
Photographed on a world tour in December 1955, the de Havilland Comet 3 was the first passenger jet airliner. It had sleek wings containing powerful jet engines.

optimism of the jet age was epitomized by Eero Saarinen's 1962 TWA Terminal at what later became JFK Airport. Air carrier Braniff employed advertising executive Mary Wells, who touted the airline as "The End of the Plain Plane", and introduced futuristic flight attendant uniforms designed by Emilio Pucci. While hot meals had been introduced in the 1930s, in-flight meals as we know them – eaten from a fold-down seatback tray – became standard in the 1960s.

▷ **Come fly with me**
"Just say the words, and we'll beat the birds down to Acapulco Bay," sang Frank Sinatra in 1958's *Come Fly with Me*. It was an invitation for millions to take to the air.

The bestselling book of 1968 was Arthur Hailey's *Airport*, which spent 30 weeks at number one on the *New York Times* bestseller list.

◁ **TWA cabin crew**
Air stewardesses initially became glamorous poster girls for flying, although nowadays both women and men serve as pilots and cabin crew.

Airlines come of age

Flying was an adventure, and one that increasing numbers of people could share. The world's first "tourist" fare, later to become "economy", was introduced in the 1950s, and by 1959, more people were crossing the Atlantic by air than by sea. Airlines were also quick to exploit the advantages of jet travel for businesses. By 1965, the annual number of US air passengers reached 100 million, double the figure for 1958. The next turning point was the launch of the Boeing 737 in 1968. This soon became the most numerous jetliner. The Boeing 747, nicknamed *Jumbo Jet*, followed in 1970. Its vast size and efficiency allowed fares to decrease further. By the end of the century, there were around 1,250 747s in the air at any time – which meant that one was either landing or departing somewhere in the world every five seconds. By this time, the word "jetliner" had fallen out of use. Jet technology had replaced the propeller-driven planes of old to such an extent that the new machines were now simply known as airliners.

▽ **Fast food**
In-flight pre-packaged airline meals on trays were introduced as standard in the 1960s, just one example of the many innovations of the jet age.

WRIGHT FLYER, US, 1903

TRAVEL AIR 4000, US, 1929

SARO A.19 CLOUD, UK, 1930

STITS SA-2A SKY BABY, US, 1952

AVRO 652A ANSON C19, SERIES 2, UK, 1946

Planes

As soon as the Wright brothers achieved take-off in 1903, engineers set about producing larger and ever-faster aircraft.

The initial goal was to give aeroplanes more power. The greater the thrust, the greater the lift, which made it possible to build larger planes that could carry heavier loads. That heavier load was sometimes fuel, which enabled aviators to make record-breaking long flights, such as Charles Lindbergh's crossing of the Atlantic in the *Spirit of St Louis*. Larger planes were developed to carry passengers. One of them was the Boeing 247, an early airliner with many features that later became standard, including an all-metal airframe, cantilevered wings, and retractable landing gear. Aircraft were subsequently designed in all shapes and sizes, from the world's smallest plane, the 1952 Sky Baby, to Howard Hughes's giant flying boat, the H-4 Hercules of 1947.

The jet engine, introduced in the 1950s, revolutionized flight. Its extraordinary thrust made passenger air travel truly viable for the first time. Today, ever-larger aircraft carry increasing numbers of passengers with greater fuel efficiency, as exemplified by the Airbus A380.

BLÉRIOT XI, FRANCE, 1909

RYAN NYP *SPIRIT OF ST LOUIS*, US, 1927

BOEING 247, US, 1933

DE HAVILLAND DH87B HORNET MOTH, UK, 1934

MORAVAN NÁRODNÍ PODNIK ZLÍN Z.226T, CZECHOSLOVAKIA, 1956

BOEING 727-200, US, 1967

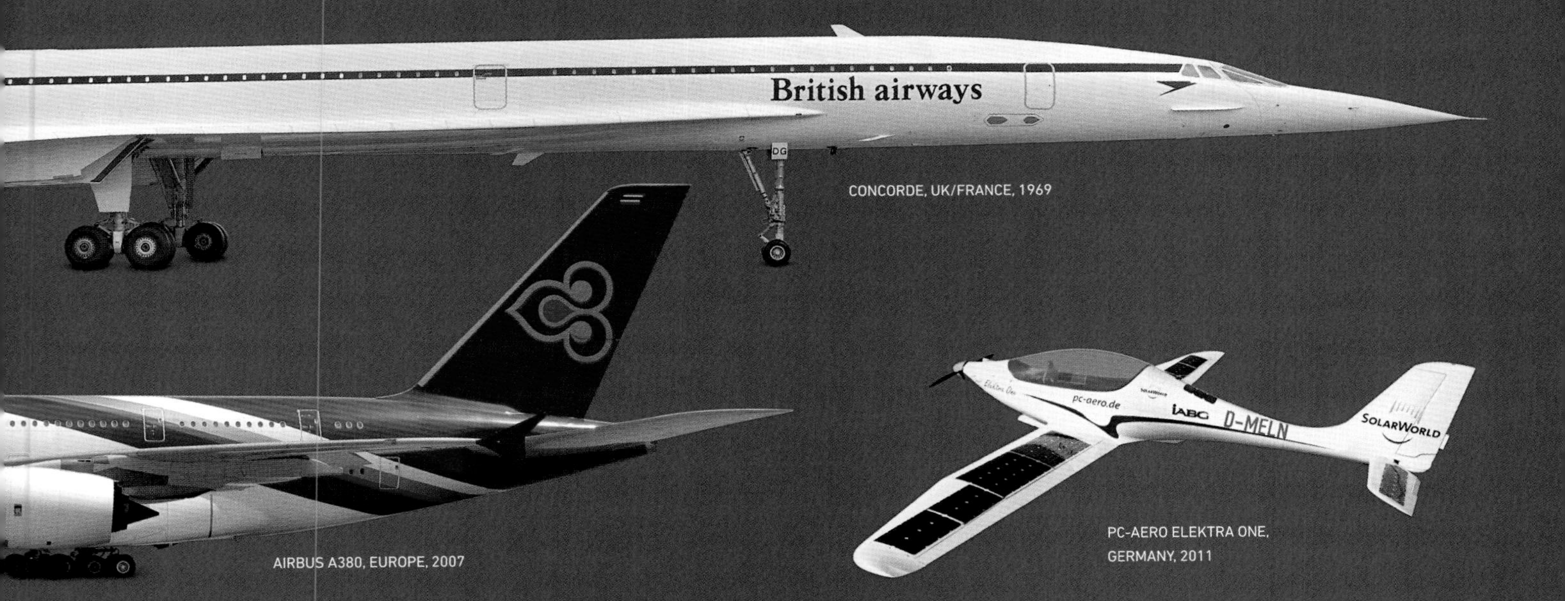

CONCORDE, UK/FRANCE, 1969

AIRBUS A380, EUROPE, 2007

PC-AERO ELEKTRA ONE, GERMANY, 2011

The roof of the world

In May 1953, Edmund Hillary and Tenzing Norgay became the first people to stand on the summit of Mount Everest. Their achievement was only possible thanks to those who had failed before them.

In 1856, the Great Trigonometric Survey of India determined that an obscure Himalayan mountain, known as Peak XV, was the highest mountain in the world (see pp.242–43). It was named Everest shortly afterwards, but attempts to climb it could not begin until 65 years later, when the forbidden kingdom of Tibet first opened its borders to outsiders.

In 1921, the British led an expedition to scout a northern approach to the mountain. The following year, they returned for a proper attempt on the summit, reaching a record altitude of 8,326m (27,316ft). The next expedition came in 1924. After an initial aborted attempt, two British climbers, George Mallory and Andrew Irvine, made another push for the summit. They were last seen disappearing into the clouds that perpetually swirl around the mountain. No-one knows if they reached the top. Mallory's frozen body was found 75 years later, in 1999. Irvine's has yet to be located.

Such are the dangers of Everest. At 8,848m (29,029ft), its peak is hostile to life. At that height, the air contains only a third of the oxygen found at sea level, increasing the chance of fatal cerebral edemas, which occur when the oxygen-starved brain swells up. In such conditions, non-essential body functions shut down, so digestion and sleep are impossible. The temperature at the summit is -36°C (-33°F), making frostbite and hypothermia likely. There are also constant threats of avalanches, crevasses, and storms. And yet, when explaining his reasons for climbing, Mallory said: "What we get from this adventure is just sheer joy. And joy is, after all, the end of life. We do not live to eat and make money. We eat and make money to be able to live."

▽ **Ascending Everest**
This photograph, from 1953, shows one Sherpa using ropes to guide another across a log bridge over a crevasse in the Western Cwm. Both wear crampons – metal plates with spikes fixed to boots.

The southern route

In all, there were seven attempts to climb Everest before World War II halted expeditions. They were all carried out by the British, who deliberately used their position of power in India and Tibet to deny other nations the chance to climb the mountain. Access to Everest was closed in 1950, after China invaded Tibet, but by that time Nepal was open, after being closed to foreigners for 100 years. In 1950, an exploratory expedition was made from the south, along the route that has now become the standard approach. In 1952, a Swiss expedition following this route reached a new, record height of 8,595m (28,199ft). The following year, a ninth British expedition travelled to Nepal. The first pair of climbers failed to reach the summit, but planted backup caches of oxygen for a second pair, New Zealand beekeeper Edmund Hillary and Nepali Sherpa Tenzing Norgay. At 11:30am local time on 29 May 1953, they became the first people to reach the summit. They took photographs, buried a few sweets and a cross in the snow, and then began their descent.

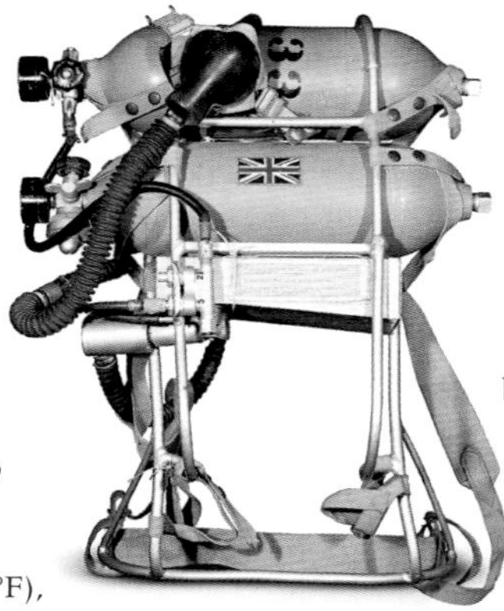

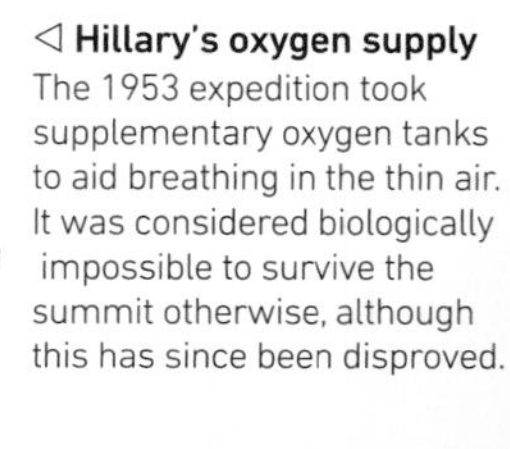

◁ **Hillary's oxygen supply**
The 1953 expedition took supplementary oxygen tanks to aid breathing in the thin air. It was considered biologically impossible to survive the summit otherwise, although this has since been disproved.

Reporting the achievement, Britain's *Manchester Guardian* concluded that the mountain "is in its nature a terminal point… It is doubtful whether anyone will ever try to climb Everest again." Of course, far from signalling an end, the ascent opened a whole new chapter in mountaineering history – one in which hundreds of climbers flock to the mountain each year.

“ **Nobody** climbs for **scientific reasons**. You really **climb** for the **hell of it**. ”

EDMUND HILLARY, MOUNTAINEER

▽ **On the southeast ridge**
This image shows Edmund Hillary (left) and Tenzing Norgay preparing to make for the final camp on their way up Everest, or *Chomolungma* to use its Tibetan name. The following day, they became the first people to stand on the summit.

IN CONTEXT
Climbing Everest today

Since Hillary and Tenzing conquered Everest, the summit has been reached over 7,600 times by around 4,460 climbers. The numbers have rocketed recently, partly because fixed ropes and ladders have made the climb easier. In just one day in 2016 (19 May), 209 climbers scaled the mountain, which is more than the number who made it in the 33 years after the first ascent.

In 1990, just 18 per cent of attempts on the summit were successful, but in 2012 that figure was 56 per cent. However, the climb is not necessarily any safer. 282 people have died on Everest since 1921, and 18 died in 2015 alone, following a massive earthquake. Nor is it cheap. The British Mountaineering Council puts the minimum spend for an Everest climb at $35,000, but climbers can pay as much as $65,000.

PERMANENTLY FIXED ROPES AND LADDERS HAVE MADE EVEREST EASIER TO CLIMB

The open road

The mass manufacture of cars after World War II made it possible for ever more families to take to the road on holiday. In the US, the road trip became synonymous with freedom and romance.

Chevy puts the purr in performance!

△ **Car ownership**
After World War II, American manufacturing switched back to consumer goods, and soon many people in work could afford a car.

The great American road trip was born in 1903, when former doctor Horatio Nelson Jackson, mechanic Sewall Crocker, and a dog called Bud set out in a red Winton touring car across the US. At the time, there were fewer than 240 km (150 miles) of paved road in the entire country. A friend bet Jackson $50 that it would take at least 90 days to drive from San Francisco to New York. In fact, the 7,242 km (4,500 miles) took 63 days. Jackson won his bet, but the trip cost him $8,000, including the price of the car.

Sixteen years later, a military convoy led by Major Dwight D. Eisenhower travelled from Washington, DC to San Francisco, at an average speed of only 8 kph (5 mph). Improving the highways became a key issue for Eisenhower as president, and his response was to create the Federal Aid Highway Act of 1956 and to construct the Interstate Highway System.

By the 1950s, one in six Americans was employed either directly or indirectly in the car business. The US was the largest car manufacturer in the world, but also its greatest buyer. There were 25 million registered cars on the road in 1950, but 67 million by 1958, tempted by the stylish cars of the time, such as the iconic '57 Chevy, and by the new multi-lane highways inviting Americans to take to the open road.

For Black travellers, however, the open road was far more complicated, as they faced racial discrimination and restrictions on where they could safely stay, eat, or repair and refuel their vehicles. To counter such dangers, an African American mailman, Victor Hugo Green, compiled *The Negro Motorist Green-Book*, listing those establishments friendly to non-white travellers. Regularly updated between 1936 and 1966, it became "the bible of Black travel" until the Civil Rights Act of 1964 outlawed racial discrimination.

Drivers and drive-ins

Meanwhile, in Europe, tyre manufacturers André Michelin and his brother Edouard had been publishing a guide for French motorists since 1900, and Britain's AA (Automobile Association) had a million

▽ **Roadside manuscript**
The cross-country road trip is quintessentially American, and no book captures it better than Jack Kerouac's *On the Road*. He taped his manuscript pages into a long scroll of paper.

▷ **Quick break**
Car culture led to the drive-in restaurants designed to catch the eye of speeding passers-by.

members by 1950. As the US was developing its highway system, Europe was becoming veined with motorways, autobahns, autoroutes, and autostrada. However, Europe's multiple borders and passport controls meant that motoring freedom was more restricted.

In the US, businesses emerged to cater specifically for motorists, such as roadside motels, drive-in restaurants, and drive-in cinemas. In the 1950s, with backing from the car industry, the first Holiday Inn opened (by 1959 there were 100 of them), as did the first Howard Johnson's Motor Lodge, and the first franchises of a roadside hamburger restaurant called McDonald's.

◁ **Michelin men**
In Europe, the French Michelin brothers built a successful tyre business. They promoted it with rubber men who showed motorists where to find the best places to eat.

The road in art

As a metaphor for freedom, the great American road trip was, and still is, a source of inspiration for writers and filmmakers. The year after Eisenhower's Highways Act, *On the Road* (1957) was published, Jack Kerouac's iconic celebration of youth and a journey to find America. Around the same time, celebrated novelist John Steinbeck came to a profound conclusion: "I discovered I did not know my own country." Aged 58, he and his pet poodle, Charley, hit the road, publishing his journey of discovery as *Travels with Charley: In Search of America* (1962). Meanwhile, that most American of cultural media, the cinema, has glorified the open road repeatedly, particularly in the iconic *Easy Rider* (1969).

▷ ***Easy Rider***
The cult 1969 film reinvented the road trip, replacing the car with motorbikes, but still celebrating the timeless American urge to hit the road and head out west.

“ The cross-country trip is the **supreme example** of the **journey** as the **destination**. ”

PAUL THEROUX, TRAVEL WRITER

Route 66

Stretching from Chicago to Los Angeles, Route 66 is the quintessential American highway. Covering more than 3,900 km (2,451 miles), it is rightly considered the mother of all road trips.

In 1946, Bobby Troup set off on a cross-country drive from Pennsylvania to California, where he wanted to try his hand as a Hollywood songwriter. The trip began on highway US 40, and he thought about writing a song about the road. However, inspiration did not strike until the route reached US 66, which, Troup's wife noted, rhymed with "Get your kicks". The resulting song, *(Get your kicks on) Route 66*, became a hit for Nat King Cole later that year, and has since been recorded by artists including the Rolling Stones, Bruce Springsteen, and Depeche Mode.

However, although the road was now enshrined in popular culture, Route 66 was famous long before Troup turned his key in the ignition and headed west.

▽ Historic route
When Route 66 opened in 1926, its purpose was to link hundreds of rural communities in Illinois, Missouri, and Kansas to Chicago. During the Depression, it became the route that many took to seek better lives in the west.

Road of dreams

Route 66 was not the first US trans-continental highway – that was the Lincoln Highway, which was dedicated in 1913, and ran from New York to San Francisco. Established 13 years later, in 1926, US 66 followed a different route, starting in Chicago, and then turning southwest, down into the heartlands of the Midwest, traversing eight states and three time zones to end at Los Angeles. The route was deliberately intended to give many small towns their first access to a major road.

▽ Route of freedom
No other road has symbolized the American spirit and captured the popular imagination like Route 66. Nearly a century after its construction, it still exerts a pull on Americans and foreigners alike.

> “ 66 is the **mother road**, the **road of flight**. ”
>
> JOHN STEINBECK, *THE GRAPES OF WRATH*

One of the main supporters of the highway was a businessman from Oklahoma called Cyrus Avery, who was determined that the route should pass through his home state and deliver the economic benefits of connectivity.

Ironically, a highway that was meant to bring wealth to middle America ended up becoming famous as an escape route. It was completed just in time for the families who were forced to leave their homes and farms in the Midwest during the Great Depression and the Dust Bowl crisis of the 1930s. Route 66 was the road they hoped would take them to better lives out west. Almost from the start it was a "road of dreams".

Route 66 reached the height of its popularity in the 1950s, when the boom in car ownership led to great numbers of holidaymakers hitting the road to find out what the rest of the US looked like. And so, a little later than originally planned, the highway finally brought prosperity to businesses all the way along it.

Mother Road

Ironically, it was the very legislation that encouraged long-distance car travel in the US – President Eisenhower's 1956 Federal Highway Act – that led to the demise of Route 66. The new four-lane interstate system bypassed it. Towns suffered from the loss of through traffic, and businesses closed down as parts of the route were abandoned altogether. By 1985, Route 66 was decommissioned and officially ceased to exist.

However, the role Route 66 played during the Depression has given it iconic status. It is sometimes referred to as the Mother Road. Thus, in recent times, non-profit organizations and the US National Park Service have mobilized support and provided funding to conserve what is left of the route. Stretches of the former highway are now promoted as a heritage site. Once again, people come from all over the US and from around the world to drive along this historic road and experience a slice of Americana.

△ **Route 66 today**
An extensive nostalgia industry has grown up around US 66. Visitors can stay in old-style motels all along the route, such as the Blue Swallow Motel in Tucumari, New Mexico.

No-frills flying

The jet engine made passenger flights fast and convenient, but it was not until the prices of airline tickets were dramatically slashed that flying became affordable for most people.

△ **Peanut airlines**
Southwest was the first of a new generation of airlines to undercut established rivals by doing away with the "frills". It served snacks instead of full meals, which earned it the nickname "peanut airlines".

In 1977, Freddie Laker, a former teaboy for the flying-boat builders Short Brothers of Rochester, made airline history by launching Skytrain, a pioneering cheap daily flight between the US and Britain.

Until that time, international flights were largely the preserve of the rich. After World War II, it was thought that competition between airlines would compromise passenger safety, so commercial aviation was strictly regulated by the IATA (International Air Transport Association). This left state airlines free to run monopolies, offering identical services at high prices.

There were exceptions to the rule. The Icelandic airline Loftleiðir declined to join IATA, and in the 1960s it offered cut-price fares between the US and Europe via Reykjavik, earning it the nickname "the hippie airline". Charter flights were also available, but generally they only operated on package holidays to Spain and other places where there were resorts.

Skytrain and Southwest Airlines

Freddie Laker proposed a system in which passengers who wanted cheap flights could queue for tickets at the airport, as at a train or bus station.

▽ **Freddie Laker**
The launch of Laker's London-to-New York Skytrain in September 1977 was the first low-cost operation of its type. Other airlines lowered their fares in response.

△ **Grabbing the headlines**
To stand out from the competition, Southwest had stewardesses who wore orange hot pants and white boots, and served drinks called "love potions." The airline's slogan was "Long legs and short nights".

After six years of negotiations with the British and US governments, the first Skytrain took off for New York in September 1977. The service had no frills (such as meals), but its fares were a third of those of rival airlines.

Another budget-fare pioneer was Texas-based Southwest Airlines, which was launched in 1971. It promoted its cheap flights with sensational posters featuring leggy air hostesses in hot pants and white boots. Its success convinced the Civil Aeronautics Board to relax its regulations on airfares, and more competitive pricing was introduced in 1978. Airfares fell by a third, and traffic more than doubled in the 1980s as more low-cost American carriers were launched. Ironically, having championed cheap fares, Skytrain went out of business when British Airways and Pan Am slashed their transatlantic fares.

Ryanair and easyJet

The success of the budget American airlines caught the attention of aspiring aviation executives across the Atlantic.

In 1990, a small loss-making Irish airline called Ryanair reinvented itself as a no-frills airline offering cheap flights to secondary European airports. A few years later, a similar business began in a cramped office at Luton Airport in the UK. Known as easyJet, it began advertising flights to Edinburgh and Glasgow, in Scotland, that were "as cheap as a pair of jeans". These two airlines, and the many copycat businesses that followed, revolutionized European air travel.

By stripping their product to the bare minimum – having a single cabin in which all the seats were the same, charging for food, and dispensing with free baggage allowance – low-cost carriers offered flights that everyone could afford. In just over 25 years, low-cost airlines became not just a fixture of the travel industry, but also a vital part of contemporary life in Europe, the US, and around the world.

◁ **Global revolution**
"Now everyone can fly" is the slogan for Air Asia. It is just one of the many budget airlines seen in this photograph taken at an airport in Kuala Lumpur, Malaysia.

Hidden costs

Cheap fares, frequent schedules, and flights to places where no-one had flown before made it feasible for people to commute between their homes and businesses in different parts of the world. Ever more flights at ever cheaper rates created an unprecedented freedom of movement – at least, until the COVID-19 pandemic, exacerbated by that very freedom of movement, brought air travel almost to a standstill. As the travel industry rebuilds and recovers, it will also need to address another, greater challenge – that of the climate crisis and the cost to the planet of the carbon footprint generated by those "cheap" flights.

△ **Low fares model**
Inspired by the success of Southwest Airlines, Ireland's Ryanair lowered fares by cutting costs wherever possible. It crammed in more flights each day, used smaller airports with cheaper landing charges, and operated just one aircraft model, the Boeing 737.

> " What railways have **done for nations**, airways will **do for the world**. "
>
> CLAUDE GRAHAME-WHITE, AVIATION PIONEER, 1914

Into the abyss

People have spent centuries exploring the surface of the Earth, but it was only in the 20th century that advances in technology made it possible to investigate the oceans.

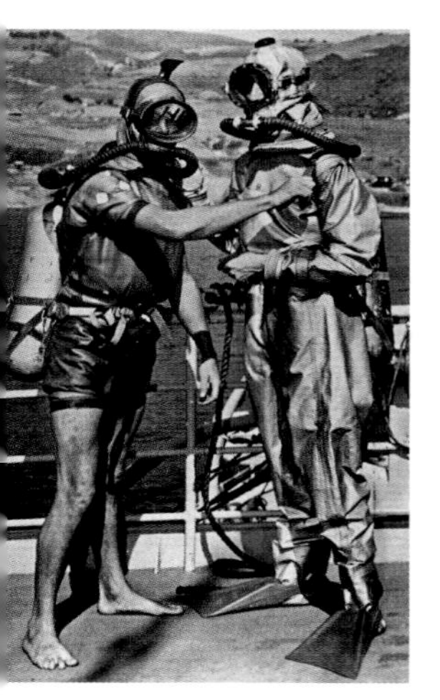

△ **New technology**
Divers Jacques Cousteau (right) and Terry Young prepare to dive with Aqua-Lungs strapped to their backs. The apparatus was soon renamed the SCUBA system.

Rarely has one man dominated a field of study so completely as Jacques-Yves Cousteau. Not only was he a pioneering marine explorer, he also invented the technology that allowed divers to swim underwater.

The Aqua-Lung

Experiments in breathing underwater date back as far as the 17th century, when English physicist Edmond Halley (of Halley's Comet fame) patented a design for a diving bell. In the 1930s, American naturalist and undersea explorer William Beebe used the newly invented bathysphere to explore the oceans to what was then a record depth of 804 m (2,638 ft). At the time, Jacques Cousteau was only in his early twenties. He had originally joined the French Naval Academy to become a pilot, but a car accident put an early end to this ambition, and instead he began to explore the underwater world of the Mediterranean from the beaches around the naval base at Toulon in France.

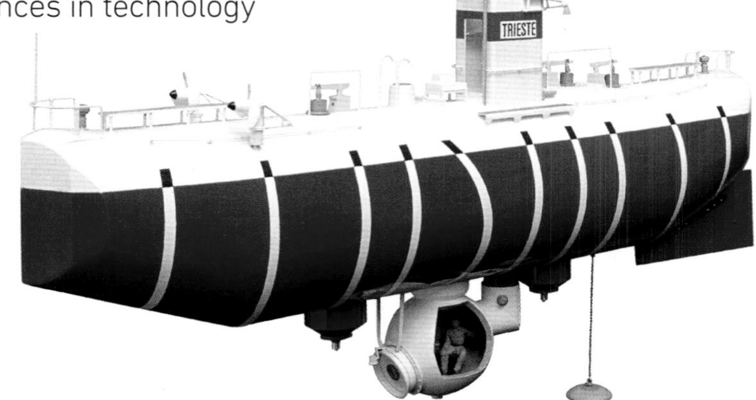

△ **Bathyscaphe *Trieste***
On 23 January 1960, Jacques Piccard and Don Walsh descended to the bottom of the Mariana Trench in a bathyscaphe called *Trieste*. A window cracked on the way down, but they still spent 20 minutes on the sea floor.

At this time, diving suits were heavy and cumbersome, and the diver received oxygen through an air line connected to a ship above, which greatly restricted movement. With help from his colleagues, Cousteau developed an alternative, which was a canister of compressed air strapped to the diver's back, with a regulator to control air flow. So successful was this Aqua-Lung, as he called it, that the US Navy bought it and renamed it the SCUBA (Self-Contained Underwater Breathing Apparatus) system.

In the 1950s, Cousteau adapted the first of several ships, each called *Calypso*, to operate as a research station, creating a base from which he could dive. A few years later, he invented a diving saucer for two people, and in 1962 built an experimental underwater living capsule, which he called the "Conshelf". He documented all of these inventions in books and television films that made him a household name worldwide.

The origin of life

Another key figure in oceanography was Jacques Piccard, a Swiss engineer who developed underwater vehicles for studying ocean currents. In 1960, aboard the bathyscaphe *Trieste*, Piccard and a companion became the first people to reach the deepest location on Earth – the floor of the Mariana Trench, in the Pacific Ocean, which lies at 10,911 m (35,797 ft) below sea level.

◁ **Hydrothermal vent**
Hydrothermal vents are fissures in the Earth's crust, from which water erupts due to volcanic activity. These vents are so rich in various life forms that many believe that they were crucial to the origin of life on Earth.

In 1969, Piccard also joined a team that spent 30 days drifting 2,414 km (1,500 miles) in the *Ben Franklin PX-15* subsurface research vessel. They were investigating the path of the Gulf Stream, a current in the Atlantic Ocean, but they also studied how they reacted to being cooped up for so long – something NASA was very interested in at the time (see pp.332–33).

In 1977, oceanographers made a discovery that overturned one of our most cherished beliefs – that the Sun played a key role in the origin of life. When the submersible ALVIN descended 2,100 m (6,890 ft) to the East Pacific Ocean floor, its cameras revealed chimney-like structures channelling super-heated water and minerals up from the mantle beneath. Large communities of sea creatures were thriving there, in utter darkness, using heat for energy and minerals for food. Many now believe that life on Earth began in such conditions.

IN CONTEXT

Underwater archaeology

Robert Ballard discovered the *Titanic*. In 1985, using a small unmanned submersible called *Argo*, he found wreckage on the floor of the Atlantic Ocean and followed it until he reached the infamous liner. Since then, entire historic cities have been uncovered on the seabed. In 2000, for example, Franck Goddio, diving off the coast of Egypt, discovered the ruins of Thonis-Heracleion, (and Canopus in 1997). Built at the mouth of the River Nile in the 8th century BCE, Thonis-Heracleion was submerged by geological and cataclysmic phenomena some 1,200 years ago. During the Late Period of ancient Egypt, it was Egypt's principal port of international trade.

A DIVER INSPECTS A BARNACLED STATUE IN THE SUNKEN CITY OF HERACLEION

" The sea, **once it casts its spell**, holds one in its **net of wonder forever**. "

JACQUES COUSTEAU, MARINE EXPLORER

▽ **Jacques Cousteau**
Aided by his many inventions, Cousteau became a pioneering underwater explorer and conservationist. His adventures were televised in the series *The Undersea World of Jacques Cousteau* (1968–75).

Flight to the Moon

In 1903, humans took to the air for the first time in a flight that lasted 12 seconds. Less than 70 years later, aviation had become space travel, and astronauts were walking on the Moon.

△ **Test launch**
An early rocket designed by Robert H. Goddard is winched onto its launch pad in Roswell, New Mexico, in 1935.

Leonardo da Vinci sketched flying machine designs back in the 15th century, but it took more than 400 years for the Italian's dreams to become reality. Yet, from the first experimental flights with liquid-fuelled rockets, it took just 33 years to reach the Moon.

Space race

American scientist Robert H. Goddard, with support from the Smithsonian Institute, ushered in the era of space flight when he successfully launched a rocket on 16 March 1926. He went on to launch 33 more, but, in 1941, Nazi Germany took the lead in rocket research. Hitler wanted to make the rocket into a weapon, and the result was the V-2, the world's first long-range guided ballistic missile. The V-2 also became the first man-made object to leave the Earth's atmosphere when it was launched on 20 June 1944.

During the Cold War, the V-2 became the model for both American and Soviet rocket designs, which in turn became the basis of their space exploration programmes. The two powers competed to be the first into space, the Americans employing captured German scientists, such as Wernher von Braun.

▷ **One step beyond**
A Russian poster celebrates Yuri Gagarin's historic achievement, becoming the first human in space. "Cosmonautics Day USSR", it reads, with the date of the flight in the rocket's exhaust plume: "12, IV, 1961".

In orbit

The first goal was to launch a satellite. The Soviets got there first: on 4 October 1957, they received the distinctive "beep… beep… beep…" from radio transmitters signalling *Sputnik I* was in orbit. Two years later, in 1959, the Soviets became the first to land an unmanned craft, the *Luna 2*, on the Moon. Later that year, the *Luna 3* took pictures of the far side of the Moon.

△ **Space dog**
Before sending a man into space, the Soviets sent a dog called Laika. A stray from the streets of Moscow, she orbited the Earth aboard *Sputnik 2*, but soon died from overheating.

On 12 April 1961, the Soviets finally launched a human into orbit, in a craft called *Vostok I*. The cosmonaut on board was Yuri Gagarin, a former fighter pilot, who initiated the launch with a shout of "*Poyekhali!*" ("Let's go!"). This momentous first space journey took just 108 minutes, which was just as well for Gagarin, since he was crammed into a compartment just 2.3 m (7½ft) in diameter. In that short time, however, he circumnavigated the Earth.

Almost a year later, on 20 February 1962, astronaut John Glenn became the first American to orbit the Earth. Both nations now stepped up their efforts to be the first to land a man on the Moon. However, in 1967, both the Apollo (American) and Soyuz (Russian) programmes suffered fatal disasters. The Americans lost the crewmen of Apollo 1 when a fire swept their spacecraft cabin during a ground test, and the Russians lost a Soyuz 1 cosmonaut when his capsule crashed due to a parachute failure.

Man on the moon

The Soviets took 18 months to recover from their disaster. The US, on the other hand, rallied quicker, and on 16 July 1969, the crew of Apollo 11 – mission commander Neil Armstrong, Michael Collins, and pilot Edwin "Buzz" Aldrin – entered their capsule at the

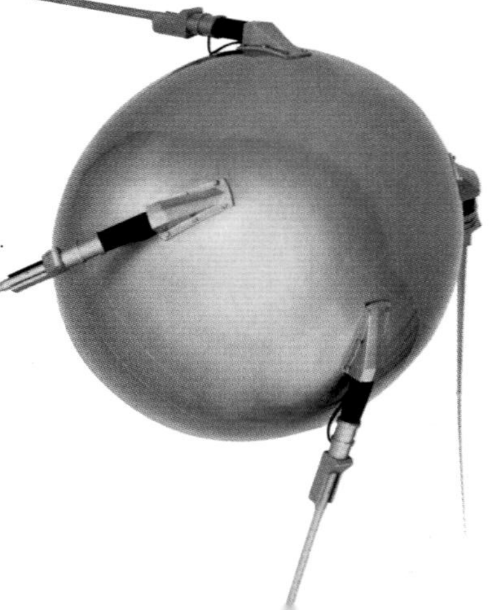

▷ ***Sputnik I***
Launched by Russia on 4 October 1957, *Sputnik 1* was the first artificial Earth satellite. Its position was broadcast by its four radio antennae.

top of a colossal Saturn V rocket. At 9:32am, the rocket took off from Kennedy Space Center in Florida, beginning its epic space flight.

Four days later, Collins remained in the command module as Armstrong and Aldrin embarked in the *Eagle*, the lunar-landing craft, and touched down on the Moon on 20 July. "Houston, Tranquility Base here, the *Eagle* has landed," reported Armstrong. Shortly afterwards, he took his first tentative step onto the surface and said the immortal words: "That's one small step for a man, one giant leap for mankind".

Aldrin joined Armstrong, and the two explored their surroundings, staying within a short distance of the *Eagle*. They took soil and rock samples, and left an American flag and a Soviet medal in honour of Yuri Gagarin. After some 21 hours on the Moon's surface, they returned to the landing craft and started on the long journey home.

◁ **Reaching Pluto**
On 19 January 2006, NASA's *New Horizons* spacecraft took off from Cape Canaveral Air Force Station, Florida. It travelled 7.5 billion km (4.67 billion miles) to Pluto, arriving on 14 July 2015.

Last man on the Moon
Apollo 11 was followed by six other manned flights to the Moon, culminating in Apollo 17, which landed in 1972. Eugene Cernan, pictured here, was the last man to stand on the lunar surface.

The Hippie Trail

In the 1960s, hoards of idealistic youths took off from cities across Europe to hitchhike or bus their way through central Asia to India and beyond, in search of peace, love, and enlightenment.

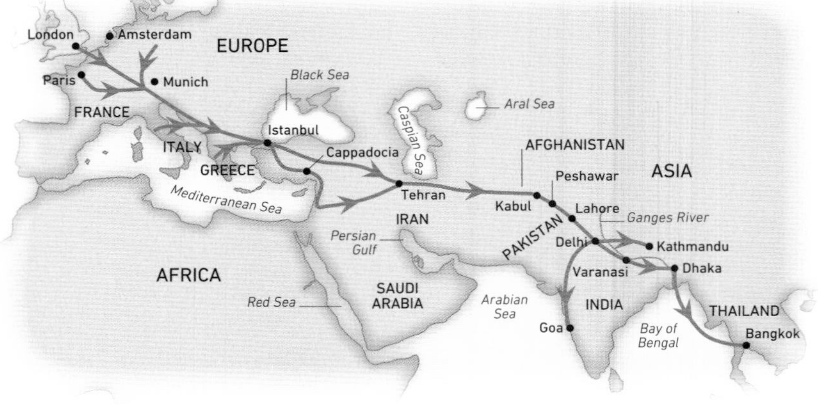

▽ **Hippie Trail route**
The point of the Hippie Trail was not just the destination (e.g. India), but the journey itself. Unlike travelling by boat or plane, the overland route exposed travellers to numerous local cultures, and gave them time to explore.

For two decades, in the second half of the 20th century, the spirit of the Grand Tour was briefly revived (see pp.180–83). But where once wealthy young aristocrats had roamed Europe to bolster their education, the young people who took to the road in the Sixties, on what became known as the Hippie Trail, travelled in search of spiritual enlightenment.

The travellers' inspiration came from figures like the American author Jack Kerouac and his fellow adventurers of the "alternative" scene. Kerouac had travelled the US in a quest for fulfilment, and turned the experience into a bestselling novel, *On the Road* (1957). The Hippie Trail was the European counterpart to the great American road trip. Its end point was India, the home of ancient eastern philosophy. "The east was not only a country and something geographical but it was the home and youth of the soul," wrote German author Hermann Hesse in *Journey to the East* (published in English in 1956).

Hitting the road

From starting points in major cities across Europe, the path to India followed the Silk Road of old, through Istanbul and on into Iran, Afghanistan, and Pakistan (see pp.86–87). After India, many went further, through Southeast Asia to Bangkok, and even to Australia.

▽ **Kombi mural**
Various companies ran buses along the Hippie Trail from Europe to India. Many travellers used their own transport, the most popular being the VW Kombi van.

One of the key elements of the trip was to travel as cheaply as possible, mainly to extend the length of time away from home. Cheap, private buses provided transport. According to Rory Maclean, author of *Magic Bus: On the Hippie Trail from Istanbul to India* (2006), the first European tourist bus to follow the route took 16 passengers from Paris to Bombay in the spring of 1956. The following year, Irishman Paddy Garrow-Fisher established the first regular coach service from Europe to the subcontinent. For almost a decade, Garrow-Fisher's Indiaman Tours operated the world's longest bus route from King's Cross, London, to Calcutta. Dozens of similar operators followed in his footsteps. Travellers also hitchhiked, or drove their own transport, typically trucks, minibuses, or camper vans that doubled up as accommodation.

△ **Backpacking**
The modern phenomenon of backpacking – travelling with everything you need in a backpack – began with the Hippie Trail to India.

A support network grew up along the Trail, and certain cafes, restaurants, and hotels became marshalling points where travellers wearing Apache headbands, paisley shirts, or Afghan coats congregated. These also served as pre-internet forums, with noticeboards pinned full of information on travel, offers of lifts, or hook-ups along the lines of: "Gentle deviant, 21, seeks guitar-playing chick ready to set out for mystical East". In Istanbul, the Pudding Shop became the place to meet; in Kabul, it was Sigi's on Chicken Street; in Tehran, it was the Amir Kabir Hotel; and in Kathmandu, it was an entire road, nicknamed "Freak Street" because of the thousands of hippies who passed through.

Journey's end

After Istanbul, the first major stop was Cappadocia in central Turkey, an area that was unknown to western Europeans until the 1950s. Its volcanic

> " All you've got to do is **decide to go** and the **hardest part is over**. "
>
> TONY WHEELER, FOUNDER OF *LONELY PLANET*

landscapes of fantastically shaped rock chimneys riddled with caves provided pleasingly alternative accommodation for travellers, who squatted there for days or even months at a time. Iran was regarded as a repressive police state, so few lingered there. Instead, they pushed on into Afghanistan, which became something of a hippie paradise thanks to native hospitality.

Pakistan was another "passing" country that could be crossed in 48 hours – then there was India, which had all the ashrams a pilgrim could need. The most popular destinations were the holy city of Varanasi on the Ganges, Goa in the west, and Kathmandu in Nepal.

The Trail came to an end in 1979. Following the Islamic revolt, Iran's borders were closed to tourists, while a Soviet invasion had the same effect in neighbouring Afghanistan. By this time, the idea of "backpacking", as this kind of budget independent travel had become known, was well established. Some of those who had taken the Hippie Trail had written up notes on the best places to stay, things to see, and how to get around, and they published these to great success, reinventing the travel guidebook. Such writers included Tony and Maureen Wheeler, founders of the *Lonely Planet* guidebooks. The idealistic travellers on the Hippie Trail may not have changed the world, but they did create a very successful modern travel publishing industry.

▽ **Hippie Trail**
The term "hippie" was shorthand for anyone with long hair (i.e. "to the hips"), so the journey east became known as the Hippie Trail.

> " I've always thought of the Concorde as a **magical object**, a symbol, a **miracle**. "

ANDREE PUTMAN, FRENCH DESIGNER

Concorde

Once man had walked on the moon, anything seemed possible, including flying commercial passengers at twice the speed of sound.

Concorde, introduced in 1976 after decades of development by France and Britain, was unique. Although engineers in the US and the Soviet Union had worked on supersonic airliners of their own, the American Boeing 2707 never made it off the drawing board, while the Soviet Tupolev TU-144 was abandoned because of performance and safety problems.

Concorde was the only commercial airliner to fly faster than the speed of sound. A Machmeter on the bulkhead told passengers when they were hitting Mach 1, then Mach 2. At this point they would be travelling at twice the speed of sound, at about 2,180 kph (1,350 mph), compared with 780 kph (485 mph) on a commercial Boeing 737. It was the closest most people came to space flight.

Nothing else looked like Concorde, with its dart-like delta wings, and adjustable pointed nose that could be lowered to give pilots better visibility on takeoffs and landings. Inside, it was not overly luxurious. The aircraft was created by engineers who, it was said, built a narrow metal tube that flew very fast and then grudgingly bolted seats in afterwards. But it was exclusive. Only 14 of the aircraft ever entered service, seven each with Air France and British Airways. Tickets were expensive, but some international businesspeople were inclined to splurge on the fare because of the time saved; Concorde could cross the Atlantic in just three hours, compared to seven on a Boeing 747. With twice-daily services from London to New York, it was possible to fly over, do business, and be back home in time for a late dinner.

For most of its service, Concorde had an exemplary safety record. This changed on 25 July 2000, when an Air France flight burst into flames and crashed shortly after taking off. 113 people died. The damage this caused to Concorde's reputation, combined with low passenger numbers following the 9/11 terrorist attacks in the US, and escalating maintenance costs, meant that in the summer of 2003 the supersonic fleet was permanently retired.

For many, the end of Concorde was a step backwards in technology. As the English broadcaster and Concorde regular David Frost once said: "You can be in London at 10 o'clock and in New York at 10 o'clock. I have never found another way of being in two places at once." We may never have that possibility again.

◁ **Concorde air stewardesses**
Air stewardesses from a variety of airlines from around the world stand in front of a scale model of Concorde. In fact, only Britain and France had airlines that operated this supersonic service.

New horizons

In 1946, Evelyn Waugh predicted the death of travel writing. As he saw it, everywhere had been written about. Yet not only is travel literature still alive, it appears to be in remarkably good health.

In the early 1970s, a young American novelist suggested to his publisher that he write a book about a train journey. The publisher agreed and Paul Theroux set off from London's Victoria railway station, bound eventually for Tokyo Central. *The Great Railway Bazaar* was published in 1975 and went on to sell more than 1.5 million copies in 20 languages. Theroux followed up his great train ride with others in South America (*The Old Patagonia Express*, 1979) and China (*Riding the Iron Rooster*, 1988), as well as plenty of excursions by boat, bus, and car.

A rapidly growing roster of names joined Theroux on the bookshop travel shelves, notably Bruce Chatwin (who debuted with *In Patagonia*, 1977), Colin Thubron (whose breakthrough was *Among the Russians*, 1983), and Jonathan Raban (*Arabia Through the Looking Glass*, 1979), followed by many others. All achieved both commercial success and critical acclaim. The travel writing genre, which had reached its first peak in the late 19th century and its second in the 1930s, was enjoying another resurgence.

Once again, the increase in the popularity of writing about foreign places coincided with a transformation in the methods of getting to them. In this case, international air travel was just becoming widely affordable. The new travel writers addressed this growing market with books that were faster paced, and more inventive than those that had gone before.

◁ ***A Walk in the Woods***
Nick Nolte (shown here as hiker Stephen Katz) stars alongside Robert Redford and Emma Thompson in *A Walk in the Woods*, the 2015 film that was based on Bill Bryson's 1998 memoir of the same name.

The endearingly grumpy Theroux filled his books with spiky encounters and diplomacy-be-damned honesty. Chatwin, a former art auctioneer, created high-concept travelogues that read like slightly surreal fiction: *In Patagonia* centres on a quest to find a piece of a brontosaurus. It has been said that people fall into Chatwin's travel stories, and stay there.

Breaking boundaries

Other new authors pushed travel literature in different directions, not just geographically. The naturalist Redmond O'Hanlon introduced black humour into the Amazon in his book *In Trouble Again: A Journey Between the Orinoco and the Amazon* (1988), gleefully detailing episodes when he shared hallucinogens with tribesmen and the effects of invasive parasites. Bill Bryson, an expatriate journalist living in the UK, went back home to poke fun at America ("I come from Des Moines. Somebody had to") for *Lost Continent* (1989). He then used the same approach for the UK, Europe, and Australia, becoming a bestselling author as a result.

Writers such as the American Tim Cahill (*Jaguars Ripped My Flesh*, 1987; *A Wolverine is Eating My Leg*, 1989) put a sense of adventure back into travel writing, often in an extreme manner, recounting exhilarating tales of harvesting poisonous sea snakes in the Philippines and dining on baked turtle dung in the Australian Outback. For *Holidays in Hell* (1988), P.J. O'Rourke travelled the world's blackspots from war-torn Lebanon to Heritage USA, a vast theme park for Christians run by famous television evangelists.

△ ***In Patagonia***
Unforgettable and influential, Bruce Chatwin's account of his journey across "the uttermost part of the earth" became an instant classic.

> " As long as there are **writers**, there will be travel writing **worth reading**. "
>
> SAMANTH SUBRAMANIAN, AUTHOR OF *THIS DIVIDED ISLAND: STORIES FROM THE SRI LANKAN WAR*

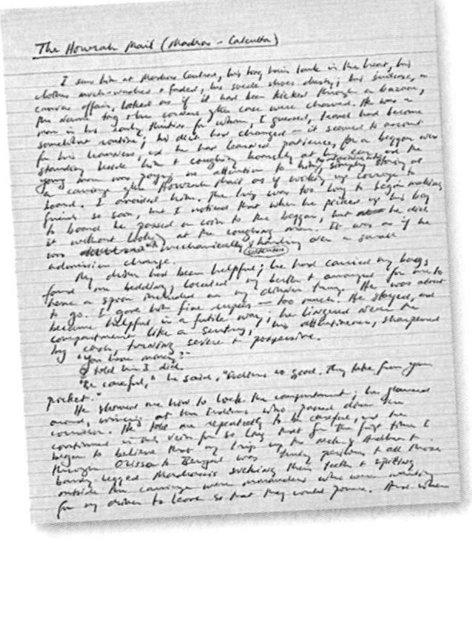

◁ **Notes for *The Great Railway Bazaar***
As well as notes for *The Great Railway Bazaar*, the Paul Theroux Collection at the Huntington Library in California contains correspondence from V.S. Naipaul and many other writers.

Fresh perspectives

For a long time, V.S. Naipaul, the Trinidadian author of the Indian travelogues *An Area of Darkness* (1962), *A Wounded Civilization* (1977), and *A Million Mutinies Now* (1990), was a solitary non-Western voice in the field of travel literature. Then, in 1983, Indian author Vikram Seth published *From Heaven Lake: Travels Through Sinkiang and Tibet*, and Indian novelist Amitav Ghosh wrote an affectionate portrait of the time that he spent in an Egyptian village, which was published in 1992 as *In an Antique Land*. More recently, the Indian-American Suketu Mehta returned to the city of his youth to write *Maximum City: Bombay Lost and Found* (2004).

These titles suggest possibilities for a rich new form of travel writing. In the last 150 years, we have grown accustomed to reading about the journeys of Westerners who have ventured out into remote corners of the globe and then reported back. Now, it will be fascinating to read the travel accounts of the millions of migrants who travel in the other direction and head to Europe and America each year.

◁ **V.S. Naipaul**
Nobel Prize-winning writer V.S. Naipaul (right) and "Rolling Stone" magazine editor Hunter S. Thompson (left) report on Grenada after US involvement in 1983.

▽ **Steam locomotive, India**
In a scene that may have been familiar to V.S. Naipaul, railway workers move a steam locomotive at the railway yard near the Taj Mahal in Agra, India, in 1983.

Exploration today

In the 21st century, it may seem that there is nothing left to explore, but adventurers are still finding new places to discover and learning things about Earth that were not known before.

People have visited the North and South Poles, stood on the highest peaks, and looked down on Earth from space, but there are still caves and mountains that have not been explored. Above all, the oceans are yet to be fully charted. According to US deep-sea explorer Robert Ballard, humans have seen only "one tenth of one per cent" of what is below the sea.

Deep beneath the Earth

In 2012, film director James Cameron, of *Titanic* fame, put his fascination with wrecks on the ocean floor to scientific use when he became only the third person to descend to the Challenger Deep, the deepest known point on Earth, in the western Pacific's Mariana Trench. This was part of an ongoing survey of the planet's most remote spot to find new animal species (using a "slurp gun" to suck up small creatures) and bring back images of rocks between the two tectonic plates that could further understanding of how tsunamis start.

Perhaps the most extreme example of underground mapping is that of Krubera Cave – a chasm on the edge of the Black Sea some 2,197 m (7,208 ft) below sea level, and thus the deepest known cave on Earth. Ukrainian diver Gennady Samokhin explored Krubera in 2007. He plunged into its aptly named "terminal sump", reaching record-breaking depths, and found out that the cave was 46 m (150 ft) deeper

▷ **Record depth**
On 26 March 2012, Canadian film director James Cameron descended to the Challenger Deep, 10,898 m (35,756 ft) below sea level, aboard *Deepsea Challenger*, a single-person vessel. He trawled the sea floor, videoing and collecting samples of sediment.

> "That there is an **inner urge** [to do this] is **undeniable**."
>
> SIR RANULPH FIENNES, ON CROSSING THE ANTARCTIC IN WINTER

◁ **Old techniques**
Today, explorers have all kinds of state-of-the-art equipment, including light, breathable clothing. Physical challenges remain tough, however. Crevasses are still crossed the hard way – using small, precarious-looking ladders.

than previously imagined. Three years later, new species were discovered, one of which – the tiny, eyeless *Plutomurus ortobalaganensis*, which had adapted to living in total darkness – was the deepest terrestrial animal ever found.

In 2012, Samokhin returned to Krubera and descended a further 52m (170ft), creating a second world record. Future explorers still have a great deal to find beneath the Earth – as Robert Ballard said: "The next generation of kids will probably explore more of Earth than all previous generations combined."

Retracing journeys

Not all of today's journeys plumb new depths – many put a 21st-century slant on places that have been visited in the past. Briton Tim Severin, for example, has retraced the steps of historic figures such as Genghis Khan and Marco Polo in an attempt to understand the past. Many individuals have set new records. American Matt Rutherford became the first person to sail solo through the Northwest Passage, which was first crossed by Roald Amundsen in 1906. In 2006, New Zealander Mark Inglis was the first double amputee to scale Mount Everest. In 2015, Briton John Beeden became the first person to row solo non-stop across the Pacific Ocean, from San Francisco, US, to Cairns in Australia. Charitable causes also spur on explorers. Ranulph Fiennes is only part of the way through his Global Reach Challenge, which aims to raise money for the Marie Curie organization that helps those with terminal illnesses. His goal is to become the first person to cross both polar ice caps and to climb the highest mountain on each continent. Turkish-American Erden Eruç began a similar adventure in 2003, when he embarked on his Six Summits Project, a bid to climb the highest peaks on each continent (except Antarctica) after reaching them by human power alone. In pursuit of this goal, Eruç made the first human-powered circumnavigation of the globe in 2012.

Advances in technology have shed a new light on places that are inaccessible or hard to see. Satellites have revealed traces of ancient civilizations beneath the deserts of Arabia, and drones make it possible for ecologists to study the impenetrable canopies of rainforests. And then, of course, there is space.

◁ **The eternal adventurer**
British adventurer Ranulph Fiennes has set numerous exploration records. He was the first person to visit the North and South poles by surface means only (1979–82), and the first to completely cross Antarctica on foot.

▽ **Reliving the past**
Tim Severin has recreated several historic or legendary journeys, such as the voyages of Sinbad the Sailor in a handmade replica of a medieval Arab boat in 1980–81.

New frontiers

The last man may have set foot on the Moon many decades ago, but plans to colonize Mars and dreams of superfast flight show that the desire to travel into space is as strong as ever.

△ **International Space Station (ISS)**
The ISS is a space station orbiting around 330–435 km (205–270 miles) above the Earth. It is the largest artificial object currently in orbit.

Although NASA was the first agency to succeed in landing astronauts on the Moon in 1969, its last manned mission there was in 1972. Recently, however, its Artemis programme stated an ambition to create a crewed lunar base within this decade, and to create an opportunity for the first female and first person of colour to set foot on the Moon.

Another focus of resources is on the International Space Station (ISS), which has had rotating crews drawn from 10 different nations in constant residence since November 2000. Mars is also a focus of attention, and NASA is trying to find out whether it might be habitable. Missions such as the Mars Rover programme – the robotic *Curiosity* rover that has been exploring the planet since August 2012, and *Perseverance*, which landed in 2021 – are providing important information about the geography and atmospheric conditions of the planet, alongside a Chinese rover, *Zhurong*, which landed in 2020.

Colonizing Mars

In May 2012, a privately funded Dutch company called Mars One announced a plan to establish a human settlement on Mars. More than 200,000 people initially applied, despite the fact that the ticket was one-way only, but the company went bankrupt in 2019. Meanwhile, NASA is recruiting to send humans to Mars by the 2030s, and China has a stated aim of establishing a manned presence on Mars within a similar timeframe.

Also in the race to Mars is Space X, an American company founded in 2002

IN CONTEXT
Outer space

When the Voyager 1 and 2 probes were launched back in 1977, Jimmy Carter had just been sworn in as US president, and the first *Star Wars* film was still in the cinemas. Four decades later, the probes are still exploring space. Both of them have flown by the giant planets, Jupiter and Saturn, and Voyager 2 has also made its way to Uranus and Neptune. The probes continue to monitor conditions in the outer reaches of the solar system, and are expected to be able to do so until around 2025. Often called the greatest feat of human exploration, the Voyager programme is a wonderful example of just how far technology can travel.

AN ARTIST'S IMPRESSION OF VOYAGER 2

▷ **Travel to Mars**
This poster, produced by NASA in 2016, imagines a future in which Mars is colonized. It also looks back at the milestones of Mars exploration, celebrated as "historic sites".

" Where **we're going**, we **don't need roads**. "

DR EMMETT BROWN, *BACK TO THE FUTURE*, 1985

by entrepreneur Elon Musk. The Space X programme envisions spaceships capable of carrying more than 100 passengers. Unlike Mars One, these spaceships would be able to take off from Mars and return to Earth again. However, Musk hopes that anyone who travels to the far-off planet will stay there, as he wants to establish a settlement. "I think there is a strong humanitarian argument for making life multi-planetary," Musk has said, "in order to safeguard the existence of humanity in the event that something catastrophic were to happen."

Travel on Earth

Musk has also explored the concept of the Hyperloop, a vehicle that could radically speed up travel on Earth. Pod-like vehicles would be propelled through a vacuum-like tube at the speed of a jet aircraft. Musk claims that the Hyperloop could transport passengers the 560 km (350 miles) from San Francisco to Los Angeles in just 35 minutes, at an average speed of 970 kph (600 mph).

Sub-orbital aircraft

New technology could also drastically reduce long-distance flight times in the future. Several private companies around the world are investing in prototypes of sub-orbital aircraft. Part-rocket and part-aircraft, these would be launched up to the top of the atmosphere, around 100 km (62 miles) above Earth. But rather than going into orbit around the planet, the aircraft would then glide down to their destinations. A vehicle like this would be capable of taking passengers from Europe to Australia in 90 minutes, or from Europe to California in an hour.

Such a journey sounds exciting, but it would cause high-level environmental damage and be extremely expensive. That has not deterred a few private companies, such as Blue Origin and Virgin Galactic, from exploring sub-orbital spaceflight commercially, and indeed successfully launching crewed sub-orbital flights in 2021. As yet, however, space tourism is reserved for the super-wealthy.

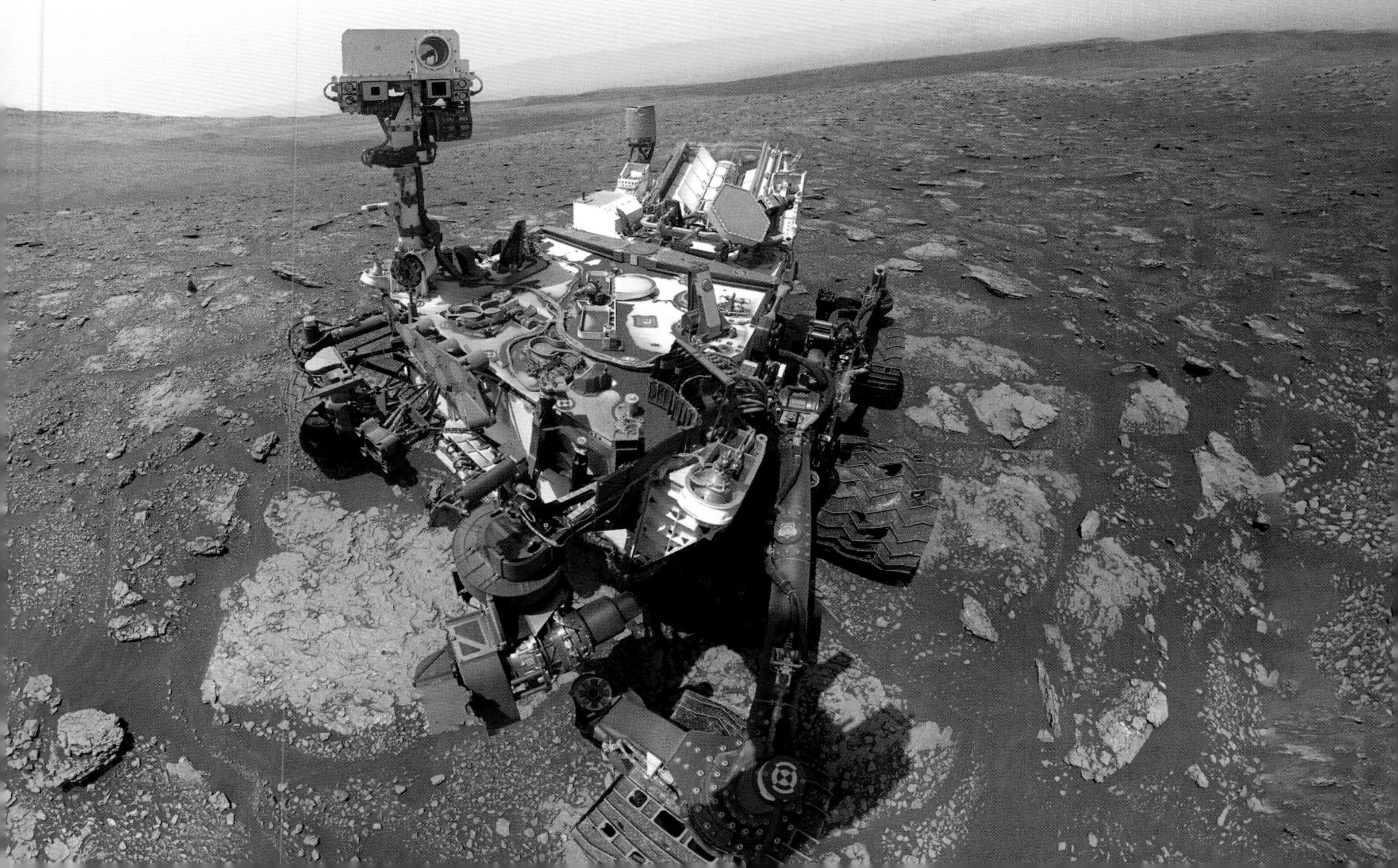

▽ **A "selfie" from Mars**
NASA's *Perseverance* rover took this selfie on 10 September 2021. The image is actually 57 individual images beamed back to Earth and then stitched together.

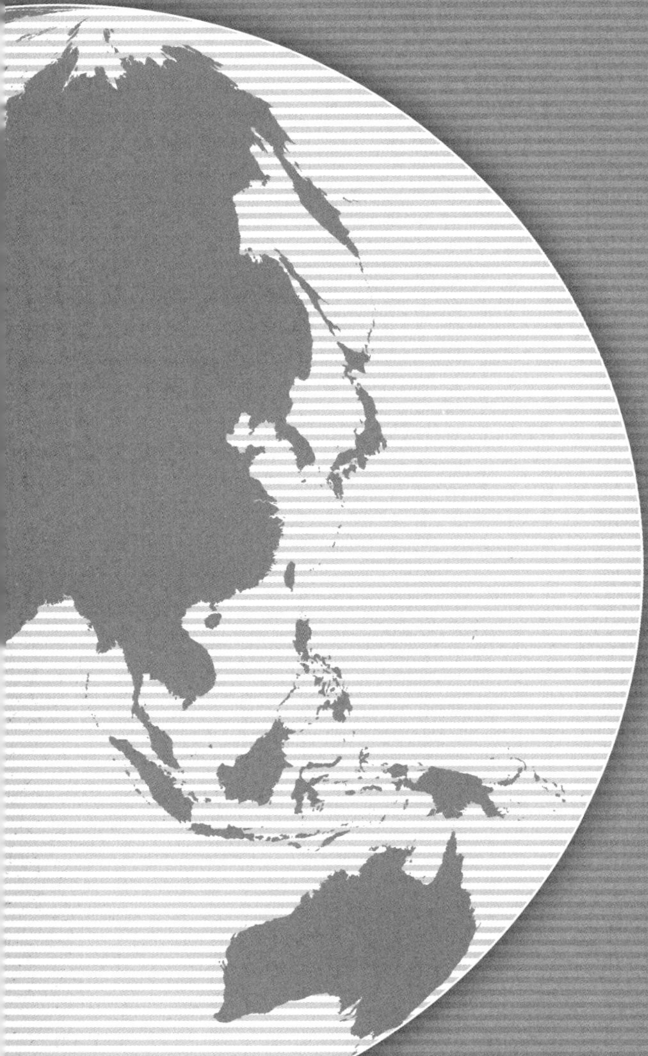

INDEX

A

B

C

D

F

G

N

P

R

T

PICTURE CREDITS

Dorling Kindersley would like to thank the following for their work on the book: Contributor: Phil Wilkinson. Indexer: Helen Peters. Editorial assistance: Victoria Heyworth-Dunne, Sam Kennedy, Devangana Ojha, Nandini Desiraju. Design assistance: Vikas Chauhan, Sourabh Challariya. Chapter openers and design assistance: Phil Gamble. Jacket: Priyanka Sharma. DTP assistance: Vishal Bhatia.

The publisher would like to thank the following for their kind permission to reproduce their photographs:

Key: a-above; b-below/bottom; c-centre; f-far; l-left; r-right; t-top

1 Dorling Kindersley: James Stevenson / National Maritime Museum, London. **2-3 Bridgeman Images:** Ira Block / National Geographic Creative. **4 Getty Images:** DEA / G. Nimtallah / De Agostini (tr). **5 akg-images:** Pictures From History (tl). **Alamy Stock Photo:** Lebrecht Music and Arts Photo Library (bc). **6 akg-images. Alamy Stock Photo:** Mary Evans Picture Library (br). **Getty Images:** MPI / Stringer (tr). **Rijksmuseum. 7 Alamy Stock Photo:** Contraband Collection (bl). **Getty Images:** Sky Noir Photography by Bill Dickinson (tr). **Library of Congress, Washington, D.C.** (tl). **NASA:** JPL-Caltech / MSSS (br). **8-9 Getty Images:** Robbie Shone / National Geographic (t). **11 Bridgeman Images:** British Library, London, UK / © British Library Board. All Rights Reserved. **12 Getty Images:** DEA / G. Nimtallah / De Agostini (l); Science & Society Picture Library (c); DEA / G. Dagli Orti (r). **13 Alamy Stock Photo:** Uber Bilder (l). **Getty Images:** DEA / G. Dagli Orti (c); Marc Hoberman (cr). **14 Alamy Stock Photo:** kpzfoto (bc). **Getty Images:** DEA / G. Dagli Orti (bl); VCG (br). **15 Bridgeman Images:** Pictures from History (br). **Getty Images:** Fine Art Images / Heritage Images (bc); Too Labra (bl). **16 Alamy Stock Photo:** INTERFOTO (bl). **16-17 Alamy Stock Photo:** Peter Barritt (t). **17 akg-images:** Erich Lessing (clb). **Alamy Stock Photo:** Classic Image (br). **18-19 Getty Images:** DEA / G. Nimtallah / De Agostini (t). **19 Alamy Stock Photo:** INTERFOTO (c). **Dorling Kindersley:** Graham Rae / Hellenic Maritime Museum (bl). **Dreamstime.com:** Denis Kelly (br). **20 akg-images:** Hervé Champollion (cl). **20-21 Getty Images:** Leemage / Corbis (b). **21 Bridgeman Images:** Egyptian National Museum, Cairo, Egypt (cr). **Getty Images:** DEA / G. Dagli Orti (tc). **22-23 Getty Images:** CM Dixon / Print Collector. **24-25 Getty Images:** Science & Society Picture Library (b). **24 Science Photo Library:** Library of Congress, Geography and Map Division (tr). **25 Alamy Stock Photo:** The Natural History Museum (clb). **Getty Images:** Science & Society Picture Library (br). **26 akg-images:** Erich Lessing (bc). **Bridgeman Images:** Pictures from History (cla). **26-27 Getty Images:** DEA / G. Dagli Orti(t). **27 Alamy Stock Photo:** Anka Agency International (br). **28 akg-images:** British Museum, London. **29 Alamy Stock Photo:** imageBROKER (tc). **Getty Images:** DEA Picture Library / De Agostini (fbr, br). **30-31 Bridgeman Images:** Private Collection / Photo © Ken Welsh. **31 Alamy Stock Photo:** Science History Images (bc). **Dreamstime.com:** Olimpiu Alexa-pop (cra). **32-33 Dorling Kindersley:** Graham Rae / Hellenic Maritime Museum (b). **32 Alamy Stock Photo:** Charles O. Cecil (bl); North Wind Picture Archives (tr). **Getty Images:** DEA / G. Dagli Orti (cla). **33 Getty Images:** Marc Hoberman (cla). **American School of Classical Studies at Athens, Agora Excavations:** (cra). **34 Getty Images:** DEA / G. Dagli Orti / De Agostini (bl); Heritage Images (cla). **34-35 Photo Scala, Florence:** White Images (b). **35 Alamy Stock Photo:** kpzfoto (tl). **Getty Images:** Universal History Archive / UIG (crb). **36 Getty Images:** DEA / M. CARRIERI / De Agostini. **37 123RF.com:** Juan Aunin (bc). **Getty Images:** Universal History Archive (br). **Rex Shutterstock:** The Art Archive (cr). **38 Western Han Dynasty Museum of the South Vietnamese:** (c). **38-39 Getty Images:** VCG (t). **39 Bridgeman Images:** People's Republic of China (bc). **40-41 Getty Images:** DEA / G. Dagli Orti. **42 Alamy Stock Photo:** Granger Historical Picture Archive (cra). **iStockphoto.com:** sculpies (bl). **43 Alamy Stock Photo:** Granger Historical Picture Archive (cr); The Granger Collection (l). **44 Alamy Stock Photo:** Kenneth Taylor (clb). **Bridgeman Images:** The Israel Museum, Jerusalem, Israel / The Ridgefield Foundation, New York, in memory of Henry J. and Erna D. Leir (ca). **44-45 Alamy Stock Photo:** adam eastland. **46 Getty Images:** DEA / G. Dagli Orti (tc). **46-47 Dreamstime.com:** Robert Zehetmayer (b). **47 Dreamstime.com:** Axel2001 (crb). **Getty Images:** PHAS (ca). **48-49 Getty Images:** Photo12 / UIG. **50-51 Bridgeman Images:** Royal Geographical Society, London, UK (t). **50 Alamy Stock Photo:** Ken Welsh (c). **51 Dreamstime.com:** Marilyn Barbone (bc). **Getty Images:** Science & Society Picture Library (br). **52 Getty Images:** Fine Art Images / Heritage Images (t). **53 Getty Images:** Ann Ronan Pictures / Print Collector (cr); Fine Art Images / Heritage Images (tl); Universal History Archive (bl). **54 akg-images:** Album / Oronoz (c). **Alamy Stock Photo:** Granger Historical Picture Archive (r). **Getty Images:** Heritage Images (l). **55 Dorling Kindersley:** National Maritime Museum, London (cr). **Getty Images:** Stefano Bianchetti (c); Print Collector (l). **56 Getty Images:** Arne Hodalic / Corbis (bl); Kazuyoshi Nomachi (bc). **Science Photo Library:** NYPL / Science Source (br). **57 Alamy Stock Photo:** GL Archive (bc). **Getty Images:** Leemage / Corbis (br); Leemage (bl). **58 Photo Scala, Florence:** The British Library Board (t). **59 Bridgeman Images:** Pictures from History (bc). **Imaginechina:** (tl). **Rex Shutterstock:** Sipa Press (cra). **60-61 akg-images:** Pictures From History. **62 Alamy Stock Photo:** ART Collection (t). **Getty Images:** Heritage Images (crb). **63 Alamy Stock Photo:** www.BibleLandPictures.com (cl). **Dreamstime.com:** Viacheslav Belyaev (br). **Getty Images:** Heritage Images (cra). **64 Alamy Stock Photo:** Niels Poulsen mus (tr). **64-65 Getty Images:** Ullstein Bild (b). **65 Alamy Stock Photo:** Science History Images (cra); Sklifas Steven (br). **Bridgeman Images:** Ashmolean Museum, University of Oxford, UK (cla). **66 Getty Images:** Arne Hodalic / CORBIS (bl). **66-67 Alamy Stock Photo:** World History Archive (t). **67 Bridgeman Images:** Private Collection (br). **Rex Shutterstock:** Alfredo Dagli Orti (tr). **68 Dorling Kindersley:** National Maritime Museum, London (r). **Getty Images:** Bettmann (clb). **69 Alamy Stock Photo:** Charles O. Cecil (tl). **Getty Images:** Leemage (clb); Print Collector (br). **70 Bridgeman Images:** British Library, London, UK / © British Library Board (bc). **71 123RF.com:** Nickolay Stanev (bc). **Alamy Stock Photo:** Heritage Image Partnership Ltd (crb). **Getty Images:** Stefano Bianchetti (t). **72 Alamy Stock Photo:** ART Collection (tc). **72-73 Alamy Stock Photo:** Hemis. **73 Alamy Stock Photo:** Arctic Images (tc). **Getty Images:** Russ Heinl (cr); Universal History Archive (clb). **74 akg-images:** Jérôme da Cunha. **75 Alamy Stock Photo:** Josse Christophel (cr). **76 akg-images:** Album / Oronoz (b). **Getty Images:** Leemage (tc). **77 Alamy Stock Photo:** INTERFOTO (cl); robertharding (tl). **Getty Images:** Heritage Images (cra, br). **78-79 Alamy Stock Photo:** Granger Historical Picture Archive. **80 Rex Shutterstock:** Alfredo Dagli Orti (t). **81 Alamy Stock Photo:** The Granger Collection (cr); robertharding (clb); GM Photo Images (b). **82 akg-images:** Fototeca Gilardi (tr). **Getty Images:** Angelo Hornak / Corbis (cla); Fine Art Images / Heritage Images (br). **83 Getty Images:** Fine Art Images / Heritage Images. **84 Rex Shutterstock:** British Library / Robana (ca). **Science Photo Library:** NYPL / Science Source (bc). **85 Bridgeman Images:** British Library, London, UK (tr). **Getty Images:** DEA / M. Seemuller (tl). **86 Alamy Stock Photo:** GL Archive (bc). **Getty Images:** Leemage / Corbis (t). **87 Alamy Stock Photo:** Niday Picture Library (bl). **Bridgeman Images:** Pictures from History / David Henley (tr). **Getty Images:** Martin Moos (bc). **88 Alamy Stock Photo:** The Granger Collection (cla). **88-89 Getty Images:** Print Collector (b). **89 Alamy Stock Photo:** Pictorial Press Ltd (bc). **Bridgeman Images:** Private Collection / Pictures from History (tr). **90 Alamy Stock Photo:** D. Hurst (tr). **Dorling Kindersley:** Courtesy of Deutsches Fahrradmuseum, Germany (tc). **Rex Shutterstock:** Bournemouth News (cr). **90-91 123RF.com:** Richard Thomas (c). **Dorling Kindersley:** Matthew Ward (b).

91 Dorling Kindersley: A Coldwell (crb); National Motor Museum, Beaulieu (tl); Jerry Young (tc); R. Florio (tr); Jonathan Sneath (cb). **Nissan Motor (GB) Limited:** (br). **92-93 akg-images:** Roland and Sabrina Michaud. **92 akg-images:** Gerard Degeorge (bl). **94 Alamy Stock Photo:** The Granger Collection (ca). **94-95 AWL Images:** Nigel Pavitt (b). **95 4Corners:** Tim Mannakee (cl). **Alamy Stock Photo:** Ian Nellist (tr). **96 Bridgeman Images:** Private Collection / Archives Charmet (bl). **97 Alamy Stock Photo:** dbimages (crb). **Getty Images:** Heritage Images (l). **Württembergische Landesbibliothek Stuttgart:** (cr). **98-99 Bridgeman Images:** Bibliotheque Nationale, Paris, France / Index (t). **98 Getty Images:** Heritage Images (br); Universal History Archive (cl). **99 Library of Congress, Washington, D.C.:** G5672.M4P5 1559 .P7 (bc). **100 Alamy Stock Photo:** Stuart Forster (l); Visual Arts Resource (r). **Bridgeman Images:** Private Collection / Index (c). **101 Alamy Stock Photo:** Alberto Masnovo (cr); North Wind Picture Archives (l); Ken Welsh (c). **102 akg-images. Alamy Stock Photo:** North Wind Picture Archives (bl). **Getty Images:** Bjorn Landstrom (br). **103 Alamy Stock Photo:** Lebrecht Music and Arts Photo Library (bl). **Getty Images:** DEA / G. DAGLI ORTI(bc); UniversalImagesGroup (br). **104 Alamy Stock Photo:** Stuart Forster. **105 Alamy Stock Photo:** Chris Hellier (bl). **Bridgeman Images:** Pictures from History (cr). **Getty Images:** Chris Hellier (br). **106-107 SD Model Makers:** (b). **106 Getty Images:** DEA / A. Dagli Orti / Contributor / De Agostini (cla). **106-107 Alamy Stock Photo:** Pat Eyre (cb). **Getty Images:** Science & Society Picture Library / Contributor / SSPL (bc). **107 Getty Images:** DEA / A. Dagli Orti / Contributor / De Agostini (ca). **107 Getty Images:** Science & Society Picture Library (tc) **108 Getty Images:** Culture Club (ca). **Rex Shutterstock:** Gianni Dagli Orti (cla). **SuperStock:** Phil Robinson / age fotostock (br). **108-109 Photoshot:** Atlas Photo Archive (t). **109 Getty Images:** G&M Therin-Weise (bc). **110 Alamy Stock Photo:** North Wind Picture Archives. **111 Alamy Stock Photo:** Chronicle (clb); musk (br). **112 Alamy Stock Photo:** Granger Historical Picture Archive (tl). **Getty Images:** Leemage (tr). **112-113 Bridgeman Images:** Photo © Tallandier (b). **113 Alamy Stock Photo:** Mary Evans Picture Library (tc); The Granger Collection (crb). **114 akg-images. Alamy Stock Photo:** Granger Historical Picture Archive (br). **Getty Images:** Fine Art Images / Heritage Images (c). **115 Alamy Stock Photo:** Chronicle (b); Prisma Archivo (tr). **116-117 Photo Scala, Florence:** bpk, Bildagentur fuer Kunst, Kultur und Geschichte, Berlin. **118 Bridgeman Images:** Private Collection / Index (t). **119 Alamy Stock Photo:** Colport (c); LMR Group (tr). **Getty Images:** DEA / G. Dagli Orti(tl). **120-121 Alamy Stock Photo:** The Granger Collection (b). **120 Alamy Stock Photo:** Chronicle (bc); The Granger Collection (tc). **Getty Images:** Bjorn Landstrom (cla). **121 Alamy Stock Photo:** robertharding (tc). **122 Alamy Stock Photo:** Lebrecht Music and Arts Photo Library (t). **Getty Images:** Stock Montage (crb). **123 Alamy Stock Photo:** Alberto Masnovo (cra). **Bridgeman Images:** Newberry Library, Chicago, Illinois, USA (bc). **124 Alamy Stock Photo:** Peter Horree (cra). **Getty Images:** DEA / G. Dagli Orti (cla). **124-125 Alamy Stock Photo:** Graham Prentice (b). **125 Alamy Stock Photo:** Granger Historical Picture Archive (tl). **Getty Images:** PHAS (cra). **126-127 akg-images:** Album / Oronoz (b). **127 Alamy Stock Photo:** Peter Horree (br); The Granger Collection (tc, c). **128-129 Getty Images:** Stock Montage. **130-131 Getty Images:** UniversalImagesGroup. **131 Alamy Stock Photo:** The Granger Collection (c). **132 Alamy Stock Photo:** Hi-Story (cl); Stock Montage, Inc. (tr). **132-133 Getty Images:** Leemage (b). **133 Alamy Stock Photo:** North Wind Picture Archives (crb). **Library of Congress, Washington, D.C.:** 02616u (ca). **134 Alamy Stock Photo:** North Wind Picture Archives (clb); The Protected Art Archive (cr). **135 akg-images. Alamy Stock Photo:** The Granger Collection (br). **Bridgeman Images:** American Antiquarian Society, Worcester, Massachusetts, USA (cr). **136 Getty Images:** DEA / M. Seemuller (bc); Robert B. Goodman (cra). **Mary Evans Picture Library:** (cla). **137 Getty Images:** Stefano Bianchetti. **138-139 Royal Geographical Society:** Jodocus Hondius Snr (b). **139 Getty Images:** De Agostini Picture Library (br). **Rex Shutterstock:** Harper Collins Publishers (cra). **140 akg-images. Şermin Ciddi**. **Getty Images:** Ullstein Bild (c). **141 Alamy Stock Photo:** Lebrecht Music and Arts Photo Library. **Bridgeman Images. 142 Bridgeman Images:** Pictures from History (bl). **Getty Images:** Historical Picture Archive / Corbis (bc); Universal Images Group (br). **143 akg-images:** Fototeca Gilardi (bc). **Alamy Stock Photo:** North Wind Picture Archives (bl). **Getty Images:** (br). **144-145 Rijksmuseum:** (t). **144 Rijksmuseum. 145 Bridgeman Images:** Pictures from History (br). **146-147 akg-images:** Rabatti & Domingie. **148 Bridgeman Images:** Dutch School, (17th century) / Private Collection (cl). **Dorling Kindersley:** Mike Row / The Trustees of the British Museum (ca). **148-149 Nationaal Archief, Den Haag:** (t). **149 Wikipedia:** Bibliothèque François Mitterrand / TCY / CC license: wiki / File:Globe_Coronelli_Map_of_New_Holland (br). **150 Alamy Stock Photo:** Photo 12 (t). **151 Alamy Stock Photo:** Granger Historical Picture Archive (bl). **Getty Images:** MPI / Stringer (tc). **iStockphoto.com:** CatLane (br). **152 Alamy Stock Photo:** ClassicStock (bl). **Nationaal Archief, Den Haag. Pilgrim Hall Museum:** (br). **152-153 Getty Images:** The New York Historical Society (t). **154 Şermin Ciddi**. **155 Getty Images:** DeAgostini (bl); DeAgostini (br). **The Walters Art Museum, Baltimore:** (cl). **156-157 Getty Images:** Historical Picture Archive / Corbis. **158 akg-images. Bridgeman Images:** Peter Newark American Pictures (cr). **159 Getty Images:** SSPL / Florilegius (cr). **Wilberforce House, Hull City Museums:** (b). **160 akg-images. 161 Alamy Stock Photo:** Granger Historical Picture Archive (clb); Uber Bilder (cr). **Bridgeman Images:** Peter Newark Pictures (cl). **162 AF Fotografie. Alamy Stock Photo:** The Natural History Museum (bl). **163 Getty Images:** Historical Picture Archive / Corbis (r). **Royal Pavilion & Museums, Brighton & Hove:** (cl). **164-165 Bridgeman Images:** Bonhams, London, UK. **166-167 Getty Images:** Anton Petrus (b). **167 Alamy Stock Photo:** AF Fotografie (tl). **Getty Images:** Fine Art Images / Heritage Images (br). **168 Alamy Stock Photo:** INTERFOTO (cl). **168-169 Alamy Stock Photo:** North Wind Picture Archives (b). **169 Alamy Stock Photo:** AF Fotografie (tl); INTERFOTO (br). **Dorling Kindersley:** Harry Taylor / Natural History Museum, London (tr). **170-171 Getty Images:** DeAgostini (b). **170 AF Fotografie:** (cl). **Getty Images:** Science & Society Picture Library (tr). **171 Alamy Stock Photo:** Granger Historical Picture Archive (tl); Granger Historical Picture Archive (tr). **172 Getty Images:** Universal Images Group (cla). **Getty Images:** Science & Society Picture Library / Contributor / SSPL (ca). **173 The Trustees of the British Museum:** (tl). **Photo Scala, Florence:** White Images (b). **174 Getty Images:** Universal History Archive (tr). **Mary Evans Picture Library:** Natural History Museum (tl). **174-175 akg-images:** Private collection. **175 Alamy Stock Photo:** Granger Historical Picture Archive (cr). **176 Alamy Stock Photo:** The Natural History Museum (bl). **177 Alamy Stock Photo:** Florilegius (tl); The Natural History Museum (tr). **Mary Evans Picture Library:** Natural History Museum (bl). **178-179 Bridgeman Images:** Royal Collection Trust © Her Majesty Queen Elizabeth II, 2017. **180 Bridgeman Images:** English School, (19th century) (after) / Private Collection (t). **181 AF Fotografie. Bridgeman Images:** English School, (19th century) / Private Collection (br). **182 Bridgeman Images:** Civico Museo Sartorio, Trieste, Italy / De Agostini Picture Library / A. Dagli Orti (tr). **182-183 Bridgeman Images:** Christie's Images (b). **183 akg-images:** Fototeca Gilardi (tc). **184 Bridgeman Images:** Archives Charmet (bl). **Getty Images:** (cl). **184-185 Alamy Stock Photo:** Mary Evans Picture Library (t). **185 Bridgeman Images:** SZ Photo / Scherl (br). **186 Alamy Stock Photo:** Chronicle (cl); The Natural History Museum (cr). **186-187 State Library Of New South Wales:** (b). **187 National Library of Australia:** (tr). **National Archives of Australia:** (br). **State Library Of New South Wales:** Bequest of Sir William Dixson, 1952 (cl). **188 Bridgeman Images:** Alte Nationalgalerie, Berlin, Germany / De Agostini Picture Library (c). **Getty Images:** Science & Society Picture Library (l); Science & Society Picture Library (r). **189 akg-images:** ullstein bild (l). **Dorling Kindersley:** Gary Ombler / Christian Goldschagg (cr). **TopFoto.co.uk:** Roger-Viollet. **190 Getty Images:** Archive Photos / Smith Collection / Gado (bl); Hulton Fine Art Images / Heritage Images (br). **The Stapleton Collection:** (bc). **191 akg-images:** Universal Images Group / Underwood Archives (bc). **Bridgeman Images:** Keats-Shelley Memorial House, Rome, Italy (bl). **Thomas Cook Archives:**

(br). **192 Bridgeman Images:** Alte Nationalgalerie, Berlin, Germany / De Agostini Picture Library. **193 Getty Images:** De Agostini Picture Library (br); iStock / andreaskrappweis (bc). **Missouri Botanical Garden:** Peter H. Raven Library (cr). **Zentralbibliothek Zurich:** (ca). **194 Alamy Stock Photo:** Hemis (tr). **Getty Images:** De Agostini / G. Dagli Orti(cl). **194-195 Bridgeman Images:** Pictures from History. **195 Bridgeman Images:** The Stapleton Collection (cr). **Getty Images:** Fotosearch (tc). **196-197 Getty Images:** Fine Art Photographic Library / Corbis. **198 Alamy Stock Photo:** North Wind Picture Archives (cla). **198-199 Alamy Stock Photo:** North Wind Picture Archives (t). **199 Alamy Stock Photo:** Don Smetzer (br); Granger Historical Picture Archive (cr). **Bridgeman Images:** Private Collection (bl). **200-201 Getty Images:** Fotosearch / Stringer (b). **200 Benton County Historical Society Museum in Warsaw, MO:** (bl). **201 Getty Images:** James L. Amos / National Geographic (br); MPI / Stringer (tl). **202 Getty Images:** Science & Society Picture Library (bl). **202-203 Getty Images:** Archive Photos / Smith Collection / Gado (t). **203 Getty Images:** Bettmann (br). **Library of Congress, Washington, D.C.:** (tr). **204 Alamy Stock Photo:** V&A Images (tr). **Getty Images:** Dea / G. Dagli Orti/ De Agostini (b). **205 Alamy Stock Photo:** John Baran (tl). **Bridgeman Images:** Keats-Shelley Memorial House, Rome, Italy (bc). **Wikipedia:** I.H. Jones (tr). **206 Getty Images:** GraphicaArtis (cl). **206-207 Bridgeman Images:** Historic England (b). **207 Alamy Stock Photo:** The Natural History Museum (cl). **Getty Images:** Science & Society Picture Library (br). **208 Historic England Photo Library:** (b). **208-209 Getty Images:** SSPL (tc). **209 Boston Rare Maps Incorporated, Southampton, Mass., USA:** (bc). **Getty Images:** Bettmann (br). **Penrodas Collection:** (tr). **210 Getty Images:** Hulton-Deutsch Collection / Corbis (cl). **210-211 akg-images:** ullstein bild (b). **211 Boston Public Library:** (tl). **Library of Congress, Washington, D.C.. 212-213 Alamy Stock Photo:** The Granger Collection (b). **212 David Rumsey Map Collection www.davidrumsey.com:** (tr). **Getty Images:** Time Life Pictures / Mansell / The LIFE Picture Collection (c); Universal History Archive / UIG (bl). **213 Bridgeman Images:** Pictures from History (tc). **214 Getty Images:** Photo12 / UIG (bl). **214-215 Bridgeman Images:** Royal Geographical Society, London, UK (t). **215 Alamy Stock Photo:** Mary Evans Picture Library (br). **Bridgeman Images:** Royal Geographical Society, London, UK (clb). **Getty Images:** The Print Collector (tr). **216 Getty Images:** Photo12 / UIG (tr); Science & Society Picture Library (b). **217 colour-rail.com:** (br). **Getty Images:** Bettmann (tc). **218-219 Dorling Kindersley:** Gary Ombler / Didcot Railway Centre (c); Mike Dunning / National Railway Museum, York (cb). **218 colour-rail.com. Dorling Kindersley:** Gary Ombler / The National Railway Museum, York (tr). **Getty Images:** Bettmann (cla). **Science & Society Picture Library:** National Railway Museum (br). **219 colour-rail.com. Dorling Kindersley:** Gary Ombler / B&O Railroad Museum (tl); Gary Ombler / Virginia Museum of Transportation (cb). **Getty Images:** SSPL (tr). **Vossloh AG:** (br). **220 akg-images:** Universal Images Group / Underwood Archives (b). **Alamy Stock Photo:** E.R. Degginger (tr). **221 Alamy Stock Photo:** Chronicle (bl); Granger Historical Picture Archive (br). **Getty Images:** GraphicaArtis (tr). **222 Bridgeman Images:** (bl). **Thomas Cook Archives. 223 Getty Images:** Hulton Archive / Stringer (l). **Thomas Cook Archives. 224-225 Getty Images:** Culture Club. **226 Bridgeman Images:** Bibliotheque des Arts Decoratifs, Paris, France / Archives Charmet (cra); Ken Walsh (bl). **227 Alamy Stock Photo:** Antiqua Print Gallery (tr). **Bridgeman Images:** Look and Learn / Barbara Loe Collection (br). **Getty Images:** Ullstein Bild (bl). **228 Alamy Stock Photo:** Amoret Tanner (bl). **Bridgeman Images:** The Geffrye Museum of the Home, London, UK (bc); Private Collection / Christie's Images (cla). **Dorling Kindersley:** Gary Ombler / The University of Aberdeen (cra); Jacob Termansen and Pia Marie Molbech / Peter Keim (tr). **Getty Images:** De Agostini Picture Library / De Agostini / G. Dagli Orti(tl); De Agostini / DEA / L. DOUGLAS (cb); Jason Loucas (crb). **National Museum of American History / Smithsonian Institution:** (tc). **Wellcome Images http://creativecommons.org/licenses/by/4.0/:** (ca). **229 Alamy Stock Photo:** Basement Stock (tr); Chronicle (tc); Caroline Goetze (cra); INTERFOTO (cla). **Getty Images:** Chicago History Museum (tl); Photolibrary / Peter Ardito (br).

230 Bridgeman Images: Look and Learn / Peter Jackson Collection (tr). **Getty Images:** The Print Collector (cl). **230-231 Getty Images:** Chris Hellier / Corbis (b). **231 Getty Images:** Chicago History Museum (br); Science & Society Picture Library (t). **232 Getty Images:** Universal Images Group (b). **NYCviaRachel:** (tr). **233 Bridgeman Images. Getty Images:** Don Arnold (tr); Joe Scherschel / National Geographic (tl). **234 Bridgeman Images:** Luca Tettoni (b). **235 Bridgeman Images:** Pictures from History (bc). **RMN:** Thierry Ollivier (cl). **236-237 Beinecke Rare Book And Manuscript Library/yale University**. **237 Bridgeman Images:** Peter Newark American Pictures (br). **238 Getty Images:** Daniel Mcinnes / Corbis (cr). **Mary Evans Picture Library:** SZ Photo / Scherl (cl). **238-239 Bridgeman Images:** Tallandier (b). **239 akg-images:** (tr). **Bridgeman Images:** City of Westminster Archive Centre, London, UK (cr). **Mary Evans Picture Library:** Pharcide (tl). **240-241 Tom Schifanella:** (all images). **242 Getty Images:** Science & Society Picture Library (l). **243 Bridgeman Images:** Royal Geographical Society, London, UK (c). **Getty Images:** Best View Stock (br). **Royal Geographical Society:** (tl). **244-245 Photo Scala, Florence:** White Images (b). **245 akg-images:** Fototeca Gilardi (tl). **Alpine Club Photo Library, London:** (tr). **Getty Images:** Kean Collection / Archive Photos (b). **246 Alamy Stock Photo:** Universal Art Archive (bc). **Getty Images:** Swim Ink 2, LLC / Corbis (cl). **247 The J. Paul Getty Museum, Los Angeles:** William Henry Jackson; I.W. Taber (tr). **248 Alamy Stock Photo:** Art Collection 2 (cl). **Thomas Cook Archives. 248-249 Alamy Stock Photo:** Old Paper Studios. **249 Bridgeman Images:** Cauer Collection, Germany (br). **Getty Images:** LL / Roger Viollet (c). **250 Getty Images:** DeAgostini (bc); Time Life Pictures / Mansell / The LIFE Picture Collection (cr). **Science Photo Library:** Natural History Museum, London (cl). **250-251 Getty Images:** Time Life Pictures / Mansell / The LIFE Picture Collection (b). **251 Alamy Stock Photo:** AF Fotografie (bc). **Getty Images:** Science & Society Picture Library (tc). **252-253 Alamy Stock Photo:** Vintage Archives. **254 akg-images. Getty Images:** General Photographic Agency (r). **255 akg-images. Alamy Stock Photo:** Lordprice Collection (l). **Dorling Kindersley:** Gary Ombler / Jonathan Sneath (cr). **256 Getty Images:** Bettmann (br); Cincinnati Museum Center (bl); Herbert Ponting / Scott Polar Research Institute, University of Cambridge (bc). **257 Alamy Stock Photo:** Contraband Collection (br). **Getty Images:** Time Life Pictures / Mansell / The LIFE Picture Collection (bc); Topical Press Agency / Stringer (bl). **258 Bridgeman Images:** United Archives / Carl Simon (bl). **The Trustees of the British Museum:** (c). **Getty Images:** Universal Images Group (cl). **259 akg-images. Museum of Ethnography, Sweden:** Sven Hedin Foundation (tl). **260 Alamy Stock Photo:** Lordprice Collection (br). **Getty Images:** Thinkstock (c). **The National Library of Norway:** Siems & Lindegaard (bl). **261 Getty Images:** Uriel Sinai (tr). **The National Library of Norway**. **Skimuseet i Holmenkollen:** Silja Axelsen (tl). **262 Dorling Kindersley:** Gary Ombler / Jonathan Sneath (tl). **Getty Images:** Cincinnati Museum Center (b). **263 AF Fotografie. Alamy Stock Photo:** Mary Evans Picture Library (bl); Universal Art Archive (tr). **Getty Images:** Bettmann (cl). **264 Alamy Stock Photo:** Marc Tielemans (tr). **The Camping and Caravanning Club:** (cla). **264-265 Country Life Picture Library:** (b). **265 Alamy Stock Photo:** AF Fotografie (tc). **Bridgeman Images:** Christie's Images (br). **266-267 Bridgeman Images:** Look and Learn. **267 Getty Images:** Sovfoto (tr). **Mary Evans Picture Library:** Illustrated London News Ltd (br). **268 Getty Images:** Print Collector / Contributor / Hulton Archive (crb). **Getty Images:** Wolfgang Steiner (tl). **269 Alamy Stock Photo:** Chronicle (tl). **Bridgeman Images:** The Advertising Archives (b). **PENGUIN and the Penguin logo are trademarks of Penguin Books Ltd:** (tr). **270-271 Alamy Stock Photo:** Contraband Collection. **270 Alamy Stock Photo:** Universal Art Archive (tr). **271 akg-images. Alamy Stock Photo:** Contraband Collection (bl). **272 Getty Images:** Bettmann (tl). **Library of Congress, Washington, D.C.. 272-273 Library of Congress, Washington, D.C.. 273 Alamy Stock Photo:** Glyn Genin (br). **274 Getty Images:** Bettmann (cl); Roger Viollet (bc); VCG Wilson / Corbis (c). **275 Bridgeman Images:** Cauer Collection, Germany (r). **Photo Scala, Florence:** (tl). **276 Getty Images:** Bettmann (cl). **276-277 Alamy**

Stock Photo: Universal Art Archive (b). **277 Getty Images:** Bettmann (tc); Universal History Archive (br). **278 Getty Images:** Herbert Ponting / Scott Polar Research Institute, University of Cambridge. **279 Alamy Stock Photo:** Uber Bilder (c). **Bridgeman Images:** Granger / Herbert Ponting (bc). **280 akg-images:** Sputnik (cr). **Getty Images:** Universal History Archive (bl). **281 Alamy Stock Photo:** Interfoto (cr). **Getty Images:** Bettmann (l); Time Life Pictures / Mansell / The LIFE Picture Collection (br, br); Time Life Pictures / Mansell / The LIFE Picture Collection (br, br). **282-283 Getty Images:** Bettmann. **284 Alamy Stock Photo:** Granger Historical Picture Archive (tr). **284-285 4Corners:** Susanne Kremer. **285 Alamy Stock Photo:** Granger Historical Picture Archive (tr). **Rex Shutterstock:** Eduardo Alegria / EPA (br). **286-287 Getty Images:** Bettmann (t). **286 Getty Images:** Topical Press Agency / Stringer (br). **287 Alamy Stock Photo:** John Astor (br). **Getty Images:** Bettmann (tr). **288-289 Getty Images:** Kirby / Topical Press Agenc (b). **289 Alamy Stock Photo:** Danita Delimont (tl). **Getty Images:** Bettmann (cr). **Mary Evans Picture Library:** John Frost Newspapers (br). **290-291 Royal Geographical Society. 291 Getty Images:** arabianEye / Eric Lafforgue (tr); Universal Images Group (tl). **Royal Geographical Society. 292 Getty Images:** Roger Viollet. **293 Mary Evans Picture Library. 294-295 Alamy Stock Photo:** Royal Geographical Society (b). **294 AF Fotografie. Beinecke Rare Book And Manuscript Library/yale University. 295 Getty Images:** Roger Viollet (tr). **296 Affiliated Auctions:** (cra). **Getty Images:** Photo12 / UIG (bl). **297 Getty Images:** Bettmann. **The Granger Collection, New York:** National Geographic Stock: Vintage Collection (br). **Mary Evans Picture Library:** SZ Photo / Scherl (cr). **298-299 Alamy Stock Photo:** Contraband Collection. **300-301 Getty Images:** SSPL. **300 Getty Images:** SSPL (tr). **301 Getty Images:** General Photographic Agency (tr). **302 AF Fotografie. Getty Images:** Corbis / swim ink 2 llc (bl). **Mary Evans Picture Library:** Everett Collection (ftr); Onslow Auctions Limited (tr). **303 1stdibs, Inc:** (tc). **Alamy Stock Photo:** Vintage Archives (tr). **Canadian Pacific Railway**. **Getty Images:** Corbis Historical. **Mary Evans Picture Library:** Onslow Auctions Limited (tl). **304-305 Alamy Stock Photo:** INTERFOTO. **306 Alamy Stock Photo:** Dimitry Bobroff. **Getty Images:** Popperfoto (l). **Rex Shutterstock:** Turner Network Television (r). **307 Alamy Stock Photo:** A. T. Willett (cr). **Getty Images:** SSPL / Daily Herald Archive (l). **NASA. 308 Bridgeman Images:** Pitt Rivers Museum, Oxford, UK (bl). **Getty Images:** Fotosearch (bc). **Royal Geographical Society. 309 Getty Images:** Hawaiian Legacy Archive (bl). **NASA. 310 Getty Images:** Margaret Bourke-White / The LIFE Images Collection. **311 Getty Images:** Corbis Historical / Hulton Deutsch (tl); Stringer / AFP (cr); Photo12 / UIG (br). **Press Association Images:** (c). **312-313 Getty Images:** SSPL / Daily Herald Archive. **314 Getty Images:** Archive Photos (cl). **314-315 The Kon-tiki Museum, Oslo, Norway**. **315 Getty Images:** ullstein bild (br). **316 Bridgeman Images:** Pitt Rivers Museum, Oxford, UK (bl). **317 Bridgeman Images:** Pitt Rivers Museum, Oxford, UK (br); Pitt Rivers Museum, Oxford, UK (l). **Roland Smithies / luped.com:** (cr). **318 Getty Images:** Hulton Archive (bl); David Pollack / Corbis (cr). **318-319 Getty Images:** Museum of Flight / Corbis (t). **319 Alamy Stock Photo:** Collection 68 (bl). **Getty Images:** Bettmann (br). **Roland Smithies / luped.com:** (cr). **320-321 Alamy Stock Photo:** Tristar Photos (b). **Dorling Kindersley:** Gary Ombler / Brooklands Museum (cb). **320 Cody Images:** (cra). **Kristi DeCourcy:** (cr). **Dorling Kindersley:** Gary Ombler / Paul Stone / BAE Systems (clb); Martin Cameron / The Shuttleworth Collection, Bedfordshire (t). **321 aviation-images. com. aviationpictures.com:** (br). **Dorling Kindersley:** Peter Cook / Golden Age Air Museum, Bethel, Pennsylvania (crb); Gary Ombler / The Real Aeroplane Company (clb); Gary Ombler / De Havilland Aircraft Heritage Centre (cra); Dave King (tl). **National Air and Space Museum, Smithsonian Institution:** (tr). **322 Royal Geographical Society. 322-323 Royal Geographical Society. 323 Getty Images:** Moment Select / Jason Maehl (cr, c). **324-325 Getty Images:** Corbis / Hulton Deutsch. **324 Bridgeman Images:** Christie's Images (bl). **Getty Images:** Fotosearch (tr/1). **325 Getty Images:** Alfred Eisenstaedt / The LIFE Picture Collection (tc); Moviepix / Silver Screen Collection (br). **326-327 Getty Images:** Sky Noir Photography by Bill Dickinson. **327 Getty Images:** Lonely Planet Images / Kylie McLaughlin (tr). **328 Getty Images:** Robert Alexander (tr); Alan Band / Keystone (cr); John Williams / Evening Standard (bl). **328-329 Getty Images:** AFP Photo / Philippe Huguen (b). **329 Dreamstime.com:** Tan Kian Yong (tc). **330 Getty Images:** Corbis Documentary / Ralph White (bc); Haynes Archive / Popperfoto (cla); SSPL (tr). **331 Rex Shutterstock:** Turner Network Television (bc). **World Wide First Limited:** Christoph Gerigk (c) Franck Goddio / Hilti Foundation (tr). **332 Getty Images:** Bettmann (tl); SSPL (tr); SSPL (bl); Blank Archives (crb). **333 Getty Images:** Business Wire (tr). **NASA:** JSC. **334 Alamy Stock Photo:** (bl); Profimedia. CZ a.s. (crb). **335 Getty Images:** Jack Garofalo / Paris Matc (b). **336-337 Getty Images:** Keystone / Hulton Archive. **338 Alamy Stock Photo:** Atlaspix (cra). **PENGUIN and the Penguin logo are trademarks of Penguin Books Ltd:** (cl). **Paul Theroux:** (bc). **339 Getty Images:** The LIFE Images Collection / Matthew Naythons (tr). **Magnum Photos:** Steve McCurry (b). **340-341 Getty Images:** CANOVAS Alvaro. **340 National Geographic Creative:** Mark Thiessen (bl). **341 Alamy Stock Photo:** (cr); Royal Geographical Society. **342 Alamy Stock Photo:** Futuras Fotos (l). **342-343 NASA:** JPL-Caltech / MSSS. **343 Getty Images:** Corbis (cr).

Endpapers: Emigrants Farewell: 20th August 1912: Emigrants wave their last goodbyes as the emigrant ship 'Monrovian' leaves Tilbury in Essex, bound for Australia. (Photo by Topical Press Agency/Getty Images) Credit: Getty Images / Topical Press Agency / Stringer

Cover images: Back: **akg-images:** Pictures From History fcl; **Alamy Stock Photo:** Granger Historical Picture Archive cl; **Getty Images:** Sky Noir Photography by Bill Dickinson fcr; **Photo Scala, Florence:** Christie's Images, London cr

All other images © Dorling Kindersley
For further information see: **www.dkimages.com**

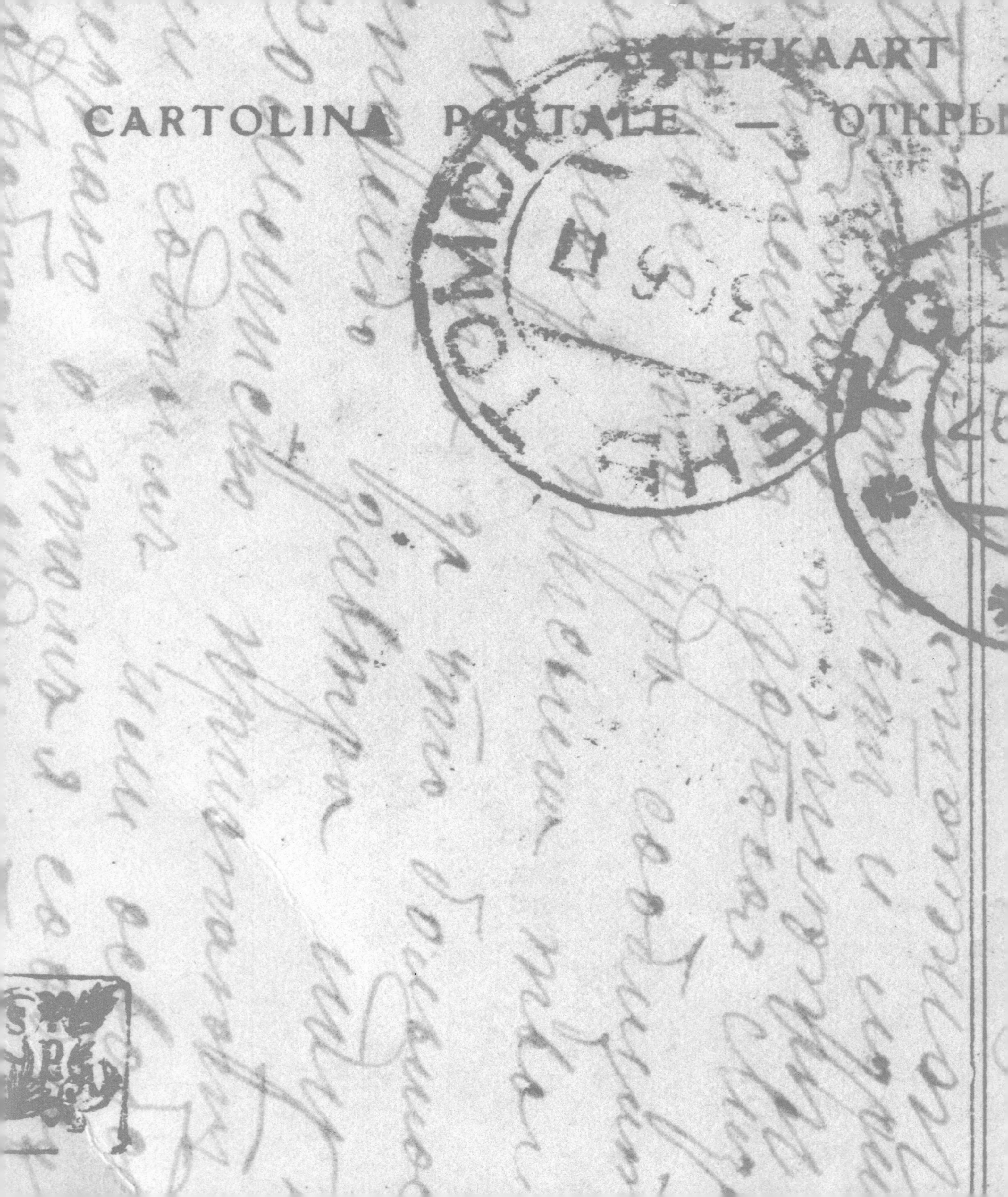

BRIEFKAART

CARTOLINA POSTALE — OTKPЫ